DECISION MAKING FOR COMPLEX SOCIO-TECHNICAL SYSTEMS
Robustness from Lessons Learned in Long-Term Radioactive Waste Governance

ENVIRONMENT & POLICY

VOLUME 42

The titles published in this series are listed at the end of this volume.

Decision Making for Complex Socio-Technical Systems

Robustness from Lessons Learned in Long-Term Radioactive Waste Governance

by

Thomas Flüeler

*Institute of Human-Environment Systems,
Institute of Natural and Social Science Interface (NSSI/UNS) ETH, Zurich,
Swiss Federal Institute of Technology ETH, Zurich,
and
Independent Environmental Consultant,
Umweltrecherchen & -gutachten,
Hausen AG, Switzerland*

A C.I.P. Catalogue record for this book is available from the Library of Congress.

ISBN-10 1-4020-3480-6 (HB)
ISBN-13 978-1-4020-3480-0 (HB)
ISBN-10 1-4020-3529-2 (e-book)
ISBN-13 978-1-4020-3529-6 (e-book)

Published by Springer,
P.O. Box 17, 3300 AA Dordrecht, The Netherlands.

www.springer.com

Printed on acid-free paper

All Rights Reserved
© 2006 Springer
No part of this work may be reproduced, stored in a retrieval system, or transmitted
in any form or by any means, electronic, mechanical, photocopying, microfilming, recording
or otherwise, without written permission from the Publisher, with the exception
of any material supplied specifically for the purpose of being entered
and executed on a computer system, for exclusive use by the purchaser of the work.

Printed in the Netherlands.

*Our people have to be able
to follow up
on the technical progress
in their minds.*

*Willi Ritschard,
Swiss Minister of Energy,
in the parliamentarian debate
on Kaiseraugst and [other]
nuclear power plants,
17 June 1975*

Preface

Waste management can be considered as a mirror of society. In its (our) fluctuating definition, besides conserving a supply of valuable materials and resources, it acts as a fascinating expression for our throw-away mentality. Waste is inevitable, but it nevertheless provides an opportunity, a need and an incentive for society to pursue a more careful management of it. If this is successfully carried out, society may well find it possible to manage itself in an overall more careful and universal way.

The management, or governance, of radioactive waste particularly is determined by a complex constellation of individuals, social aggregates, and institutions. A wealth of interconnected parts forms a complex and volatile system that is highly technical and explosively political in nature. The problems inherent in this system must be managed so as to reach a safe, acceptable, responsible, feasible, and sustainable "solution". It is a system with an objective long-term dimension, with a hazard potential of, in part, hundreds of thousands of years; and it has a long-term institutional dimension insofar as its implementation must last for decades. The very notion of risk extends from the long life of radionuclides, over the perception of dread to the decisional risk of whether, when and how to implement an agreed upon "solution".

Such a complexity makes it understandable, and perhaps even necessary, that the controversial "radwaste" issue be protracted. It is futile and unproductive to allocate "guilt" to those on whatever side of the issue for the slow pace of formulating a solution. In this contribution to the dialogue, my purpose is to examine the major mechanisms of radioactive waste governance, and to compare the efforts and ideas from all sides of the problem.

As to the quest for transparency, it is necessary to lay open my own record. Until 1990 I was the Director of the leading Swiss environmental organisations in energy issues, the Swiss Energy Foundation (SES), for that matter anti-nuclear. From then on I have been an environmental consultant, to the NGOs, but also to the International Atomic Energy Agency (IAEA). From 1992 to 2004 I was a member of the Swiss Federal Nuclear Safety Commission (KSA), an advisory body to the Swiss Government. Until its dismissal in 2002, I also served on the Cantonal Expert Group Wellenberg (KFW) consulting a potential host canton in Central Switzerland. I am involved in several international projects as well.

As far as it is possible for an actor in the arena of radioactive waste, I attempt to cast a "second look" at the issue, after Goeudevert, in hopes perhaps for some "discovery of the other" [*ibid.*M32:57-62]. Sometimes it is only feasible to crystallise contradictions … and leave them there (as to the "final" proposal for "the" solution).

This study, hence, is some sort of an experiment, and in a contentious environment. It is a study of an ongoing process by an actor with insights into the arena from diverse perspectives. It is a risk, with both components: danger and opportunity; danger of being biased and single-eyed; opportunity for shedding light on aspects, different than usual, on a different background. With relation to research, the transdisciplinary character of the study leads to two different major challenges and insights, respectively: Firstly, various disciplines had to be sufficiently understood and to be profitably focussed in the research questions. Secondly, frictional energy had to be overcome, so to speak against the research fields, because the cross-cutting focus was, as a necessity, not "at the front" of the specific disciplines.

In view of the driving forces behind radioactive waste and its Cinderella position in the nuclear debate, I hope to, at least, add some aid for reflexivity and reflection in the indispensable discourse to come. At any rate, it applies what Bechmann 1995 said with regard to risk and uncertainty, key aspects of radioactive waste: "It is not a matter of banning uncertainty, but dealing with uncertainty has to become a societal learning process. Only by taking care of risk may one avoid failing because of it".

June 2004

Summary

Worldwide long-term governance[1] of radioactive waste continues to be an unresolved major complex socio-technical issue. In 1998, the International Atomic Energy Agency (IAEA), stated in "Topical Issues on Nuclear, Radiation and Radioactive Waste Safety": "The key principle involved is the concept of defence in depth corresponding to multiple barriers which provide a combination of lines of defence against potential challenges to the safety of the disposal system. One of the important aspects is the evaluation of the robustness of the repository system …. One of the main issues is obtaining a better understanding of the meaning of principles such as defence in depth in the context of waste disposal" [G114:244]. The Nuclear Energy Agency (NEA) 1999 identified the need to analyse long-term management options as a whole and, specifically, reversibility and retrievability in the final disposal[2] concept. The Sixth European Framework Programme for Research and Technological Development launched in 2002 states that "[r]esearch alone cannot ensure societal acceptance; however, it is needed in order to … promote basic scientific understanding relating to safety and safety assessment methods, and to develop decision processes that are perceived as fair and equitable by the stakeholders involved" [G38:187*passim*].

The present study attempts to give a novel, empirically based and specific response to these issues. It proposes an integrated overall system robustness

---

[1] Governance is more than management and denotes, adapted according to the corresponding European White paper, "rules, processes and behaviour that affect the way in which powers are exercised … particularly as regards openness, participation, accountability, effectiveness and coherence" [M12:8]. This is in line with the seminal TRUSTNET Framework [M18:1][G88] to emphasise a broader scope including issues of justification, participation, definitory power, *etc*. It is not used where traditional project management is meant.

[2] For precise definition of disposition concepts see footnote 4.

in a sustainable radioactive waste governance concept facilitating a combination of technical design issues, analysis methods and institutional backup within a dynamic procedure. It strives for a continuous improvement of the risk analysis and risk management methodology and, hence, follows recent developments to deal with risks in an integrated way, along the "New Approach"/"Global Approach of the European Communities 1999/2000" [M10] [M11]. In accordance with the search for a thorough stakeholder[3] involvement, a well defined process orientation is of paramount importance.

A survey of international literature reveals that a multitude of technical, social, institutional and ethical analyses exist. To date though, no attempt has been made to examine the various aspects collectively, in a systematic and integrated manner, in order to identify the obstacles on the way to a sustainable governance of radioactive waste. On behalf of the stakeholders – proponents, implementers, authorities, experts and the public – fundamental data and insights shall be presented for an integrated, long-term and transparent decision-making strategy, which will allow society to satisfactorily cope with the complex problem of safe disposition[4] of radioactive waste.

The system characteristics of radioactive waste lead to the fact that long-term safety of waste disposal intrinsically cannot be mathematically demonstrated. The hypothesis is investigated that a "convincing set of arguments" (NEA 1999b, [G182:11]) can only be given in an extensive stepwise process and consists of technical elements (barriers), scientific methods (*e. g.*, uncertainty analyses) and comprehensive procedural aspects. This inter-relationship between technical and non-technical aspects is the topic of the study presented here. Its objectives are threefold and on three levels:

- A reconstruction of the decision-making processes in radioactive waste management and governance on the basis of two Swiss case studies (siting procedures for a final disposal of high- and intermediate-level as well as low-level radioactive waste, respectively);
- A presentation of decision-making fundamentals for disposition options – from surface storage to final disposal in the deep geological underground – on the ground of the principle of sustainability;
- The development of a concept of robustness of the socio-technical overall "radioactive waste system" with regard to an integrated, long-term, and transparent decision-making strategy.

[3] The pragmatic definition by NEA 2000 for "stakeholders" is adopted: "all persons or groups with an 'interest' in the project," "somebody with a role to play in the process" [G186:-118]. Similar notions are, in the present context, "actors" or "players".

[4] The generic term "disposition" is to denote very long-term management of radioactive waste, in line with the US Board on Radioactive Waste Management 2001 [G21]. "(Final) disposal" refers to this activity "without the intention of retrieval", "repository" denotes final disposal in the deep geological underground whereas indefinite control is named "storage". See footnote 48 and *Figure 12-4*.

The reconstruction is achieved on level 1 by means of an (empirical) content or document analysis with the help of criteria from both risk perception and decision science research. The proposed layout of disposition options (level 2) is normatively based on a specified principle of sustainability and empirically on the reconstruction of the decision process under investigation. The approach using an overall system robustness facilitates – on the basis of 1 and 2 as well as the risk analysis literature – a combination of technical design aspects, suitable analysis methods and institutional backups in a dynamic procedure. It is meant to be an extension of the existing risk analysis and management methodology (level 3).

For the formation of categories in the content analysis, the study tackles the issue from two theoretical angles: The first perspective originates from the concerned stakeholders. It adopts insights from risk perception research postulating that institutional decision-making anomalies of proponents and authorities may be diminished through an enlarged rationality and risk model, which integrates the risk perception of concerned parties. The models are to be expanded by aspects of the risk notion, the hazard properties and the social context.

The second perspective starts out from institutional decision makers. It utilises insight from normative, as well as empirical, decision science and systems theory, saying that complex situations of decision can only be met adequately if certain preconditions are given which fulfil the complexity of the respective issue: Understand the system sufficiently, recognise perception and communication aspects (see first perspective), avoid fallacies and biases, consider and adapt problem structures, decompose into subsystems and re-integrate, explore goal relations, treat different levels adequately, utilise latency periods as a chance for learning.

The content analysis over a lapse of 40 years substantiates that the radioactive waste proponent's, partly also the safety authorities' course of action, was not transparent nor traceable for many years, even so for experts. The reasons for the initial over-optimism are manifold but three main aspects are paramount:

- Well-known structural engineering was prioritised whereas the decisive issue of long-term safety was initially underestimated;
- Political considerations predominated the fact-based ones whereby the waste issue was instrumentalised in energy politics by all parties;
- Procedural issues of such a complex disposal programme were not duly followed, and if they were, they were dealt with as "political" aspects under point 2.

Specifically, it proved that the implementation delay of more than three decades after the establishment of the radioactive waste proponent, Nagra, was not merely a "political" problem as often dogmatised. The (technical) decision basis is in close interconnection with the (political) policy and consequential decisions. In the "grey range" of this interrelationship there are aspects such as the definition of criteria, transparency, duty of publication, traceability of arguments, plausibility, reviewing process, and step-by-step procedure. Durable political decisions in the socio-technical system called "long-term disposal of radioactive waste" are, consequently, based on solid technical grounds, which in turn can only be erected if corresponding pre-decisions leave them a sufficient framework (including resources).

The "inner" circle of decision makers has been gradually expanded, strengthened, and professionalised; increasingly relevant arguments are being used in a well-informed discourse. Herewith the chances rise that an "extended final disposal" of radioactive waste may be achieved whose primary goal is passive long-term safety, while allowing for control mechanisms to validate the performance assessments and to enhance confidence and securing a broad political backup (the specified notion of sustainability favours passive safety without abandoning technical and institutional control).

The empirical content analysis and literature studies suggest that controllability, retrievability and procedural issues like transparency, traceability of arguments, and stakeholder involvement are key elements in the safety assessment of the system "radioactive waste". A corresponding decision is well supported if it integrates relevant parts of both the problem and solution ranges of the main stakeholders. Therefore, the principle issues have to be thoroughly and broadly discussed. The aspect of control is one example of how in a complex factual field the dimensions are often debated in reverse order, especially pertaining to a technological constraint like waste: first technical and commercial, then political and economical, afterwards social and, last and least, under ethical aspects. Theoretically, it should be the other way around: First, one should have a broad debate and decision on political principles over ethical guidelines, this should in turn lead to the selection of the corresponding optimum technical variant, in consideration of ecology, economy and society, the "magic" triangle of sustainability. To reach this stage, an extended "control" by third parties has to be implemented – through strengthening the safety authorities, intensifying the review process and involving stakeholders hitherto excluded or not judged equivalent. As a counter to it, all partners have to proactively discuss eventual contradictions and inconsistencies and to duly consider time dimensions with respect to the construction of disposal facilities and system impacts. Ultimately, the generations taking profit from nuclear power have to assume responsible ownership of its related by-product, the radioactive waste.

Table of Contents (condensed)

Preface — vii
Summary — ix
Introduction — xxix

PART I: OVERALL ISSUE AND METHODOLOGY
Chapter 1: Setting and topics at issue — 3
Chapter 2: Objectives and aim — 21
Chapter 3: Research political embedding — 25
Chapter 4: Issues under investigation, evidence and validation — 29

PART II: PERSPECTIVE "FROM BELOW": RISK PERCEPTION OF THE PUBLIC
Chapter 5: Insights from risk perception research — 43
Chapter 6: Risk perception in radioactive waste issues — 51

PART III: PERSPECTIVE "FROM ABOVE": DECISION PROCESSES
Chapter 7: Insights from decision research — 101
Chapter 8: Development of decision making in technical systems — 113
Chapter 9: Decisions in radioactive waste governance — 123
Chapter 10: Siting as an example of sub-optimum decision making — 151

PART IV: CONCLUSIONS AND FURTHER DEVELOPMENT
Chapter 11: Patterns of arguments in radioactive waste governance — 175
Chapter 12: Fundamentals of a comparison of disposition options — 201
Chapter 13: Proposal for a concept of an overall system robustness — 231

References — 283

Contents

Preface	vii
Summary	ix
Table of Contents (condensed)	xiii
Table of Contents	xv
List of figures	xxi
List of tables	xxiii
Abbreviations	xxv

Introduction   xxix

PART I: OVERALL ISSUE AND METHODOLOGY

Chapter 1: Setting and topics at issue   3

1. Historical milestones   3
2. Radioactive waste in society as a political topic and reflection of value   9
3. Radioactive waste in science and technology: long-term dimension as a challenge to risk analysis   13
4. Complex socio-technical "radioactive waste system"   18

Chapter 2: Objectives and aim   21

Chapter 3: Research political embedding   25

Chapter 4: Issues under investigation, evidence and validation — 29

1. Theoretical basis — 29
1.1 Analysis "from below": risk perception of the public — 29
1.2 Analysis "from above": multi-dimensional decision strategy — 29

2. Concrete issues — 30

3. Methodical basis — 31
3.1 Principles — 31
3.2 Methods: historical-critical source analysis and content analysis — 32

4. Operationalisation of issues and utilisation of theoretical constructs — 33

5. Data gathering, evidence — 36

6. Validation — 37
6.1 Reliability — 38
6.2 Validity — 38
6.3 Open issues — 40

PART II: PERSPECTIVE "FROM BELOW": RISK PERCEPTION OF THE PUBLIC

Chapter 5: Insights from risk perception research — 43

Chapter 6: Risk perception in radioactive waste issues — 51

1. Transfer of criteria to radioactive waste issues — 51

2. Risk notion — 54
2.1 Risk definition — 54
2.2 Risk analysis — 55
2.3 Risk concern — 57
2.4 Empirical findings on risk — 59

3. Type of hazard — 61
3.1 (Amount of) damage potential — 61
3.2 Emergence/onset of impacts, time dimensions — 64
3.3 Scientific uncertainties/controversies — 64
3.4 Experience with danger — 67
3.5 Voluntariness and inevitability — 69
3.6 Individual controllability, damage defence — 70
3.7 Reversibility of actions — 71
3.8 "Commonplace character", familiarity — 72
3.9 Perceivability to the senses — 72
3.10 Empirical findings on hazard — 72

4. Societal context — 80
4.1 Benefit and responsibility: Possible strategies of main stakeholders — 80
4.2 Spatial and temporal risk distribution, concern, participation in procedure — 83
4.3 Degree of information, understanding, knowledge — 86
4.4 Credibility, trust — 87
4.5 Empirical findings on social context — 90

PART III: PERSPECTIVE "FROM ABOVE": DECISION PROCESSES

Chapter 7: Insights from decision research — 101
1. Preliminary remarks — 101
2. Decision, problem, information, and uncertainty — 102
3. Ill-defined problems — 105
4. Criteria of decision science — 106

Chapter 8: Development of decision making in technical systems — 113
1. Early days (until 1960s) — 114
2. Emergence of (probabilistic) risk analysis (1970s) — 116
3. Social rationality — 118

Chapter 9: Decisions in radioactive waste governance — 123
1. Transfer of criteria to radioactive waste issues — 123
2. System understanding — 125
2.1 Empirical findings on system understanding — 127
3. Avoidance of logical fallacies and biases — 128
3.1 Empirical findings on logical fallacies and biases — 129
4. Consideration and adaptation of problem structures — 130
4.1 Empirical findings on problem structures — 132
5. Decomposition into subsystems and re-integration — 136
5.1 Empirical findings on decomposition and re-integration — 137
6. Goal relations investigation, complex goals — 138
6.1 Empirical findings on goal relations — 140

7. Adequate management of diverse levels	142
7.1 Empirical findings on adequate-level management	142
8. Decisions under uncertainty	144
8.1 Empirical findings on decisions under uncertainity	147
9. Co-operation problem	148
9.1 Empirical findings on co-operation problems	149
10. Utilisation of latency periods as opportunities	149
10.1 Empirical findings on latency periods	150

Chapter 10: Siting as an example of sub-optimum decision making — 151

1. General	151
2. Siting of a low-level radioactive waste disposal facility	154
3. Siting of a high-level radioactive waste disposal facility	161
4. Conclusions with regard to siting procedure	167

PART IV: CONCLUSIONS AND FURTHER DEVELOPMENT

Chapter 11: Patterns of arguments in radioactive waste governance — 175

1. Document analysis (Objective 1a)	175
2. Patterns of arguments (Objective 1b)	175
2.1 Compilation of arguments	176
2.2 Findings	177
2.2.1 Decision-making process	177
2.2.2 Improvement of system	181
3. Comparison with decision strategies in other countries	185
3.1 Analysis by country	185
3.2 International organisations, research and development	192

Chapter 12: Fundamentals of a comparison of disposition options — 201

1. Variants as bases of decisions	201
2. Ecological, temporal and spatial dimensions	203
3. Ethical dimension	204
4. Societal and political dimensions	209
5. Economic dimension	215

6. Technical dimension: implementation	218
7. Discussion: on the search for a well-supported decision	222

Chapter 13: Integrated risk analysis: outline of an overall system robustness — **231**

1. Need for an integration of aspects: General remarks on dealing with dissenting views	231
2. Technical robustness	233
3. Societal robustness owing to an extended decision model	243
4. Approach to robustness of the overall "radioactive waste system"	264
5. Special role of the regulatory bodies	269
6. Requirements in view of an inclusive technical-scientific and societal discourse	274
7. Conclusion	276

Acknowledgements — **281**

References — **283**

1. Primary literature	283
2. Secondary literature	297
2.1 Methods (M)	297
2.2 Decision science, learning organisations, institutional aspects (D)	302
2.3 Risk perception (R)	306
2.4 Radioactive Waste: General, international (G)	311
2.5 Radioactive waste: Meta-analyses of decision processes (MA)	328

Index — **333**

List of figures

Figure 1-1	Perspective 1: political stalemate	10
Figure 1-2	Perspective 2: complex decisional situation	11
Figure 1-3	Perspective 3: mental models	12
Figure 1-4	Waste streams and main actors in the Swiss disposal concept	14
Figure 1-5	System properties of highly toxic waste disposal	17
Figure 2-1	Approach to an overall system robustness	23
Figure 4-1	Perspectives of the issues under study for the content analysis	30
Figure 4-2	Theoretical frame with constructions and auxiliary models	35
Figure 6-1	Retrievability with goal conflicts	79
Figure 7-1	Decisions in the feedback model of an institution	103
Figure 7-2	General model of the strategic decision process	104
Figure 8-1	Linear decision making	114
Figure 8-2	Sophisticated linear decision making including risk communication	117
Figure 9-1	Phases of the Swiss decision-making process	126
Figure 9-2	Divergent system units in problem definition	131
Figure 9-3	Perspectives of stakeholders: power utilities shareholders	133
Figure 9-4	Perspectives of stakeholders: proponent or waste implementer	134
Figure 9-5	Perspectives of stakeholders: safety authority	135
Figure 9-6	Perspectives of stakeholders: general public	135
Figure 9-7	Time dimensions for duties of disposal and liability	137

Figure 9-8 Sustainability based on protection and intervention 140
Figure 9-9 Classification of the key notion of "uncertainty" 146
Figure 10-1 Locations of potential sites in Switzerland 155
Figure 10-2 Approaches and mechanism of a siting selection procedure 168
Figure 11-1 Integration of relevant aspects into the official Swiss concept 185
Figure 11-2 "Learning curve" in participation 198
Figure 12-1 Sustainability of disposition systems: trade-off among eight dimensions.... 202
Figure 12-2 Waste management with a long-term safety and project character 204
Figure 12-3 Major stakeholders in the Swiss radioactive waste system 210
Figure 12-4 Recommended options to comply with sustainability 219
Figure 13-1 Technical robustness 235
Figure 13-2 Key elements of a "monitored long-term geological disposal" (EKRA) ..237
Figure 13-3 Decision process according to the current minimum legal procedure 242
Figure 13-4 Proposal of an integral and recursive decision-making process 243
Figure 13-5 Dynamic and pluralistic decision making 250
Figure 13-6 Societal/institutional robustness 259
Figure 13-7 Transfer of knowledge 262
Figure 13-8 Integral robustness 265
Figure 13-9 Overview of disposition phases, data gathering and responsibilities 266
Figure 13-10 Proposal for re-structuring the Swiss institutional setting 273

List of tables

Table 1-1	Technical and programme data base of Swiss nuclear waste production	...15
Table 5-1	General risk perception criteria	47
Table 6-1	Criteria for the perception of risk in the radioactive waste field	51
Table 6-2	Time notions for planning horizons in radioactive waste management	85
Table 7-1	Criteria to assess decisions and decision-making processes	106
Table 9-1	Criteria for decision making in the radioactive waste field	123
Table 9-2	Definition of the system and problem ranges	132
Table 11-1	Patterns and strategies of decision in radioactive waste governance	179
Table 12-1	Relations (and hierarchy) of consensus and dissent at diverse levels	214
Table 12-2	Disposition conceptions compared: implications of objectives	220
Table 13-1	Corner stones of third party involvement	245
Table 13-2	Complex waste governance: stakeholders, functions and responsibilities	258

List of abbreviations

AGNEB	Interdepartmental Working Group on Radioactive Waste Management (established in 1978)
AECL	(former) Atomic Energy of Canada Ltd. (proponent)
ANDRA	Agence nationale pour la gestion des déchets radioactifs (French waste implementing organisation)
ANS	American Nuclear Society
ASK	Abteilung für die Sicherheit der Kernanlagen (from 1983 HSK)
BaZ	Basler Zeitung (daily newspaper from Basle)
BBr	Brückenbauer (weekly magazine)
BNFL	British Nuclear Fuels Ltd. (reprocessing company)
BT	Badener Tagblatt (daily newspaper from Baden)
CC	Waste principle of "concentrate and confine" treatment
CEC	Commission of the European Community
CERN	Multinational nuclear research centre by Geneva
Cogéma	French reprocessing company
CORE	Swiss Federal Energy Research Commission
COWAM	Community Waste Management (EU projects, 1 and 2)
D	References pertaining to decision science, *etc.*, References, 2.2
DD	Waste principle of "dilute and disperse" treatment
DETEC	Federal Department of the Environment, Transport, Energy and Communications (UVEK, formerly EVED; Ministry for the Environment)
EKRA	Expert Group on Disposal Concepts for Radioactive Waste (national conceptual group advisory to the Federal Government, 1999–2002)
EU	European Union
FDI	Federal Department of the Interior

Federal	Usually: Swiss, pertaining to the Confederation
FOE	Federal Office of Energy (BFE, formerly BEW)
FOPH	Federal Office of Public Health (BAG)
FSC	NEA's Forum on Stakeholder Confidence (as to radioactive waste)
G	References pertaining to general literature on radioactive waste, References, 2.4
GBP	British pound
GNW	Cooperative for Nuclear Waste Management, Wellenberg (proponent, dissolved in 2003)
HLW	High-level radioactive waste (usually short for HLLW)
HLLW	High-level and long-lived intermediate-level radioactive waste
IAEA	International Atomic Energy Agency, Vienna
ICHLRWM	International Conference on High-Level Radioactive Waste Management, Las Vegas
ICRP	International Commission on Radiological Protection
LLILW	Long-lived intermediate-level radioactive waste (TRU, transuranic)
KARA	Koordinationsausschuss Radioaktive Abfälle (first conceptual body, 1975–1977)
KASAM	Swedish National Council for Nuclear Waste
KFW	Kantonale Fachgruppe Wellenberg (advisory committee to the potential host canton of Nidwalden, 2000–2002)
KNE	Kommission Nukleare Entsorgung (current geoscientific advisory body, from 1989)
KORA	"Conflict-solving group radioactive wastes" (1st mediation attempt in Swiss radioactive waste governance, 1991–1992)
KSA	Swiss Federal Nuclear Safety Commission (permanent Federal advisory body)
LLW	Low-level and short-lived intermediate-level radioactive waste (L/ILW)
M	References pertaining to methods, *etc.,* References, 2.1
MA	References pertaining to meta-analyses, References, 2.5
MCHF	Million Swiss francs (around 650,000 EUR)
MEUR	Million Euros (around 1.2 MUSD)
MIR	Radioactive waste from medicine, industry and research
MNA	Komitee für die Mitsprache des Nidwaldner Volkes bei Atomanlagen (regional citizens' group)
MSEK	Million Swedish crowns (around 100,000 EUR)
mSv	Milli-Sievert (unit for absorbed and weighted radiation dose to humans, 1 Sv = 1 Joule/kilogram)
Nagra	National Cooperative for the Disposal of Radioactive Waste (Swiss radioactive waste proponent, waste institution), Wettingen
NEA	Nuclear Energy Agency of the OECD, Paris

NGO	Non-governmental organisation
NPP	Nuclear power plant
NRC	US Nuclear Regulatory Commission, or: National Research Council
NZZ	Neue Zürcher Zeitung (daily newspaper from Zurich)
OECD	Organisation for Economic Co-operation and Development, Paris
P	References pertaining to primary literature, References, 1
PSI	Paul Scherrer Institute (Swiss nuclear research centre), Würenlingen
R	References pertaining to risk perception, References, 2.3
R&D (&D)	Research &Development (& Demonstration)
SAEFL	Swiss Agency for the Environment, Forests and Landscape
SES	Swiss Energy Foundation, Zurich (national anti-nuclear NGO)
SF	Spent fuel
SKB	Swedish Nuclear Fuel and Waste Management Company (implementer)
SKi	Swedish Nuclear Power Inspectorate
SoZ	Sonntags-Zeitung (weekly newspaper from Zurich)
SSI	Swedish Institute for Radiation Protection
STS	Science and technology studies
STUK	Finnish supervisory body
Subgroup	Subgroup Geology (Untergruppe Geologie) of AGNEB (1st geoscientific advisory body, 1980–1988)
SVA	Swiss Association for Atomic Energy, Berne
TA	Tages-Anzeiger (daily newspaper from Zurich)
tf	Comment by myself
USD	US dollar (around .84 EUR)
US DOE	United States Department of Energy
UVEK	See DETEC
WIPP	US Waste Isolation Pilot Plant
y/yr	Year
ZWILAG	Central Interim Storage Facility, Würenlingen

Introduction

Regarding the issue of radioactive waste governance[5] in Switzerland, apart from regulatory experts' reports, there have been prepared a multitude of technical, partly institutional analyses of the disposal concepts valid at the time[6]. Polls and surveys have been covering the field[7]. Intensive research revealed, though, that up to this study it has never been attempted to examine various and diverse ethical, technical, social and institutional aspects concurrently – in an integrated way, in order to explore the stumbling blocks on the way to a sustainable governance of radioactive waste. This is all the more remarkable since the Swiss Government has already alluded in their message of 1957 to an Article on Atomic Energy to be incorporated in the Federal Constitution: "Perhaps the answer to the question whether the atomic ashes [*sic!*] may be rendered harmless in a technically sound way will decide on the future utilisation of atomic energy" [P249:1142].

The "integrated" approach of this study shall not be equated to an "integral", let alone a "holistic" one (see Chapter 3); such an endeavour would be impossible under the given conditions anyhow. The voluminous reference body shall enable the reader, though, to trace the diverse arguments laid out and, if necessary, to pursue them further. Indeed, in the spirit of a "Sustainable Development of Switzerland" [P144][P145], there is not only need for research but for action as well [P145:55*passim*]. This need has lately also been identified internationally [G182:17-18][G193:3].

---

[5] See footnote 1 on page ix.
[6] [P41][P47][P48][P71][P43][P44][P165]. Numbers in brackets denote References (p. 270 *passim*) subdivided into specialised fields. Numerals following colons represent pagination. For abbreviations see List of abbreviations.
[7] [P103][P104][P101][P56][P142][P105].

PART I:
OVERALL ISSUE AND METHODOLOGY

Chapter 1

SETTING AND TOPICS AT ISSUE

1. HISTORICAL MILESTONES

Long-term or final disposal of radioactive waste faces major difficulties in Switzerland and internationally. The following milestones are noteworthy in the Swiss context but generically also typical for other countries[8]:

1957	The Federal Government statutes in their Message to the Atomic Energy Article in the Federal Constitution: "A task of eminent relevance lies in the formulation of guidelines about the disposition of unusable fission products (so-called radioactive waste, also called 'atomic ashes')" [P249:1152]. Corresponding regulations, though, are not issued in the Atomic Energy Act of 1959 (not until the Federal Decree of 1978).
1950s/60s	Waste from medicine, industry and research (so-called MIR waste) are "disposed of" via refuse collection, landfills or sewage. Dumping of solid waste is practised until the 1980s.
1963	The Federal Office of Public Health (FOPH) organises the first centralised collection. No account of their radioactivity is given until 1973.
1967	Local protest is raised against the first interim-storage project, called "definitive storehouse" (Lossy in Western Switzerland).
1968	The owners of NPPs under construction, Beznau and Mühleberg, enter into reprocessing with Cogéma (F) and BNFL (UK). No waste has to be taken back by the producers.
1969	After a major reactor incident at the Pilot Atomic Energy Plant of Lucens in the Western Canton of Vaud the project of a central storage in

---

[8] Precise referencing for each statement is undertaken in Flüeler 2002e, esp. Vol. II [G72].

	the underground reactor cavern is launched. In 1972 it is dropped due to both technical difficulties and severe opposition by the population and the Cantonal Government (State Council).
1969–82	Swiss radioactive waste is dumped into the Northern Atlantic Ocean.
1969	The first nuclear power plant, Beznau I, goes into operation.
1971/72	The Swiss nuclear community declares: "The waste problem has found solutions" (Swiss Association for Atomic Energy, SVA); "The radioactive waste problem is not of a technical type but above all of political and social nature" [P38].
1972	The National Cooperative for the Disposal of Radioactive Waste (Nagra) is founded, originally with regard to bearing an interim storage at Lucens. For the Confederation the Statute is signed by the Federal Office of Energy (FOE)[9] (which later on – in the spirit of a separation of promotion and protection in nuclear technology – was superseded by the Office of Public Health, FOPH).
from 1973	Extensive investigations are carried out, first with the aim of a sub-surface storage for low-level radioactive waste (in gypsum and anhydrite), from 1975 with the aim of a deep geological repository.
1976	An editor of SVA Bulletin notes: "It is only a few years ago that not the relevance of disposal of nuclear energy [*sic!*] but the corresponding necessary technological and financial effort has been grossly underestimated" [P286].
1977	The electricity-producing industry presents a first "Concept for the nuclear disposal in Switzerland" [P282]. In 1978 it is revised: The deadline of 1985 – for the "Project Guarantee 1985" with which the NPP waste producers have to demonstrate final disposal in Switzerland – is set due to the Federal Decree of 1978.
1978	The nuclear industry protests again, in the revised disposal concept: "The technical feasibility of the disposal steps altogether is secured, *i. e.,* the nuclear disposal is technically mastered"[10].
	The Federal Council mandates the ministry in charge "to judicably draw attention to the fact that the [nuclear power] utilities have to be decommissioned if the waste issue is not resolved in a concrete manner by the mid-80s"[11]. This is the key idea underlying the above-mentioned "Project Guarantee 1985".
1978–1983	The Swiss NPP operators place baseload contracts with the reprocessing firms Cogéma and BNFL, this time on condition of returning high-

---

[9] This was criticised by Seiler 1986 [P253:16] and Buser 1988 [P43:156].
[10] This assertion virtually is a *leitmotiv* of nuclear disposal in Switzerland as is conclusively demonstrated in Flüeler 2002e [G72].
[11] A corresponding passage is added to the operating licence of every Swiss NPP albeit with the supplement that the period may be "adequately" extended.

	level waste. In 1994 they acknowledge: "To maintain the operation of the [nuclear] power plants the [fuel] elements had to be transported ... to an interim storage. Such interim storage services were offered by the reprocessors" [P37].
1979	The majority of Swiss voters favour the Federal Decree on the Atomic Energy Act, according to which "the permanent and safe final disposition and disposal of the ... radioactive wastes" have to be "guaranteed". The polluter pays (causality) principle is stated[12] as well as the extension of the general licensing procedure requirement to disposal projects. The Decree is a so-called indirect counter-proposal of the Government to the (first) anti-nuclear people's initiative, which is rejected by a close vote. With the Ordinance on Preparatory Measures [P268] legal grounds are laid for geological investigations on disposal.
1980	Nagra schedules twelve drillings for high-level radioactive waste, which are rejected by the population in all but one municipalities.
1985	Nagra submits their reports for "Project Guarantee 1985". Rudolf Rometsch, President of Nagra, asserts in the media: "There admittedly exists the bizarre situation that one has to give technical-scientific answers to psychological-political questions. This is a crazy venture, which one should never undertake. It's just that Nagra does have no other choice in their present position."
1988	The Federal Council decides on "Project Guarantee 1985": The disposal of low-level and long-lived intermediate-level waste[13] (with the potential site of Oberbauenstock, for the geographical setting see Figure 10-1, page 155) as well as the safety but not the site of high-level waste (in crystalline host rock) are viewed to be demonstrated. The Government demands extension of investigations for high-level waste to sedimentary formations.
1990	Vote on the two people's initiatives "Moratorium" (10-year ban on the construction of NPPs) and "Electricity Without Nuclear" (for a stepwise shutdown of the nuclear power stations): Three quarters of all actual voters endorse the statement that "radioactive waste cannot be safely disposed of". Even the majority (54 per cent) of the ones who reject both anti-nuclear initiatives support this view [P101:26].

---

[12] Formally this principle only applies to NPP waste; the producers of MIR waste are merely obliged to deliver their waste to the government body in charge, the Office of Public Health. Actually, MIR waste is dealt with like NPP waste.

[13] For the reason of simplification: If "LLW" is referred to, actually L/ILW is meant, *i. e.*, low-level and intermediate-level waste. "HLW" denotes high-level and long-lived intermediate-level waste. For a specification see Table 1-1. The issue of the "short-lived" character of some waste types is dealt with in Section 10.2.

1993	The ministry in charge, DETEC, states in a letter accompanying the Bill of a further revision of the Atomic Energy Act: "The disposal of radioactive waste is an eminent national task of the years to come. ... Thereby it gets clear and clearer that the construction of such a repository is rather a procedural and a political than a technical problem" [P68].
1994	Nagra selects Wellenberg in the Canton of Nidwalden as their favourite LLW site. A separate company, the Cooperative for Nuclear Waste Management Wellenberg (GNW) is established. Nagra submits HLW preparatory investigation applications for the sites of Böttstein/Leuggern (crystalline formations) and Benken (sediments: Opalinus Clay). With the prospect of taking into operation the Central Interim Storage Facility at Würenlingen "the Swiss NPP operators agree – for the time being – not to place any further reprocessing contracts" [P37]. "In such a way a long time span may be bridged without having to yield to external constraints as was previously [with reprocessing?, tf] the case" [P181].
1995	The independent geoscientific expert group KNE (Commission on Radioactive Waste Disposal) judges crystalline host rock in Northern Switzerland to be "unfavourable". The electorate of Nidwalden rejects the GNW application for LLW exploration and construction licences at Wellenberg by 52 per cent [P217:2]. A survey at GNW's behest reveals a month later that over 60 per cent would have voted in favour of a submission for an exploratory gallery only and if the general concept had included controllability and retrievability.
1996	After a debate with the Federal Nuclear Safety Inspectorate HSK and KNE, Nagra decides to dislocate the planned crystalline investigations westward (to the Mettau Valley in the Canton of Aargau) and to restrict them to seismic tests for the time being. The preparatory application for a calibration borehole and other field trials in the Opalinus clay (of the so-called Zürcher Weinland or Zurich Vineyard Region, Benken site) in turn is granted for by the Federal Council. Constructing the Central Interim Storage Facility "has markedly abated time pressure" on Nagra's disposal programme.
1997	Charles McCombie, one of Nagra's directors, states at a scientific conference: "The common assertion that 'radioactive waste disposal is purely a political problem' is not true" [P193:33].
1998	Nagra postpones their investigations in the crystalline area. Disposal costs to date amount to *ca.* 775 MCHF (500 MEUR). No repository has been built yet. The members of the cooperative "restruc-

Setting and topics at issue

ture" Nagra, staff are reduced by 15 per cent, *viz.,* from 80 in 1995 to 60 collaborators.

The Minister of Energy initiates a so-called "Energy Dialogue – Waste Disposal" among the main stakeholder groups like authorities, proponents and NGOs, to compare, *i.a.,* various disposition concepts ("controlled retrievable long-term storage" *vs.* final disposal). Conclusively the Minister appoints an expert group "on Disposal Concepts for Radioactive Waste" (EKRA) [P60].

2000 EKRA proposes the concept of "monitored long-term geological disposal", an extension of the traditional concept of final disposal by integrating controllability and retrievability (see *Figure 13-2*) [P135].

Media response: "Way out of the impasse ... suddenly an end to the ongoing thick of battle for or against disposal ... in sight ... progress in radwaste debate ... first link".

The Cantonal Government of Nidwalden installs a "Cantonal Expert Group Wellenberg" (KFW) for their advice. After debating with KFW, HSK and the regional opposition group MNA, GNW elaborates a revised application for Wellenberg: It deals solely with an exploratory gallery; the concept of EKRA shall be implemented.

The ministry (DETEC) makes another attempt to revise the Atomic Energy Act: In the Nuclear Energy Act to be set up the disposal concept shall be codified on the basis of the recommendations of EKRA.

2001 Application by GNW for a concession for an exploratory gallery.

Total expenses of Nagra: more than 850 MCHF (560 MEUR) [P225].

2002 The Nidwalden electorate refuses to grant a licence for an exploratory gallery at Wellenberg by almost 58 per cent [P231]. The Cantonal Council, predominantly in favour of the submission, concludes that "by this, the site of Wellenberg as a potential repository for low-level and intermediate-level waste is definitely dismissed". GNW abandons the potential site. In a press release of the very day of defeat it states that the "operators of the Swiss NPPs have asked the Swiss Federal Government to provide for a political and legal environment which will enable them to solve the problem. ... The problem is a purely political one".

Nagra submits the project "Entsorgungsnachweis" (demonstration of feasibility and siting of disposal) to the Federal Council. The documentation is to demonstrate how and where spent fuel (SF), high-level radioactive waste (HLW) and long-lived intermediate-level waste (LLILW/TRU) can be safely disposed of in Switzerland. Around 2006, the Federal Government, on the basis of an extensive review, will have to take a decision on the further procedure.

Total expenditure of Nagra: 886 MCHF (575 MEUR), 34 MCHF for

2003, of which 5 MCHF for public relations and documentation [P225, 2003:35].
Total expenses of the regulator HSK (for all inspection, research, *etc.*, in all domains incl. NPPs): 28 MCHF [P131:I.2], of which 2.2 MCHF for disposal activities [P132].

2003   In a national ballot the Swiss electorate rejects two new popular initiatives on phasing out nuclear power: the initiative on a phase-out by 66 per cent, the initiative on a continuation of the moratorium by 58 per cent [TA, 2003-5-19].
Total expenditure of Nagra: over 920 MCHF (600 MEUR) [P225].

The pertinent question remains: Is it indeed merely a "political" problem that three decades after Nagra was established no long-term/final disposal has been realised in Switzerland? The facts presented above suggest that the protraction came about through a mixture of factors such as

- At all levels planning was suboptimal, *e. g.*, a sufficient legal basis was not laid down before 1979;
- The disposal concept showed serious technological shortcomings – such as no seismic data available before drilling. This was only done retroactively and correspondingly rectified.
- Opinions of third-party experts were only considered in a late stage (Subgroup Geology of the Interdepartmental Working Group on Radioactive Waste Management (AGNEB), KNE);
- The period of time was grossly underrated (scheduling of "Project Guarantee 1985", also need for interim storage[14]), due to both political pressure by opponents and waste producers.
- The structure of the "system" has to be revised.

These factors shall be analysed and supplemented below. The paramount topic of this study is to explore what the stakeholders involved, proponents from nuclear industry, regulators, experts and others might contribute to an integrated problem solving, maybe even to a "solution" itself. From this action-oriented perspective "radioactive waste" shall be first analysed as a societal issue (Section 1.2), then as a scientific-technical one (1.3) and, finally, in its combination (1.4).

---

[14] Originally it was doubtlessly assumed that reprocessing would substitute final disposal, *i. e.*, that the spent fuel shipped for reprocessing would not be returned. The first contracts of the 1960s did not contain return clauses [P253:21]. In 1976 the German Federal Government went as far as to stipulate reprocessing as a "harmless recycling" (para 9a lit. 1 Atomic Energy Act, which was changed in 1994 [G16:177*passim*]). See also footnote 72.

2. RADIOACTIVE WASTE IN SOCIETY AS A POLITICAL TOPIC AND REFLECTION OF VALUE

For diverse reasons, the governance of radioactive waste has always been regarded as a particularly difficult issue. As mentioned, the Federal Council focussed on it in their 1957 Statement to the Atomic Energy and Radiation Protection Article in the Federal Constitution: "… perhaps the response to the question whether the atomic ash … will be rendered harmless or may even be utilised productively will decide upon the way how to apply atomic energy in the future" [P249:1141*passim*]. When in the 1970s for the first time the issue was dealt with more thoroughly, the Minister of Energy of that time, Federal Councillor Willi Ritschard conceded: "Our great problems are the radioactive wastes. Therefore, a commission is working on it. Resistance against disposal, against these wastes, is substantial. But we still hope to find a solution" [P250:1667, by "commission" probably meant KARA].

In 1999, D. W. North, a long-standing expert of the scene, wrote in Risk Analysis, a journal renowned beyond the nuclear community: (High-level) "nuclear waste management has the deserved reputation as one of the most intractable policy issues facing the United States and other nations using nuclear reactors for electric power generation" [MA35:751]. The issue is, thus, far away from being "closed" in the science and technology studies sense. The situation is so muddled that even the originally technocratic International Atomic Energy Agency (IAEA) focuses on the topic, *e. g.*, in the seminal Córdoba Conference in March 2000: "In almost all of the Conference's technical sessions, there was discussion of the need to involve all interested parties ['stakeholders'] in the decision-making processes related to radioactive waste management" [G118:vi]. This need has been increasingly perceived recurrent as a virtually theme in recent years as is shown below.

According to the perspective chosen on the political and societal stage, various and diverse levels and basics become apparent in determining how to judge radioactive waste.

1st reflection

On the surface level (level 1) the current situation in the radioactive waste arena in Switzerland for some is such that it is solely a "political", *i. e.*, a technically solved, issue, for others it is an unsolved, even "unsolvable" problem. There is a stalemate between implementers and regulators (who have to erect and license disposal facilities respectively) on the one side and national as well as local opposition (who raise the waste issue to the level of energy policy and/or block construction) on the other side. The actual issues are repository projects and preparatory activities at

Wellenberg[15], in Benken, or in the Mettau Valley (see *Figure 10-1*). Two "camps" face each other: the camp of the implementers (Nagra, US DOE)/ safety authorities and the camp of (regional and national) resistance (see *Figure 1-1*).

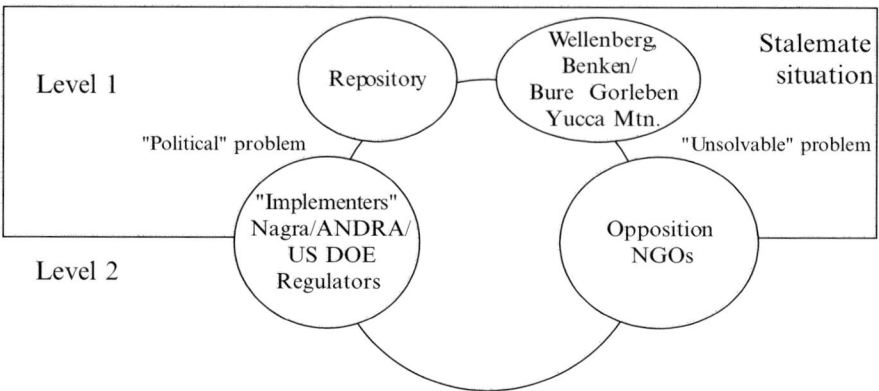

Figure 1-1. Political stalemate. In daily politics two "camps" face each other: the "implementers", above all Nagra (in the USA: Department of Energy (DOE), in France: ANDRA), and the regulators on the one side and the opposition on national and local levels on the other side. They are in a dispute over a project (repository) at an actual site (Wellenberg, Benken, or – for that a matter – Bure in France, Gorleben in Germany or Yucca Mountain in the USA).

2nd reflection

On a closer reflection one gets to Level 2: The complexity of a problem is defined by the number of its elements as well as their diversity and interrelations, which vary with time (particularised in Chapter 7). Decision-making processes as in radioactive waste governance are undoubtedly complex problem situations (see *Figure 1-2*). According to the actor model of Rohrmann 1991 [D72] and Wiedemann 1991 [D87] the implementing organisations (Nagra, once GNW) as well as the authorities, the environmental organisations and parts of the experts are institutional actors[16]; citizens' groups and other parts of the experts are non-institutionalised, social actors. Waste producers are institutional (NPP operators, indirectly Office of Public Health) and social actors (electricity consumers). The actors usually have ambi- or

---

[15] This potential site was discarded after the September 2002 negative vote.
[16] Consistent with the model the term "actor" is used here whereas later on it is replaced by "stakeholder".

Setting and topics at issue 11

even polyvalent relations; correspondingly the judgements are not plainly pro or con but there are different views on a technical project or a resource, respectively (assessments A, B ...). The consequences thereof are individual and institutional decision anomalies [D49:138*passim*], meaning – in simplified terms – ways to behave which cannot be followed by other actors (*e. g.*, regarding the principle of sustainability) (expanded in Chapter 5)[17].

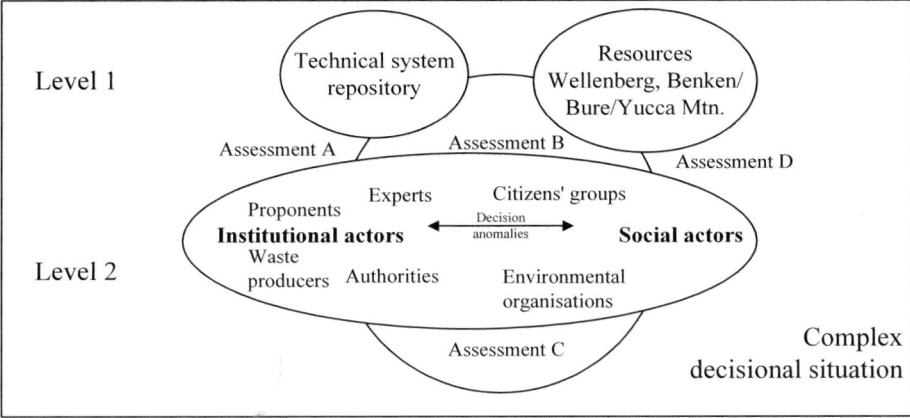

Figure 1-2. Complex decisional situation. On Level 2 a much more subtle decisional situation is displayed than was presumed according to Level 1. Both "camps" consist of diverse institutional and social actors who do not plainly accept or reject a project (generally: a technical system) at a given site (a resource). They carry out differentiated assessments (A through D), which might go wrong for traditional "camp (group) thinking".

3rd reflection

The base for decision anomalies (also for the stalemate situation in the 1st reflection) is the fact that the actors start out from diverse reality models whose grounds in turn are diverse rationality concepts[18]. Such interrelations become apparent only if Level 3, the level of "mental models" [D11], with the "basic underlying assumptions" according to Schein 1985 [D75], is considered (see *Figure 1-3)*. These assumptions make understandable how the

---

[17] Decision anomalies are "empirically observable (systematic) deviations of individual judgement and decision behaviour from standard assumptions of decision logics ..." [D49:1].

[18] Evidently there are various concepts of rationality. Rationality may be, *e. g.*, absolute, bounded (Simon 1955 [D76], Gigerenzer & Selten 2001 [D28]), social (Perrow 1984 [D67]) (see Section 8.3), communicative (Habermas 1981 [D34]), rational with regard to objectives (Weber ^{5}1963 [D84]), rational with regard to the system under scrutiny (Luhmann 1968 [D54]).

technical systems and their safety/risk analyses are erected and how they are assessed from diverse viewpoints. With respect to science a convergent approach was chosen by Kuhn 1962*passim* in analysing the social influence on scientific knowledge when he coined the term "paradigm". In this context, its sociological sense is pertinent, *viz.*, it is "the entire constellation of beliefs, values, techniques, and so on shared by the members of a given [scientific] community" [M54:175]. These considerations have a bearing on how to evaluate the uncertainty notion as well as expert and value judgments, paramount aspects in the issue under study as will be shown.

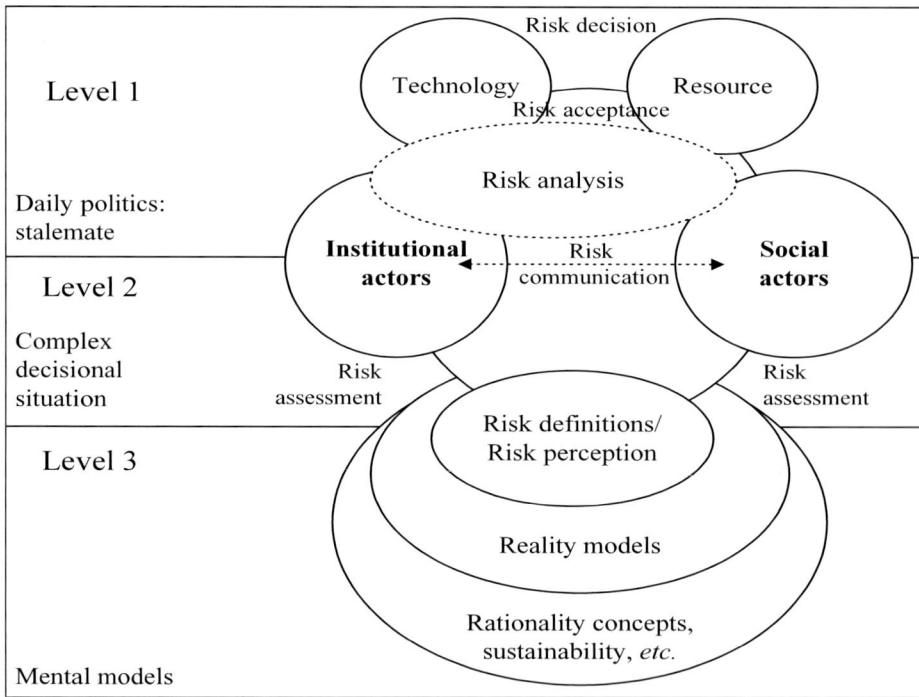

Figure 1-3. Mental models. Only Level 3 revealing the mental models of the actors (for their concepts of rationality, reality, or risk) makes it comprehensible that the concerned parties have particular perspectives resulting in different analysis, assessment, and decision-making with regard to technology, resources, and corresponding risks. Therefore, the "acceptance" of a risk, *e. g.*, due to disposal of radioactive waste, is not given. Significant divergences lead to a disturbed (risk) communication.

Out of the analysis of the diverse levels of actor perspectives one may deduce the following first working hypothesis, which is not trivial in view of the worldwide unsatisfactory governance of radioactive waste:

> Hypothesis 1: There are neither purely "technical" nor purely "political" perspectives in radioactive waste governance since its assessment necessarily consists of ethical, social, political, economical, ecological, and technical dimensions, which <u>have to</u> be integrated into an overall perspective.

The multi-dimensionality of the issue demands application of the principle of sustainability[19]. According to the Brundtland Commission 1987 "sustainable development is development that meets the needs of the present without compromising the ability of future generations to meet their own needs" [D90:43]. Sustainability, therefore, is based on two pillars: protection, *i. e.,* safety, and intervention potential, *i. e.,* the ability of today's and future generations to control (see Section 9.4). The decision on that is both scientific-technical and societal, amounting not only to a "co-production", *i. e.,* "the simultaneous production of knowledge and social order" as the STS community calls the interplay of various spheres in society [M44:393], but also to a "co-decision" on radioactive waste.

3. RADIOACTIVE WASTE IN SCIENCE AND TECHNOLOGY: LONG-TERM DIMENSION AS A CHALLENGE TO RISK ANALYSIS

The hazard potential of conventional toxic and radioactive wastes is defined by the toxicity of, in part, highly concentrated substances [G78][G156][G157]. Decisive are their potentially long-term effects [G23][G24].

For Switzerland, the waste streams are shown in *Figure 1-4* overleaf and the main data base is given in *Table 1-1* (on page 15).

---

[19] See sustainability dimensions in *Figure 12-1* and focusing of the principle in *Figure 9-8*.

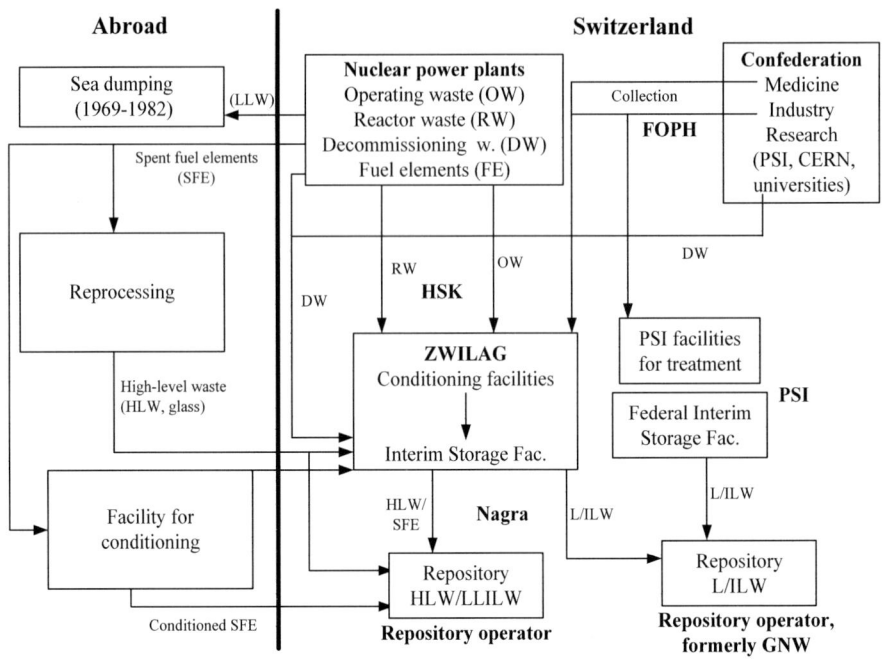

Figure 1-4. Waste streams and main **actors** in the Swiss disposal concept. The nuclear power plants generate about 80 per cent of the radioactive waste by volume, and about 99 per cent by toxicity; the remainder stems from medicine (mainly therapy and diagnosis), industry (*e. g.,* fluorescent paint, smoke detectors), and research. Spent fuel elements are either reprocessed abroad (to high-level active glass, HLW) or will be conditioned and afterwards stored in the Centralised Storage Facility ZWILAG, and then disposed of in an adequate repository. The waste from medicine, industry and research (so-called "MIR") is collected by the Federal Office of Public Health (FOPH), stored in the Federal Interim Storage at the Paul Scherrer Institute (PSI) and it is planned to be disposed of in the repository for low- and intermediate-level waste (L/ILW). Until 1982 L/ILW were dumped into the Northern Atlantic Ocean. All waste categories require a confirmation for final disposal by Nagra and a licence by the nuclear safety authority HSK. Special abbreviations are explained in Table 1-1 overleaf.

Table 1-1. Technical and programme data base of the Swiss nuclear waste production (Cont. overleaf).

Waste type (Facility type)	Disposition type	Waste origin	Dominant radio-nuclides	Activity (10^{15} Bq)	Time needed for isolation	Vol. (m³)	Res.	Gross volume (m³)	Host rocks (sites)	Political boundary conditions (el. power)
1978 (Disposal Concept by VSE *et al.* [P283])										5,000 MW$_e$ (5 times 1,000 MW during 40a)
(LLW) Waste category I	Caverns, near-surface	OW, DW, activated components	Co-60		HL <30a, some 100a	55,000		55,000	No natural isolation	
(LLW und ILW) Waste category II	Caverns, geological	Solidified waste, OW, industry	Sr-90, Cs-137		HL ~30a, 500a	<30,000		<30,000	Anhydrite, clays/marls, clay-slates	
(HLW) Waste category III	Boreholes or caverns	RW	<400a: FP >500a: Actinides		Some 1,000a, >10,000a	500		500	Rock salt, anhydrite, crystalline formations, clays/marls	
1985 ("Project Guarantee 1985" by Nagra [P201])										6,000 MW$_e$ (6 times 1,000 MW during 40a)
LLW (Type A)	Caverns	NPP decommissioning, MIR		<0.37		100,000	40,000		Alpine marls (Oberbauenstock), crystal-	

LLW/LLILW (Type B)	Caverns, mined	OW, MIR, RW 2-6 (main part)	Np-237, Ra-226	<5,550		60,000	200,000	line formations, anhydrite
HLW (Type C)	Caverns, mined	RA-1 (glass)	Np-237	<163,000	Some 10,000a	10,000 (ILW) 1,000 (HLW) [4,000 DFD]	11,200	Crystalline formations (Böttstein)
1993/1994 *passim* (after the governmental decision on "Project Guarantee 1985" of 1988 [P213][P214][P215])								3,000 MW$_e$ (today's plants during 100a at a maximum)
L/ILW	Caverns	OW, DW, MIR, RW-6	C-14	160	300-3,000a	100,000	150,000	Alpine marls (Wellenberg)
HLW LLILW	Caverns	RW-1 – RW-5, LLILW	Cs-137	30,000 200	~200,000	500 RW 100+ 4,400 DGD	500 100+ 4,400	Crystalline formations (Böttstein) Opalinus clay (Benken)

Abbreviations: a years, Actinides (plutonium, thorium, *etc*.), DFD direct final disposal/DGD direct geological disposal (spent fuel elements instead of HLW glass [RW-1] from reprocessing), DW decommissioning waste (from reactors), FP fission products (like Sr-90, Cs-137, *etc*.), HLW high-level waste, HL half-life period of radionuclides, ILW intermediate-level waste, L/ILW low-level and intermediate-level waste, LLILW long-lived intermediate-level waste, LLW low-level waste, MIR waste from medicine/industry/research, MWe Megawatt of electric power, OW operating waste, RW reprocessing waste, RW-6 (LLW from reprocessing)

Though the wastes are highly concentrated, their chemical and physical properties (no explosives, no criticality) are such that no catastrophic events have to be expected from a disposal site; in an underground (geological) site, a repository, there even are no big driving forces. The main mechanism is a low-level but long-term, chronic release into the environment; it may be described as a slow degradation of an open system with concurrent large uncertainties (see *Figure 1-5*). Such potential impacts are hard to detect with respect to location and time (except for some scenarios of human intrusion). These system characteristics make radioactive – and conventional highly toxic – waste disposal unique compared to other technical risks[20].

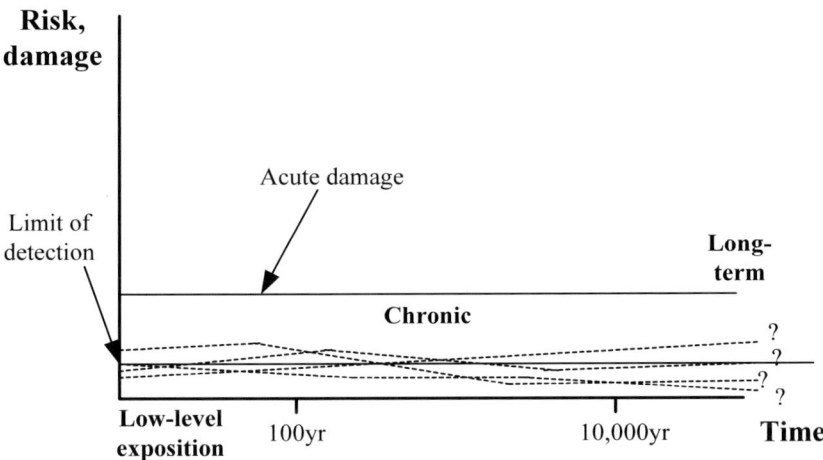

Figure 1-5. System properties of highly toxic waste disposal. The impact of the waste is potentially chronic (no peak, no acute damage), perceived to be "creeping" (badly traceable, partly below the limit of detection, low-level dose), and long-termed. Thereby, the actual low risk appears to be "smeared" over a long time span. No scale is given due to the generic character of the sketch.

These system characteristics lead to the admission that the required long-term safety "is not intended to imply a rigorous proof of safety, in a mathematical sense, but rather a convincing set of arguments that support a case for safety" [G182:11][G201:10*passim*]. Due to the required longevity of the disposal system, "[t]he aim of the performance assessment is not to predict

---

[20] Climate change risks are not dealt with. It has to be emphasised that the present study does not pretend to give a full life-cycle analysis but is focussed on indispensable fundamentals for a long-term decision with respect to the back end of the nuclear fuel cycle.

the behaviour of the system in the long term, but rather to test the robustness of the concept as regards safety criteria" [G114:245-246][21]. Following this, sophisticated disposal philosophy and design were developed, with succeeding technical barriers, review cycles and quality assurance [G129][G182].

The fact that long-term safety cannot be strictly demonstrated leads to a second working hypothesis:

> Hypothesis 2: A convincing assessment of long-term safety of a disposal system for radioactive waste can only be given in an extensive stepwise process and consists of technical elements (barriers), scientific methods (*e. g.,* uncertainty analyses) and comprehensive procedural aspects.

The present study attempts to explore both hypotheses and, in expanding the concept of robustness, to outline approaches of a corresponding decision-making process. Generally speaking, a system is robust if it is insensitive to significant parameter changes, *e. g.,* due to external influence. As a matter of course, robust procedures as defined in a narrow sense can only be achieved when the problem at hand is strictly technical[22]. Yet, the system characteristics of radioactive – and chemically toxic – waste are, as outlined, unique and technically complex. Since the long-term safety cannot be demonstrated stringently, even technical robustness cannot be treated, let alone achieved, as in "conventional" technical systems. Nevertheless, it is precisely the robust control systems that are designed to manage the mentioned uncertainties.

4. COMPLEX SOCIO-TECHNICAL "RADIOACTIVE WASTE SYSTEM"

Radioactive waste governance is characterised by all relevant features of a complex situation (of action) [D9:58-66], see Section 9.2. Many individual attributes have to be considered, often simultaneously and in an interconnected manner, *i. e.,* with all their side effects, long-distance effects of a technical, institutional and political nature. These interactions are not static

---

[21] According to the state of the art of radioactive waste risk analysis several phases of a "safety analysis" have to be covered; the procedure depends on the work stage and the integration of uncertainty analyses, from "performance assessment" via "safety assessment" to the "safety case". For details of the technical risk analysis see, *e. g.,* [G173][G174][G182:25*passim*][G183].

[22] The robustness of a system can only be tested if its parameters are clearly defined and if it is guaranteed that the system ranges within "predetermined bounds" (according to Weinmann 1991 [M86:33]).

but dynamic and extremely long-term. The system's own momentum with the technical, institutional and political subsystems is to be assessed in their development. Hereby, the situation is not transparent for the actors since they do not have complete information, they do not even know exactly which situation they are in at the moment. Uncertainties about the state of disposition increase with time, let alone about the social setting. The actors' assumptions as well as their state of knowledge on the structure of this complex socio-technical system[23] and their model of reality are likely to be defective and inaccurate to a great extent. In the following it shall be attempted to specify, describe and approach the open system of "radioactive waste", way beyond its "official" (mechanistic) definition given by the IAEA 1995[24].

In addition to the actors mentioned in *Figure 1-4* there are – as fully documented in Flüeler 2002e [G72] – a multitude of further concerned and affected parties at various stages and levels. Their institutional interconnection and functions are as follows:

Federal level

Politics
Federal Council: issues general licences, appoints commissions, decrees, ordinances.
Federal Parliament (House of Representatives, Senate): enacts laws, commissions tasks to the Federal Council and the administration.
Interest groups: nuclear industry, Swiss Association for Atomic Energy (SVA), national environmental organisations.

Administration
Department of the Environment, Transport, Energy and Communication (UVEK/DETEC): conducts licensing process, issues operating licences, position of the court of appeal.
FOE of DETEC (Federal Office of Energy): organises licensing process and prepares licences, issues regulatory guidelines.

---

[23] After Ropohl 1978/1999 a socio-technical system is "a system of action or a working system where human and technical subsystems constitute a unity" [M76:142-143]. See also Pidgeon 1991 [M68:131]. We examine the term of "complexity" in Section 9.2, into "system" in *Table 7-1*. Patterns and strategies of decision making are dealt with in PART III.

[24] "The objective ... is to ensure that within the Member State the components of a comprehensive radioactive waste management system are established. ... The use of the term 'system' does not necessarily imply a single centralized system for the Member State. Rather it is the summation of all the individual components, for example body of laws, regulatory organizations, operators, facilities, etc. that are required for the management of radioactive waste" [G108:4-5].

HSK of DETEC (Nuclear Safety Inspectorate): issues regulatory guidelines, issues clearances, technically prepares licences, supervises facilities, reviews proponents' projects.

Department of the Interior (DOI) with Office of Public Health (FOPH): is responsible for waste from medicine, industry and research (MIR), supervisory authority of the research centre of the Paul Scherrer Institute (PSI).

Interdepartmental Working Group on Radioactive Waste Management (AGNEB): is the co-ordinating body within the Federal administration.

Committees

Swiss Federal Nuclear Safety Commission (KSA): advisory committee to the Federal Council, "second opinion" to HSK

Commission on Radioactive Waste Disposal (KNE): advisory committee to the Federal Council in geoscientific matters

Cantonal and regional levels

Politics (*e. g., potential host canton or region*)

Cantonal Council (Government): grants licences under cantonal sovereignty.

Municipality Council: responsible for communal land-use planning (zoning)

Parliaments: launch laws and commission tasks to the respective Councils and administrations

Interest groups: Citizens' groups, *etc.*

(Cantonal) Administration

Responsible for legal fields under cantonal sovereignty (spatial planning, conventional protection of the environment, work safety)

Committees

KFW (Cantonal Expert Group Wellenberg): advisory body to the Cantonal Council of Nidwalden (dismissed in Sept. 2002 after the second negative vote on the potential site)

Local liaison committee(s): bodies at the potential sites (Wolfenschiessen in the case of Wellenberg, Benken in the case of the HLW programme) with representatives of the Federal Administration, the respective canton and the municipality)

According to policy science the "Swiss radioactive waste system" may be defined as a policy field [D89:21–22]. It contains specific interpretational patterns and values, instruments, a defined constellation of actors, forms of co-operation, and a knowledge system (see *Figure 11-1*).

Chapter 2

OBJECTIVES AND AIM

The mentioned interrelation of technical and non-technical aspects and the integration of knowledge in radioactive waste governance is the subject of this study. Its objectives are threefold[25]:

1. A reconstruction of the decision-making processes in radioactive waste management and governance on the basis of two Swiss case studies (siting procedures for a final disposal of high- and intermediate-level as well as low-level radioactive waste, respectively), knowledge integration of natural, technical and social systems (systems knowledge) (O1);
2. A presentation of decision-making fundamentals for disposition options – from surface storage to final disposal in the deep geological underground – on the basis of the principle of sustainability, definition of so-called target knowledge (O2);
3. The development of a concept of robustness of the socio-technical overall "system radioactive waste" with regard to an integrated, long-term, and transparent decision-making strategy, fundamentals for so-called action-related or transformation knowledge (O3).

ad 1.: The reconstruction is carried out by means of an empirical content or document analysis with the help of criteria from both risk perception and decision science research (Chapters 5 and 6, 7 to 11). The aim is not to give a full account of events but the focus lies on the analysis of the decision-making process with its obstacles, contradictions and boundary conditions.

---

[25] Originally it was intended to compare radioactive with non-radioactive toxic waste [G63]. This undertaking, however, failed due to its overcomplexity, *e. g.*, the technical and institutional boundary conditions already are completely different as was analysed in Flüeler & van Dorp 2000 [G78]. Even projects within the frame of the wide-ranging research in the European Union are restricted to (unrealistic) reference cases [G23][G24].

ad 2.: The proposed layout of disposition options (level 2) is normatively based on a specified principle of sustainability and empirically on the reconstruction of the decision-making process under investigation (Chapter 9). Fundamentals and criteria for a comparison of options are missing even internationally[26]. In this sense NEA 1999 remarked in their "Strategic areas of radioactive waste management": "... it should be helpful to examine how far the present concept of deep geological disposal would need to be modified to ensure retrievability/reversibility at several time scales" [G184:14].

ad 3.: The approach of an overall system robustness facilitates – on the basis of 1 and 2 as well as the risk analysis literature and systems theoretical insight – to combine technical design aspects, suitable analysis methods, and institutional backups in a dynamic procedure. It is meant to be an extension of the existing risk analysis and management methodology (level 3) (see Chapter 13).

The combination of the threefold approach may be graphically schematised (see *Figure 2-1* overleaf).

"Overall robustness", in a way, is a fuzzy notion but it recognises the complex socio-technical character of the issue and has the potential to stepwise and iteratively integrate structural and procedural/dynamic elements into the radioactive waste governance.

The aim is the following: Fundamentals of an integrated, long-term and transparent decision-making strategy shall be provided for the actors/stakeholders, like proponents of the radioactive waste industry, regulators, experts and otherwise concerned parties. Such basics shall come up to expectations stemming from the legal pretext of a "permanent and safe final disposition and disposal" [P39] as well as the sustainability requirement codified by the Swiss electorate in the revised 1999 Federal Constitution: "The Confederation and the Cantons strive for a permanently sustainable relationship between, on the one hand, Nature and its potential for regeneration and, on the other hand, its utilisation by man" (Art. 73) [P40].

---

[26] Attempts of multi-attributive benefit analyses are unsatisfactory [G92], also because the disposition options only distinguish between "early" and "late closure". Therefore, they do not conceptualise "extended" disposal as portrayed in Section 13.2.

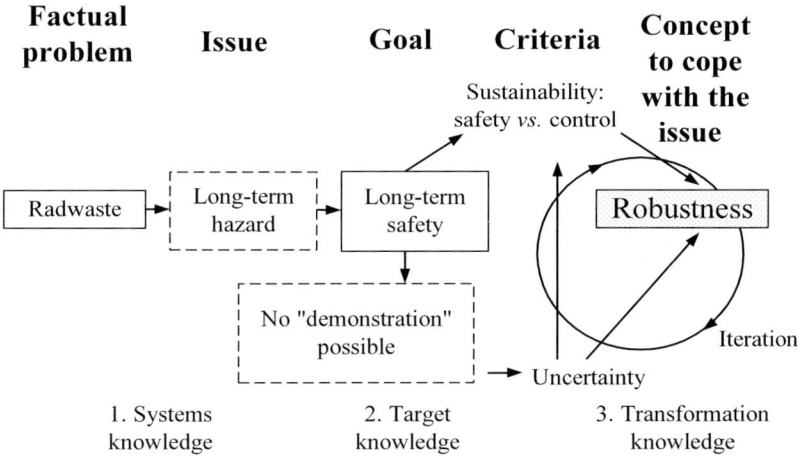

Figure 2-1. Approach to an overall system robustness. Radioactive waste is an existing problem to be solved. Disposal sites constitute a long-term hazard for man and the environment. Long-term safety cannot be strictly demonstrated because of various and diverse uncertainties resulting from the long isolation period required. The goal is a waste governance duly considering these uncertainties and complying with the principle of sustainability, *i. e.,* passive safety as well as active control. The concept of robustness attempts to fulfil the demands in a phased, iterative process and to create confidence in the analysis as well as trust in the system including the actors. It is also a matter of knowledge integration: from knowledge about systems (1st reconstruction) via expectations about goals and targets (2nd option analysis) all the way to the knowledge of how to implement the conceptions (action-related or transformation knowledge) (3rd overall robustness).

Chapter 3

RESEARCH POLITICAL EMBEDDING

"Waste", on the one side, is subject to scientific-technical analysis (with its chemical-physical properties), and on the other side it is defined as such in legal and societal terms – in contrast to valuable material or "resources". Since the study is problem (and thus action) -driven, scientific-technical approaches have to be incorporated into (social science) risk perception and decision research. This is a model, and transdisciplinary, interface of natural and social science[27]: The analysis in the environmental and waste field as to potential progress of material treatment and disposal technology is targeted at exploring the freedom of action of the proponents and other stakeholders, this via an amplification of rationality and risk models as well as institutional learning and decision research. The issue, indeed, is trans-disciplinary. And Weinberg, at that, was quite right when he termed, way back in 1972, "trans-scientific" those questions "which cannot be answered by science" [M84:209]. One would have to add: not by science alone, but also by science, and to be decided on by society.

In the end, the study should result in improved fundamentals for an adequate decision strategy. This is the core quality of this report. Waste governance is paradigmatically at the intersection between natural and (socio)technical systems. The overall goal is to generate relevant knowledge and its integration: systems knowledge (of the current factual status of the natural, technical and social aspects), effectiveness or target knowledge (what shall be achieved with systems knowledge? final state), and knowledge for action

---

[27] For that purpose I initiated, organised and directed a so-called "mutual learning session" on the topic at the International Transdisciplinarity Conference 2000 in Zurich [M24][M25]. On the definition of transdisciplinarity see Gibbons *et al.* 1994 [M31] as well as Nowotny and colleagues [M64][M65].

or transformation knowledge (how shall the objectives be met?)[28]. In the base study the disciplinary and transdisciplinary competence was assured with comprehensive communicative validation and reviewing (see Chapter 4).

Including current international research and activities[29] the approach, analyses and recommendations elaborated in Flüeler 2001a [G66], 2001b [G67], 2002 [G72], 2003b [G74], 2004 [G75], and Flüeler & Scholz 2004 [G77] are herewith transferred and focussed to an international level. The guiding principle is, *i. a.,* the specification of regulatory requirements, structures, and activities as outlined in the project idea submitted by a joint Swiss-American team within a call for anticipatory research by NRC[30] [G76].

The **Sixth Framework Programme of the EU** launched in 2002 holds in their Priority Thematic Area 2.2 "Management of Radioactive Waste" that "[t]he absence of a broadly agreed approach to waste management and disposal is one of the main impediments to the continued and future use of nuclear energy". And: "Research alone cannot ensure societal acceptance; however, it is needed in order to ... promote basic scientific understanding relating to safety and safety assessment methods, and to develop decision processes that are perceived as fair and equitable by the stakeholders involved" [G38:187*passim*]. This is the connection to the new Priority Thematic Area 1.7 "Citizens and Governance in a Knowledge-based Society". Here, "an integrated understanding" shall be created "of how a knowledge-based society can promote the societal objectives of the EU ... of sustainable development" [*ibid.*:44]. As a whole the present report also fits into the Sixth Environment Action Programme "Environment 2010: Our Future, Our Choice" [G37].

A survey of the international literature reveals that a multitude of technical, social, institutional and ethical analyses exist. Until Flüeler 2002e [G72] it was not attempted to examine the different and various aspects collectively, in a systematic and integrated manner in order to scan the obstacles on the way to a sustainable governance of radioactive waste. On behalf of the stakeholders – proponents, implementers, authorities, experts and the public – fundamental data and insights were presented for an integrated, long-term and transparent decision-making strategy which allows us to satisfactorily cope with the complex problem of safe disposal of radioactive waste. As for the intergenerational aspect, a paramount characteristic of radioactive waste

---

[28] The terminology in knowledge studies is not consolidated [R7][R11][R71][M48][M49] but the three types of knowledge are acclaimed.

[29] See Section 11.3.2.

[30] Call by the Nuclear Regulatory Commission (NRC) (delay 2002-06-01): See www.nrc.gov/what-we-do/regulatory/research/rsch-projects.html. NRC denied NRC urgency under this title by letter of 2002-10-25.

governance, the lack of systematic studies is noteworthy. A Panel of the US National Academy of Public Administration attending the topic in 1997 identified as next steps "a need for a body of case histories of specific hazardous waste sites to illustrate, categorize, and prioritize more fully the range of specific issues" [M67:16]. Other calls were raised to learn from disasters or further failures by systematically and broadly examining case studies, such as the OECD's 2003 report on emerging systemic risks [M66] and the European Environment Agency's 2001 late lessons from early warnings [M23].

If we look at the **Swiss context** in particular, the concept of sustainability was kept diffuse as mentioned. The principally laudable idea of the Energy Ministry DETEC to base its strategy on sustainability and "to thereby disclose goal conflicts and substantiate the value judgements made" gets blurred if various "sustainabilities" are mentioned [P278:18). The explanations are kept general: "Sustainability in energy matters means in detail: Ecological sustainability ... The safe disposition of nuclear waste ... economical sustainability ... The internalisation of external costs ... [*ibid.:*3,18*passim*]. No more help was the debate initiated by the Swiss Agency for the Environment, Forests and Landscape (SAEFL) on sustainability and sustainability research, respectively [P144][P252].

As a potential approach and strategy the Conference of the Swiss Scientific Academies (CASS) established productive propositions on "Research on Sustainability and Global Change – Visions in Science Policy by Swiss Researchers" [R11, emphasis original]:

> "Understanding **complex systems** requires more than an analysis of the parts." (Thesis 2)
> "Early identification and rapid understanding of unexpected environmental change require **specialized knowledge in unpredictable areas.**" (Thesis 6)
> "Sustainable development is a task to be resolved by all society, and science should supply the **systems knowledge, target knowledge and transformation knowledge.**" (Thesis 7)
> **"Our understanding of processes and interrelations,** in particular interactions between natural and man-made systems, needs to be improved." (Thesis 9)
> "Sustainable development needs knowledge which facilitates wording of concrete socio-economic **target notions.**" (proposition 12)
> Both the concept of sustainability and the image of humans and their position in nature require **ethical clarification.**" (Thesis 13)
> "Environmental research needs to study the interrelatedness of knowledge and action, and to make a greater contribution to **the application of knowledge.**" (Thesis 15)

"The public needs to be much more involved in the planning and realization of research projects – **participation** means ongoing implementation." (Thesis 16)

"Universities and institutions promoting research need to **adapt** their **structures** for sustainability research to become **inter- or transdisciplinary.**" (Thesis 17)

With respect to **disposal in Switzerland** "the issues relevant for decision" are *not* "sufficiently well covered" unlike alleged by the Federal Energy Research Commission (CORE) [P50] in their Research Concept for 2000–2003[31]. To the contrary, research topics should include technical and institutional controllability and retrievability, topics of "anticipatory" regulatory research as well[32]. Until now though, virtually the entire (technical) research concerning disposal has been left up to the proponent/implementer Nagra (this institution in turn commissioning orders primarily to the Paul Scherrer Institute). There is, up to the present, no regulatory research in this field [G95]. This lack or need has, in the meantime, been recognised by HSK[33].

---

[31] CORE even advocated a resource reduction by 50 per cent, *viz,.* from 8 down to 4 MCHF [*ibid.*]. The Swiss Federal Nuclear Safety Commission (KSA) opposed this proposal in their comment of 1999 [P179]. Indications of a need for concrete research had already been expressed a year before (KSA 1998 [P178]). See Section 9.5.1.

[32] See NEA 2001 for the distinction between "confirmatory" and "anticipatory research" [G188]. According to the US Nuclear Regulatory Commission (NRC) this latter type "arises from ... an effort to try to foresee where the NRC may need information to respond to future regulatory issues. If we wait until these potential issues become actual regulatory concerns, it may be too late to develop the technical information to respond to them in a timely fashion" [G208].

[33] In early 2002 a joint HSK/KSA project group named "Strategy on regulatory safety research" was initiated. Their proposal was put up for consultation [P289:200].

Chapter 4

ISSUES UNDER INVESTIGATION, EVIDENCE AND VALIDATION

1. THEORETICAL BASIS

1.1 Analysis "from below": risk perception of the public

The study tackles the problem from two theoretical angles: The first perspective – in PART II – originates from the concerned stakeholders, so to speak "from below". It adopts insights from risk perception research postulating that institutional decision-making anomalies of proponents and authorities may be diminished through an enlarged rationality and risk model by integrating the risk perception of concerned parties. The models are to be expanded by aspects of the risk notion, the hazard properties and the social context.

1.2 Analysis "from above": multi-dimensional decision strategy

The second perspective – in PART III – starts out from institutional decision makers, *i. e.,* comparatively "from above". It utilises insights from normative as well as empirical decision science and systems theory saying that complex decisions can only be made adequately if certain preconditions are given which fully represent the complexity of the respective issue: Understand the system sufficiently, recognise perception and communication aspects (see first perspective), avoid fallacies and biases, consider and adapt problem structures, decompose into subsystems and re-integrate, explore goal relations, treat different levels adequately, utilise latency periods as a chance for learning.

Figure 4-1 diagrams both perspectives under scrutiny ("from above", "from below"), which are chosen in radioactive waste governance, as well as the methods (risk analysis) and the goal (sustainability):

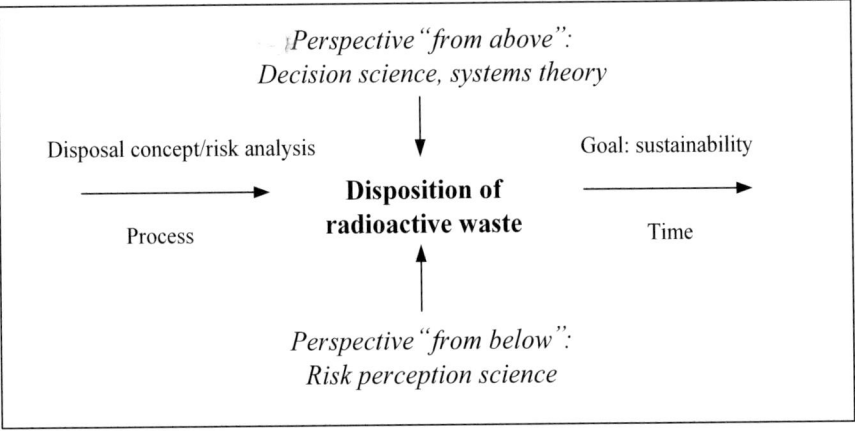

Figure 4-1. Perspectives of the issues under study for the content analysis. The approach chooses a parallel system- and process-orientation.

The theoretical constructs utilised will be presented when operationalising the issue in Section 4.4.

2. CONCRETE ISSUES

In view of the existing deadlock in the radioactive waste issue in Switzerland the assumptions derived from Section 4.1 are as follows:

1. The perspectives of the concerned parties have to date not been adequately appreciated in risk governance.
2. Applicants and authorities have largely acted according to a restricted rationality and, therefore, limited risk model.
3. The decision process has been linear and not multidimensional, *i. e.,* the problem complexity was not appropriate.

I emphasise that these preliminary assumptions are necessary but not sufficient. I do not intend to reach a false conclusion (of conversion) in which case the conclusion of a proposition would be false even if its premises were correct. To date decision science has not provided a uniform set of necessary conclusions from chosen hypotheses. Therefore, we cannot use it as guidance, particularly in the case of behavioural or situational decision processes like the one under scrutiny. This is to say: Even in case of "correct" and properly implemented recommendations the "success" of a safe and responsible governance of radioactive waste cannot be guaranteed – the process is not completed (there does not exist yet a technically and societally satisfactory disposal facility in Switzerland). My study refrains from falsifying the hypotheses as well: In view of the factual situation, falsification would be trivial if Hypothesis 1 presumed that the perspective of the concerned parties has been adequately considered. In terms of the theory of cognition it is the goal of the study only to explore and analyse the diverse arguments and their evolution in radioactive waste governance.

3. METHODICAL BASIS

3.1 Principles

There are two empirical ways to examine the assumptions under Section 4.2:

1. Analytical reconstruction of the decision process (by means of interview techniques or document analyses) or;
2. Participatory assessment and reconstruction by means of mediation techniques, *e. g.*, via focus groups or consensus conferences.

My choice was the document analysis (also called content analysis) due to several reasons:

1. Methodologically, the persons, institutions, and facts are not altered, reproducibility is in principle possible. As to the procedure I concede small chances of success for mediation at this time because of the impasse mentioned and because such techniques have been utilised several times and with little success [P169] [P168][P275][P64][P152]. I make a similar judgement on the probable outcome of other communicative methods like surveys.
2. For a content analysis there exists a solid and long-standing data base from which particular interests can be sifted with defensible

effort; if it is skilfully designed (see Section 4.3.2), it is equivalent to a survey or an observation but not critically dependent on the actors.
3. Besides directly analysed documents there is information from the surroundings permitting statements on the communication and decision process as such.
4. Additionally, the fact is challenging that to date practically no content analyses have been carried out on radioactive waste management[34]. Even technical experts propose the application of this method as did Thompson 1999 in a special issue of the Journal of Risk Analysis on the American repository of WIPP [G258:840].
5. This work is also, so to speak, a study on policy evaluation because the efficiency and effectiveness of the government and para-public course of action (authorities and implementer) is investigated[35]. According to evaluation research this study is a descriptive design of two longitudinal case analyses [D6:185*passim*].

3.2 Selection of methods: historical-critical source analysis and content analysis

Reconstruction of the argumentation in radioactive waste management requires a historical perspective. This approach heightens the awareness of time scales, concerning the disposal system itself (*Figure 1-5*) and the long project range (*Figure 9-7*). Historians work with the so-called "historical-critical method" which is qualified by three complexes of issues [M8:157*passim*][36]:

---

[34] Kraft and Clary 1993 [MA26] evaluated 1045 statements gathered from participants of hearings which were conducted by the US Department of Energy in 1986 on siting low- and intermediate-level radioactive waste in four US states. Each unit of analysis, *i. e.,* each transcribed statement, was analysed with totally 15 questions.

[35] "so to speak" is deliberately chosen because here it is not scrupulously differentiated among interventions (new policy measures) as preset by evaluation research.

[36] Source criticism with source description and text securing is of subordinate relevance in the present case because the access to – contemporary – documents is direct and the authors are interested in texts being attributable to them. The source interpretation, however, with regard to explanatory power and value of information is crucial: Intentions of the authors, situation of authors and addressees, circumstances and context have to be considered. The issue of "contemporary reality" is especially significant regarding the evaluation of site selection procedures (see Chapter 10). It seems that the science of history does not offer respective methodological instructions.

- Textual criticism: text authenticity with respect to authorship;
- Historical criticism: text relation to contemporary "reality/realities", part of this "reality", perspective of this "reality", reality and effectiveness of the text itself;
- Ideological criticism: political and world view of the author as well as the researcher.

The present study is a type of "retrospective technology assessment" which takes note of the historical dimension of technology development within technology assessment. A link between historic insights and current policy making may be constructed under certain conditions, namely according to Menkes 1977 [M59] if

1. "historical processes can be identified with or connected to policy decisions,
2. the processes in contrast to events are transfe[r]able,
3. the experience is generalizable, and
4. parties at conflict can be identified and isomorphic situations can be modelled" [*ibid.*: 324].

To guarantee transparency and comprehensibility of the reconstruction content analysis was selected as the empirical method. Methodological aspects, explanatory power and potential biases of the technique are exhaustibly dealt with in Flüeler 2002e [G72].

4. OPERATIONALISATION OF ISSUES AND UTILISATION OF THEORETICAL CONSTRUCTS

For the purpose of a theory-based formation of categories the issues in Section 4.2 are specified by the definition of variables (categories) which are deduced from two theoretical frameworks, *i. e.,* from risk perception and decision science. Theoretical criteria like perceivability or credibility are subsumed and have to be illustrated by means of indicators and anchor statements later on (see [M58]):

ad 1. and 2.: from risk perception science:
– main category "risk notion" (with the categories "risk definition", "risk analysis", "risk-'target'");
– main category "hazard" (with the categories "damage potential", "appearance of effects", "scientific uncertainties", "experience with hazard", "voluntariness", "controllability"/"damage defence", "reversibility", "'common place' character"/"familiarity", "perceivability");

– main category "social context" (with the categories "profit", "risk distribution"/"concern"/"participation in procedures", "degree of information"/"risk understanding", "confidence"/"trust").

ad 3.: from <u>decision science:</u>
– category "system understanding";
– category "recognition of perception and communication problems" (see above, set of variables regarding risk perception);
– category "fallacies and biases" (*e. g.*, the proposition of a technical solution leads to the solution of the entire problem);
– category "consideration and adaptation of problem structures" (ill-structured: not all elements are known, no clearly determinable "optimum" solution);
– category "decomposition into subsystems and re-integration" (*e. g.*, technical risk assessment/comprehensive long-term safety demonstration);
– category "exploration of goal relations" (*e. g.*, analysis of aspects of "sustainability" as a complex goal);
– category "adequate treatment of different levels" (*e. g.*, stakeholder groups);
– category "behaviour in case of delays" (*e. g.*, utilisation of latency periods as a chance of learning).

As theoretical constructs, these categories reflect the rationality concept in risk perception research and the complexity concept of decision science into which obviously the risk assessment method has to be integrated for the technical and scientific part. The socio-techno-scientific controversy of radioactive waste governance is analysed using concepts of deconstruction from science and technology studies (STS) [M14][M3]. Latour's 1987 approach of analysing "Science in action" comes in handy [M55]: The complexity and the ongoing process of the controversial issue entail looking at it "in the making", and this by its actors [*ibid.*:29,4,13,141]. The constructivist tool of "interpretative flexibility" [M13][M4] is employed to trace ways and arguments by various actors using seemingly the same terms and notions. It is investigated how far away the issue is of being "closed". According to Bijker *et al.* 1987 "closure" – in science! – occurs "when a consensus emerges that the 'truth' has been winnowed from the various interpretations" [M4:12]. To identify the actors (stakeholders) involved, the categories are sorted according to the actor model (Rohrmann 1991 [D72] and Wiedemann 1991 [D87]). Since considerations and decisions in radioactive waste governance are taken in organisations, institutional aspects receive special attention; these are analysed according to the resource concept in administration science (legislation, knowledge, time spent, human resources, organisation, financing, strategy), see Knoepfel *et al.* in Bussmann *et al.* 1997 [D6]. To explore interindividual rationalities, the concept of institutional anomalies is applied [D49]. Finally, the phase model of political science as a principle of order is applied for political processes, according to Windhoff-Héritier [D88][D89]. The concept of sustainability provides the normative central idea (see below). On the grounds of insights from risk assessment and system science the conceptions mentioned are merged to reach the study objectives (Objectives 1 to 3). Schematically the theoretical constructs as utilised to reach the three objectives present themselves as follows (*Figure 4-2*).

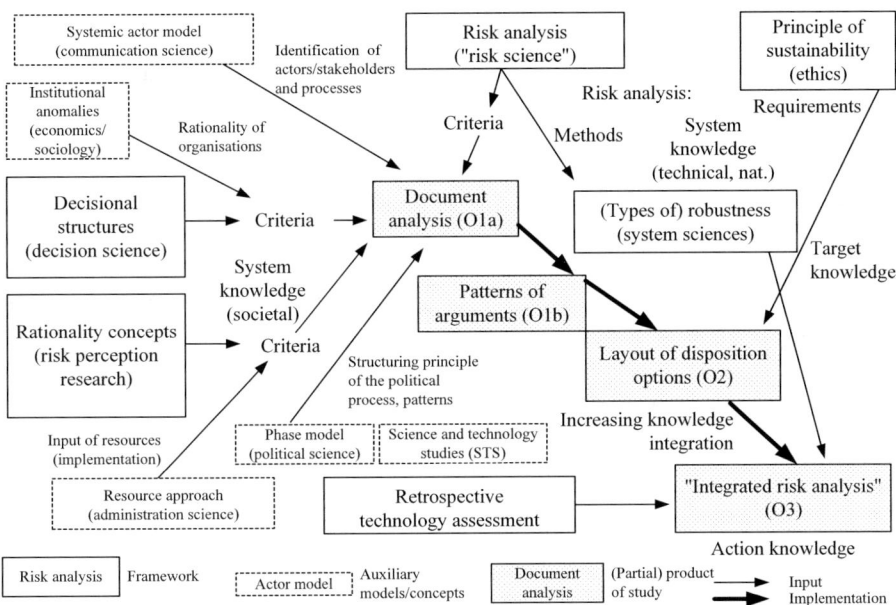

Figure 4-2. Theoretical frame with constructions and auxiliary models from different disciplines to reach the three objectives (O1 through O3). They are dealt with in depth upon the introduction of the constructs in the respective working phase.

With the exception of risk perception all theoretical constructs utilised have in common the character of being system- and process-oriented. This accommodates the approach of an "integral robustness" and makes it productive, as becomes apparent from *Figure 2-1* and as shown in Section 13.3.

5. DATA GATHERING, EVIDENCE

The research propositions determine, together with categorisation, the criteria for selecting the data:

- Documents dealing with radioactive waste management in Switzerland;
- International documents, if of interest to the Swiss case, and if constituting comparable analyses in other countries (also with respect to the twofold approach "bottom up"/"top down");
- Period in chronological sequence from 1956 to the present (with punctual reference to earlier texts);
- Authorship determined by the stakeholder origin;
- Institutional aspects also considered.

The collection of source material was phased and iterative: In a <u>first round</u> (1997/97), the primary material was selected according to the historic decision-making process in a document-oriented manner. Units of collection were safety reports, applications, experts' reports and testimonials, official enactments, licences, objections, minutes of parliamentary debates, relevant press releases, *etc.*

In a <u>second round</u> (1998), the relevant institutional and social stakeholder groups (see *Figure 12-3*) were examined for eventual additional documents for best mapping the communication processes:

Institutional actors:

- Applicants;
- Regulatory bodies;
- Decision-making authority;
- Parliaments;
- Waste producers;
- Expert bodies;
- Cantons (esp. potential siting regions);
- Local communities/municipalities (especially potential siting areas);
- Environmental organisations.

Social actors:

- Citizens' groups;
- Individual experts.

Other actors:

- Media.

In a third round (February 1999), the main stakeholders were addressed in a letter to get specific information, hints, proposals with regard to gaps, *etc.*

In a fourth round (January 2000), distinct exponents of stakeholder groups were approached for additional major documents.

Over a period of more than 40 years ((1945) 1956–2002) more than 2800 documents were sifted and appraised and over 2400 entries were made (see volume II of Flüeler 2002e [G72]). A full survey was carried out on three groups of publication series as well as parliamentary documents:

- "SVA Bulletin", the periodical of the leading pro-nuclear lobby institution, the Swiss Association for Atomic Energy (SVA), period since first publication in 1957 until mid-2002 (entries from more than 950 issues);
- "nagra informiert" (1979–1998), "nagra bulletin" (successor), "nagra aktuell" (1981–1990), "nagra report" (1991–1998), "nagra News" (from September 1998), Nagra-"FOCUS" (from 1998);
- "Energie + Umwelt", periodical of the Swiss Energy Foundation (SES), the leading anti-nuclear non-governmental organisation in Switzerland, since first publication in 1976 until mid-2002 (entries from over 80 issues);
- With the assistance of the Federal Office of Energy, nearly all inquiries in the Federal Parliament were collected, from 1957 to 2001 (over 200 motions, interpellations, questions, *etc.*).

6. VALIDATION

If content analysis is to be a method to map social reality, "inference" has to be possible, *i. e.,* the conclusion from text-internal to text-external features like the communicator, the recipient and the situation [M60:58*passim*][37]. If, in addition, the document analysis has to be restricted to the manifest text content, special importance is attached to the "basic underlying knowledge and assumptions" of the encoder: It has to be guaranteed that "latent content", *i. e.,* relations between the text and its interpreter, do not slip into the analysis [*ibid.*:56*passim*]. Furthermore, the explanatory power of inferences is limited – content analysis itself cannot conclusively reconstruct a highly complex process; interpretation has to rely on external criteria not stemming from content analysis [M29:43]. In the present case of an individual encoder

---

[37] Further considerations to methodological and cognition theoretical issues are given in Flüeler 2002e [G72:53*passim*].

– who, after all, is also acting in the process under scrutiny (see Preface) – internal (reliability) and external validity has to receive special attention.

6.1 Reliability

According to Krippendorff 1980, with sufficient reliability it is ensured that analytical results represent (an extract of) reality. It is thus a prerequisite of validity, which is to assure that the results "represent what they claim to represent" [M52:129]. With regard to principles[38], even though not by definition, in the validation discourse it is distinguished between temporal reliability (stability of the results), intersubjective reliability (intercoder reliability) and instrumental reliability (precision of the measuring instruments).

Stability is the weakest type of reliability and signifies that, *e. g.*, a coder intrasubjectively produces the same results at different times. Intersubjective reliability aims at the repeatability (duplication) of research results: Independent coders shall come to the same conclusions at different locations and times while using the same instructions for coding the same data set. With instrumental reliability, as the strictest method of demonstration, differences between the coder/measuring instrument and the (correct) standard are ascertained. Whereas Krippendorff requires reproducibility, Guba & Lincoln 1989 reject this positivistic approach and propose "dependability", a parallel criterion for the assessment of stability [D31:242]. Accordingly it has to be guaranteed that third parties are able to trace the research process as well as the coder's selection and interpretation.

Section 4.5 suggests – and Flüeler 2002e [G72] proves – that with the various "rounds" for data collection the stability criterion is met. The fact that, also out of resource reasons, I refrained from using a second coder is compensated by two counter-measures: 1. extensive citations (on over 600 pages in Volumes 1 and 2 of Flüeler 2002e), *i. e.*, a comparatively low degree of abstraction or a "closeness" to the respective authors, and 2. comprehensive, structured listing of patterns of arguments or "prototypes" by topical issues.

6.2 Validity

Krippendorff 1980 distinguishes among three major types of validation: A. data-related validity, B. product-oriented validity, and C. process-oriented validity [M52:156*passim*]:

---

[38] *e. g.,* [M52:129-268][M21:206*passim,*287*passim*][M37:198*passim*][M60:279-313[D6:220 *passim*][M58:109-115][M80:331-350][M29:171-174][M62:112-166].

A. Data-related validity: Semantical validity examines the logic of categorisation whereas sampling validity assesses the relation between sample and population serving the representativity of the data. Data-related validity primarily is to be determined in early stages of content analysis.

B. Product-oriented validity: Correlational validity depicts the degree of correlation between findings gathered by different techniques and, thus, substitutability of instruments. Predictive validity relates to the degree of agreement in the prediction carried out with different techniques.

C. Process-oriented validity: This type is also termed construct validity and assesses the relation between processes/categories and models. For a comparison, external information sources are brought in, such as results of similar constructs, experience with the context of the available data, related established theories and models, and expert opinions.

The study under question deals with validation issues as follows:

ad A.: The best possible mapping of the communication process was reached with a phased collection of data, first document-, then author-oriented. By way of an early reviewing (1997/98 for [G63] and October 1998 for [G64]) both domestic and international experts made coder-independent statements on the representativity of the raw data. For data acquisition and improvement the stakeholders were requested to indicate more documents twice (see Section 4.5). The phased abstraction should be traceable: from the raw data (documents) via the appendix (Volume 2 of Flüeler 2002e [G72]) to the chronological and issue-related patterns of arguments (Volume 1 in Flüeler 2002e [G72]).

ad B/C. Risk methodology was discussed and reviewed within the National Science Foundation project called "risk-based regulation" [G78]; English versions of the, hence enlarged, Chapters 12 and 13 were reviewed in an international journal [G66] and a conference (ESREL 2001) [G67]. With a so-called "mutual learning session" at the International Transdisciplinarity Conference 2000 various hypotheses, esp. from Chapter 9, were presented to an international range of stakeholders of diverse origin (the echo was positive throughout) [M24][M25]. In the circle of the Cantonal Expert Group Wellenberg (KFW) I got the chance to present parts of the content analysis and the patterns of arguments in the LLW field. For Section 11.3 a substantive and partially methodological comparison of international studies was carried out, with the following main question as a basis: Did the available and methodically diverse investigations lead to the same/similar conclusions (of inference)? Within KFW in 2000 to 2002 an extensive debate was held on the site selection procedure (Chapter 10) as well as the linkage of the components and subsystems waste–barriers–site–performance analysis–long-term safety demonstration (*Figure 13-3* and *Figure 13-4*). In the framework of the EU project Community Waste Management (CO-WAM) mainly the aspects of knowledge, site selection and stakeholder involvement were discussed in detail, from 2001 [MA33][MA10][G89]. By including actors, one can ensure "communicative validity" [M37:15,176*passim*,198*passim*][M58:112].

As to methodological rigour, it has to be emphasised that the technique of content analysis primarily serves the systematic examination of the documents at hand and not their rigid classification.

6.3 Open issues

By virtue of the complexity and longevity of the problem and because intercoder reliability was not tested, relevant issues remain unsolved, such as methodologically with regard to the uniformity of the category system or to eventual residual categories (which would have to be discussed in a debate on values), substantively with regard to conclusiveness and explanatory potential of interpretations and subsequent conclusions, particularly as the issue of radioactive waste governance, fiercely instrumentalised in energy policy, has not come to an end. From a STS point of view though, it is precisely this unresolved – "unclosed" – state of affairs which leaves the issue open, meaning: prone to analysis [M13].

PART II

PERSPECTIVE "FROM BELOW": RISK PERCEPTION OF THE PUBLIC

Chapter 5

INSIGHTS FROM RISK PERCEPTION RESEARCH

Serious risk analyses[39] are the basis of responsible risk management; the studies hitherto executed in this field are necessary but not sufficient. Ultimately decisive are the risk perceptions by the decision makers. These perceptions depend on various qualitative criteria as outlined below.

International risk research shows that decision anomalies are created if these criteria are not observed [D49] – the behaviour of individuals, but also of institutions, is (no longer) rational. The ones who do not agree, often – and prematurely – resort to the assumption that a particular argument was not "factual" anymore but "political". This may be, but does not have to be true.

The study attempts to, firstly, analyse reasons for this presumably purely "political" argumentation leading to individual decision anomalies and, secondly, point out approaches to a reduction of "institutional" decision anomalies (see Chapter 9).

In the course of the 1970s it was recognised that laypeople proceed heuristically when judging probability and damage, *i. e.,* they utilise intuitive "thumb rules of thinking"[40]. In this sense, an event is held all the more likely the better a similar incident can be remembered ("availability heuristics" according to Tversky & Kahneman [R89][R90]). This way *one* explanation of the differences between "objective" and "subjective" risks was found: Because of this "misperception" so-called "cognitive biases" with laypersons were detected. They had to be found with acceptance research and, in a

---

[39] Risk analysis is defined in an integrative way (*Figure 8-1*, *Figure 8-2* and *Figure 13-5*). For extensive references in risk perception research see Flüeler 2002e [G72]. For official definitions consult, *e. g.,* ISO 2002 [M42].

[40] There are diverse cognitive heuristics and biases: representativity, availability, anchoring, and adaptation. See [R89][R90][R35].

makeshift fashion, to be recovered with risk communication. An integrated risk analysis approach with an enlarged decision model is proposed in Chapter 13.

Even if the surveys in the so-called psychometric research[41] did not, as hoped, yield the stable factors of risk perception, it quickly became clear that risk assessment and appraisal – by laypersons *and* experts – is "inevitably subjective and that understanding judgmental limitations is crucial to effective decision making" [R81:182]. Neither is the term risk indisputably defined nor is there an objective risk [R58:126] nor a real "object" called risk [R34:169*passim*][42].

Availability heuristics as mentioned is, on the one hand, to be put into perspective in so far as laypersons do not estimate probabilities "worse" than experts [R46][R93], and, on the other hand, to be expanded as to both groups overestimate their own capability of judgement [R32][R22][R94].

The main difference between laypersons and technical experts lies in the fact that laypeople attribute a high value to qualitative criteria such as voluntariness of a risk taken, its controllability, distribution, hazard potential, *etc.* and, inevitably, reach different conclusions. Consequently, risk is not equalised with fatalities or frequencies of accidents [R46]; laypersons differentiate between one airplane crash with 100 fatalities as opposed to 100 car accidents with a casualty each [R40][R83]. In my view, it is a short-circuit to subsume "everything else" but the quantitative risk notion "that goes into laypeople's risk perceptions" as "outrage", as Sandman 1987 proposed [R70].

Not only general risk dimensions like voluntariness play a role but also the attitudes of those questioned towards diverse risk sources [G202] as well as towards the institutions in charge [R91]. Hence it becomes clear that risk debate cannot encompass all facets of the issue of risk acceptance or non-acceptance [G215][G260][G55]. Respective investigations gain importance with nuclear energy because this technology holds a special position in the majority of studies [R22][R45][R32][R94]. It is associated with a high, involuntary, uncontrollable, and not equitable catastrophic potential [R82]. This explains why Gamma-rays and ionising radiation from nuclear reactors are not equally perceived by laypersons (as opposed to experts) [R81:196].

---

[41] For references and critical appraisal see Flüeler 2002e [G72].
[42] This is not to connive at any type of subjectivism. Determined and defined risks are assessable in an objective and quantitative manner. Neither shall rationality as a concept of logics be brought into question; to understand other people the assertion of rationality is indispensable [R17:25], particularly with respect to action orientation as in the present case.

According to numerous empirical studies many people value the risks due to nuclear technology as unacceptably high, *e. g.,* [G55], whereas the majority of the nuclear community assess the probability of an accident with devastating consequences to be extremely low [G50:146]. Nuclear technology is coupled with economic growth, it is prototypical of a centralised large-scale industry [G260:96] and, therefore, a source of so-called "evolutionary risks" according to Krohn & Krücken 1993 [R43:21*passim*]. Evolutionary risks are difficult to compare, statistically insufficiently supported (little experience, safety scenarios are based on hypothetical assumptions), and characterised by great spans with regard to quality and extent of uncertainties. According to Luhmann 1990 a transfer takes place: from the (self-born, calculated) risk of the decision maker to the (imposed) danger for the concerned and by the decision affected [R50:150]. Additionally, it is not possible any more to distinguish between an objective risk and its perception by diverse observers – judgements are formed more and more subjectively. Evolutionary risks – Beck 1986 and 1988 termed them "large-scale hazards of late industrialism" [R5][R6:77] – are characterised as follows: They cannot be limited with regard to location, time and affected and concerned population; causality and liability cannot, in the end, be attributed to anyone; the irreversibility of potential consequences cannot be compensated [*ibid.*:76-77].

Polarisation is the consequence: On the one hand, there are the ignorant laypersons, on the other hand the arrogant, even corrupt representatives of industry and experts [R73][R19:94]. The groups remain by themselves, there is practically no exchange of ideas [*ibid.*:90]; to the contrary: Groupthink prevails [D41][D14]. One restricts information gathering to confirmatory sources from like-minded people. Krugman uses the term "incestuous amplification" which is defined by Jane's Defense Weekly as "a condition in warfare where one only listens to those who are already in lock-step agreement, reinforcing set beliefs and creating a situation ripe for miscalculation" (after [R44]). At least direct counterparts in the Swiss radioactive waste debate have to be diagnosed likewise, such as in the so-called "conflict-solving group radioactive waste (KORA)" in the beginnings of the 1990s when some mediation exercises were done after a partial victory of the anti-nuclear movement [P62][P63][P168]. This will be examined further in Chapter 6.1.

The criteria for risk perception may be classified according to Sjöberg & Drottz-Sjöberg 1994 [R76:42] into three major categories:

- Risk notion (risk definition, risk analysis, risk 'target');
- Hazard potential (damage potential and impact time, scientific uncertainties, experience with hazard, voluntariness resp. inevitability of

risk, individual controllability/damage defence, reversibility of decisions, 'common place' character/familiarity, perceivability);
– Social context (effective/perceived benefit resp. responsibility, spatial/temporal risk distribution/concernedness/participation in proceedings, understanding/knowledge, trust).

A review of the literature permitted a regrouping and an enlargement of the list. Thereby, the adaptability to radioactive waste management issues was given priority; the origin of the criteria was of secondary importance. Individual socio-economic characteristics of those questioned, like gender, education, age, income, psychic sensitivity, and personal abilities (without "risk training") in part correlate highly with risk perception; they do not modify, however, the statistical variance compared to the following three pertinent explanatory variables:

– General risk sensitivity (average of statements concerning non-ionising radiation);
– Pooled measure of the risk criteria or risk dimensions[43];
– Perceived risk of background radiation.

Sjöberg & Drottz-Sjöberg 1994 [G242:38] accordingly performed a study on risk perception of radioactive waste in Sweden. In a multiple regression it explains 65 per cent of the variance solely with these three variables. Likewise, Brehmer 1987 concludes from his results that the risk source attributes are more pertinent than the characteristics of the qualifying persons [R10]. Most of the criteria in *Table 5-1* overleaf resort to the seminal study by Fischhoff *et al.* 1978 [R23] and their numerous variations; the issue itself was originally addressed by Starr 1969 [R85][44].

---

[43] Sjöberg & Drottz-Sjöberg use the term risk "dimensions". Fischhoff and colleagues utilised the expression risk "characteristics" in their classical study of 1978 [R23:139*passim*]. The term "criteria" is chosen here because respective "properties" of the risk sources are the measures for the perceivers by which they assess the risks.

[44] As has been repeated: This is not the place to probe into the various risk perception paradigms and models. Some hints are given in Flüeler 2002e [G72:58*passim*].

Table 5-1. General risk perception criteria (Cont. to page 49).

Risk perception criterion	Elucidatory remarks[45]
1. Type of risk notion	
Risk definition	The notion is not unambiguously defined: probability, damage (also extent of precautionary measures), function of both terms, variance, semi variance, combination. For laypeople the product of 100x1 is not equal to 1x100. Contrary to experts they do not focus on mortality likelihood, all the same not regarding involuntary risks; with nuclear power the maximum damage potential is decisive. Risk is generally situative, societally defined and value-dependent. "Hypothetical" risks are indeterminate (based on the ignorance of potential risks), they are connected to the insecurity of technically facilitated catastrophic potentials. Probability is difficult to grasp by non-experts.
Risk analysis	Laypersons assess accident frequencies quite well: They are sensitive to frequentistic phenomena. Rare death causes, however, are overestimated, frequent events are under-estimated. Consequently, prudence is called for with probabilistic issues.
Risk "target" (concerns)	Interviewees judge nuclear technology to be a danger for a great number of persons whereas they do not feel threatened personally. Risks appear to be greater with personal concern (some "Not In My Back Yard, NIMBY" effect, but see below and page 119).
2. Type of hazard	
(Amount of) damage potential	The catastrophic potential was sensed to be decisive early in risk research. "Dread" [R23:133] and catastrophic potential increase the risk perception of nuclear power plants compared to chronic low-level x-rays. Managers in industry and petrol station owners judge the technical risk of liquid gas storage lower than local residents and environmentalists.
Emergence/onset of impacts	The issue is "immediacy" [*ibid.*] of the impacts after the initiating event, with or without prior warning.
Scientific uncertainties/controversies	The criterion was introduced as a risk "known to science" [*ibid.*], besides knowledge of the concerned people themselves). There is little long-term experience with industrial systems. The ways scientists see themselves leave their marks: Biologists, *e. g.,* assess risks differently than engineers. Experts may considerably differ in risk appraisals.
Experience with hazard (also history of incidents)	Direct confrontation and novelty have a negative impact on risk perception. Missing experience is substituted by conceptions of whatever type.

[45] Extensive references to every statement are given in Flüeler 2002e [G72:63*passim*].

Risk perception criterion	Elucidatory remarks
Voluntariness resp. inevitability	The fact that voluntariness positively influences risk perception is well known. Probabilistics play a role with risks voluntarily taken, the NIMBY effect does not.
(Individual) controllability, damage defence	The feeling of high personal control and controllability has a negative impact in the sense that my own actions are valued to be safer than others' ("It won't happen to me!"): "unrealistic optimism" according to Weinstein 1980 [R94] or overconfidence in one's own actions. Additional safety measures decrease the perceived risk for proponents and increase it for opponents.
Reversibility of actions	This criterion is the ultimate damage defence. It is based on responsibility ethics and sustainability. See Section 12.3.
"Commonplace character" of damage potential and familiarity of damage type	The initial studies used the term "familiarity" (as opposed to "dread"). Some researchers believe that risk perception increases with an increased coverage in the media [R12][R81], due to the mechanism of availability heuristics. This view is not concurrent [R74]. There is no clear – positive – correlation between acceptance and degree of knowledge [G260].
Perceivability to the senses	Non-perceivable issues are connected to other criteria (knowledge, information, experience, controllability) and are predominantly not familiar.
3. Social context	
Perceived benefit (incl. responsibility)	For Starr 1969 there is a direct and positive mathematical relationship between risk acceptability and benefit [R85]. In view of the risk approach it is evident that subsequent studies resulted in a lower value of benefit. The aspects gain relevance in the allocation of responsibility (polluter pays principle).
Temporal and spatial risk distribution, concern, participation in procedure	Chronic risks are judged more negatively. Spontaneous statements on the concern about future generations are rare. Personal concern partly increases criticism (regarding danger), partly makes more confident (towards risks taken voluntarily). Formal and substantive involvement in the decision-making process enhances acceptance and is demanded for ethical reasons.
Degree of information, understanding, knowledge	Personal knowledge and experience make risks appear positive. With regard to frequency and probability assessments experts and laypeople may lull themselves in a false sense of safety (see above). Both groups may be sceptical to data and investigations. Some authors question the benefit of information campaigns because these might promote overestimations of low probabilities [R12]. Others reveal lower errors of judgement for laypeople [R13] putting into perspective the "media effect" and its underlying availability heuristics. To make the media responsible for confusion in the aftermath of the Chernobyl accident seems unjustified [G226]. Affective and motivational aspects should not be ruled out in risk perception even though the

Insights from risk perception research

Risk perception criterion	Elucidatory remarks
	common explanatory models are cognitive-oriented. With nuclear power plant staff, knowledge emerges to be inversely proportional to their risk perception.
Confidence, trust	New evidence appears to be credible if it supports one's own views. Trust is also a strategy to deal with uncertainty (it is not the antonym of "mistrust"): Retrospectively it conveys safety out of a positive experience, prospectively it is associated with some sort of risky advance concession – or pre-commitment – to deal with the unknown. Trust amounts to yielding control. In this sense, it may enhance efficiency since information and transaction costs may be dropped. In long-term phenomena, mutual trust may even become a requisite for (process-based) rational action [R9:250]. Trust is looked at as a key notion in complex fields of technology because otherwise they are not tangible outside the expert community. Some view the "risk crisis" to be truly an "institutional crisis" [R38].

Chapter 6

RISK PERCEPTION IN RADIOACTIVE WASTE ISSUES

1. TRANSFER OF CRITERIA TO RADIOACTIVE WASTE ISSUES

Risk perception criteria as defined in the last Chapter are applied to the radioactive waste field in *Table 6-1*.

Table 6-1. Criteria for the perception of risk in the radioactive waste field. Possible approaches for future management are outlined. Explanations follow in Sections 6.2 to 6.4) (Cont. to page 53).

Risk perception criterion	Transfer to the radioactive waste field	Perspectives for radioactive waste governance
1. Type of risk notion		
Risk definition	Function of consequence and probability, protection goals for man and environment.	Emphasise consequences (work on high robustness of disposal design[46]).
Risk analysis	Mixed deterministic and probabilistic (long-term issues, presumably low individual doses).	Elaborate transparent consequence analyses/scenarios with probabilistic elements, sensitivity and uncertainty analyses; integrate time in risk function.

[46] For a comprehensive and systematic approach towards robustness see Chapter 13.

Risk perception criterion	Transfer to the radioactive waste field	Perspectives for radioactive waste governance
Risk-"target" (concerns)	Expand protection goals (one-self/individuals/others/population; ecosphere).	Discuss individual as well as collective doses, compare geochemical fluxes, examine biological indicators.
2. Type of hazard		
(Amount of) damage potential	(High) toxic potential extended to a very long time – the dread of radioactive waste disposal is viewed higher than the danger imposed by nuclear reactors [G248][MA22][R69]; chronic effects.	Follow rigorous CC instead of DD approach (material fluxes as short as possible); emphasise consequences (work on high robustness of disposal design); explain the notion of collective as well as individual doses; set up post-operational but pre-closure controlling programme to validate long-term safety; accumulate adequate financial funds for post-closure events; promote research in transmutation/partitioning[47].
Emergence/onset of impacts	Long-term creeping and chronic effects, albeit at a low level.	Only indirect demonstration of safe management possible (see above).
Scientific uncertainties/controversies	Unique systems, little experience about long-term behaviour and safety; certain conclusions from natural analogues and analogies.	See damage potential. Demonstrate suitability by means of external reviews.
Experience with hazard (also history of incidents)	*Do.*	*Do.*, analyse proactively contaminated sites (clarify differences and scientific progress); refer to competent performance in past activities.
Voluntariness resp. inevitability	Inter-/intragenerational issues, polluter pays principle.	Introduce option of controllability and analyse retrievability.
(Individual) controllability, damage defence	*Do.*, degree of "organised safety" [R91] in the institutions in charge.	*Do.*, strengthen safety authorities, intensify reviewing process (more "control" by third parties); discuss inconsistencies: safety/risks/resources; financial funds (see above).

[47] See Section 12.7.

Risk perception criterion	Transfer to the radioactive waste field	Perspectives for radioactive waste governance
Reversibility of actions	*Do.*	*Do.*, discuss inconsistencies as regards retrievability: safety/risks/resources
"Commonplace character" of damage potential and familiarity of damage type	Unfamiliar damage; delayed effects.	Build trust in experts and authorities: Ensure fair procedures and strive for competent waste handling in related projects/activities.
Perceivability to the senses	No perceivable risk.	*Do.*, refer to factual proof of competence and responsibility in the past (track record).
3. Social context		
Perceived benefit (incl. responsibility)	Electricity production: polluter pays principle.	Communicate message: Profiting generations have to take care; establish implementer's position as producer-independent as possible.
Temporal and spatial risk distribution, concern, participation in procedure	Inter-/intragenerational/procedural issues: Potentially affected persons are not identical with responsible persons.	Actively discuss responsibilities/potentials to intervene: political decision; discuss equity issues thoroughly, *e. g.,* geographic procedural equity: Do not narrow down site selection too early.
Degree of information, understanding, knowledge	Very complex and interdisciplinary reasoning.	Secure "complete" transparent and demonstrable information transfer, formulate explicit exclusion criteria in siting process.
Credibility, trust	Value-loaded symbolic character of the hazard potential of nuclear reactors, historic and factual relation to atomic bombs; loss of trust in conventional waste areas as well.	Establish implementer's position as producer-independent as possible (takes no stand in energy options); demand minimum and demonstrably controlled waste generation; ensure independence of regulatory body, also separate regulation from promotion (in agreement with international conventions [G103][G110]).

Just like other methodological approaches, such as technical analyses or risk communication, this one may not be utilised in a mechanistic way. But the review of literature and the passages above make it clear that this risk perception approach may be productive and, to date, has not been systematically applied to the radioactive waste field. Given the interconnection of the aspects, overlaps are preset, particularly with related criteria such as voluntariness and risk distribution or cross-cutting issues like trust and responsibility. According to the concept of social amplification of risk [R39] events and issues with signal potential influence other fields, in other locations, at different times without being directly related (*e. g.*, nuclear weapons, military legacies, chemical catastrophes, cases of corruption in the industry of public administration). See Section 6.3.4.

In the following Sections the respective issues, classified according to *Table 6-1*, are dealt with in the radioactive waste field and introduced and specified. At the respective end of the Sections 6.2, 6.3, and 6.4, the propositions are empirically backed up and illustrated by insights from the content analysis done for Switzerland [G72:Vol. II].

2. RISK NOTION

2.1 Risk definition

To show the complexity of the issue, it is useful and necessary to outline the strategy of the nuclear waste community. According to an internationally consensual technical definition, disposal is a nuclear installation for the final disposition of radioactive waste "without the intention of retrieval" [G107:18*passim*][48]. As a consequence thereof, the respective Swiss Guideline R-21, "Protection objectives for the disposal of radioactive waste", of 1993 states as follows: "The overall objective of disposal is ... that ... human health and the environment are protected in the long term ..." [P133:2]. If this aim, as in Switzerland, is to be achieved with geological disposal (= repository) the waste has to be "concentrated and confined" (CC principle) by way of a maintenance-free multi-barrier system so that in the long term no or just a small release of radioactivity is expected to comply with the protection goals. This presupposes a long-term system in full working order which, in turn, has to be demonstrated with safety performance assessments[49].

---

[48] Interestingly enough, "final disposal" or "final repository" are unused terms. By international definitions disposal extends to surface disposition such as in the case of the French Centre de l'Aube with an intended monitoring phase of 300 years [G7:21].

[49] This is not the place to cover the methodology of safety cases for radioactive waste disposal. Refer to, *e. g.*, [G174][G179][G182][G194].

In technical terms, risk is mostly defined as the multiplication of damage and probability of occurrence (per time unit). Due to the chemical and physical waste properties (no explosives, sub-criticality) and the multiple technical and geological barriers, with small driving forces, it is assumed that radioactivity eventually emitted does not cause direct acute but at the most stochastic long-term damages.

As opposed to other technical systems, an inappropriate behaviour of a (geologic) disposal site is not characterised by an instant failure of a component (except for some scenarios of human intrusion): The main mechanism is a low-level but long-term, chronic release into the environment, *i. e.,* a continuous system degradation. Special measures are not required because we have to deal with high technology or a physically critical hazard potential but because of the "unlimited" openness of the system. Consequently it has to be recognised that an annual radiological individual dose does not suffice as the one and only damage indicator, especially from the point of view of risk perception, widening the narrow notion of quantitative risk. The supposedly minimal dose value obscures the fact that its associated risk is, so to speak, spread or smeared over time[50]: The risk, admittedly low, continues to last over an unimaginably long period of time as depicted in *Figure 1-5*. This argumentation is carried on in Section 6.2.3.

2.2 Risk analysis

Such a long-term dimension entails taking into account probabilistic aspects in safety cases. The use of probability facilitates a systematic examination of uncertainties and variabilities. Geological systems are too complex though, too little predictable and unique to be tackled solely probabilistically [G48:521]. It is difficult, if not impossible, to assign probabilities of failure to specific events (like water intrusion or boring in x years from now) or components (backfilling in y years)[51]. To date only few comprehensive – "full scope" – probabilistic safety assessments for repositories have been

---

[50] This is somewhat reflected in the high scores for the exposure descriptor "Population at risk" as examined in Hohenemser *et al.* 1983 [R29]. The "Delay" hazard descriptor, though, does not admit an unambiguous interpretation [*ibid.*:379-380].

[51] It is fair to say that international radioprotection norms and recommendations are derived from risk considerations. This is also the case with the Swiss Guideline mentioned. Advantages and disadvantages of the respective deterministic and probabilistic approaches are subject of discussion in Flüeler & van Dorp 2000 [G78:7-18][G75].

carried out[52]. With this insight, most disposal proponents, including Nagra, prepare deterministic consequence analyses with an, ideally, enveloping scope of scenarios, which are supplemented by probabilistic considerations and calculations.

Such a strategy accedes the requests from risk perception: Probabilistic calculations are often hard to comprehend [M56] and comparably non-transparent because diverse types of uncertainty, variability and probability of scenarios are subsumed resulting in lost information. It is rather attempted to make plausible, with so-called "robust approaches" and conservative assumptions, that the disposal system is designed not to suffer dramatic failures. This may, as far as principally possible, satisfy the requirement of error tolerance [D83].

So-called "worst case" scenarios are damage-related. According to Elster ^{2}1997 they are "downhill" scenarios because they are confined to smooth and plausible transitions and because they do not allow "uphill" movements [D13:75*passim*]. A decision under uncertainty, like here, is more beneficial to the decision maker than "realistic" assumptions, which are difficult to comprehend and may be erroneous anyway. For the proponent and the investor behind this, "overbuilding" may be unnecessarily costly; external parties, however, welcome this approach, for traceability reasons and for the credibility of arguments. The issue is to rather make unfavourable assumptions in case of doubt and to be satisfied with a certain deterministic causality. It has to be said though, that the term "worst case" is infelicitous in so far as it suggests to have covered the worst scenario of all. If system information is so incomplete that unfavourable assumptions have to be made it is, in consequence, not possible either to localise the worst case. At any rate, an essential criterion of trade-off is the priority of pessimistic scenarios overoptimistic, or realistic, scenarios.[53]. Technical robustness is examined in Section 13.2.

Less rigorous is the utilisation of "conservativeness". Conservative assumptions are made if uncertainties in the calculation base have to be made up, such as in [P106:103*passim*]. While modelling, however, it has to be paid attention to that for all assumptions it has to be shown that they are, in the end, conservative [P116:2,13*passim*,18,37]. Conservativeness may not substitute insufficient system understanding.

Safety analyses have to be site-specific, including validations of models and data. Hence, the site selection is of eminent importance. One of the most

---

[52] The general considerations by NEA [G177][G179] and international reviews [G176][G178] of Canadian and US American performance assessments [G10][G11][G84] support this argumentation. See also SKi/HSK/SSI 1990 [G246:30*passim*].

[53] Jonas 1977/1984 termed it the "priority of the bad over the good prognosis" [M45:70*passim*].

problematic issues is that the selection procedure in Switzerland, especially for low-level waste, was not traceable. Since the careful handling of this aspect is viewed to be decisive it is discussed separately, *viz.,* in Chapter 10.

2.3 Risk concern

With regard to radioactive waste we have to deal with two crucial decisional issues:

- The "free rider" problem: Everyone wants the profit from an action or situation but have no disadvantage (this can be passed to future generations).
- A type of progressive "prisoner's dilemma": We, as beneficiaries, presumably are only marginally affected by the potential self-caused long-term contamination; we have to guarantee our descendants protection as well as freedom of action being aware of limited knowledge and communication on our side and the need for trust on theirs.

According to Luhmann 1990 the benefiting generations decide in favour of a risk, and this amounts to a danger for forthcoming generations [R50:152]: "The risk-taking behaviour of one person turns into danger to another one, and the difference of danger and risk becomes a political problem" [*ibid.*:161]. Research reveals that we have trouble exposing ourselves to danger caused by others [M77:119*passim*, 300*passim*][R81:196,205*passim*] and that we call in high protection measures [R27]. In spite of this handicap and the dilemmas mentioned we have to make a special effort on the behalf of future generations – there is no reciprocity in this case. Accordingly, it is of key importance to offensively launch an open debate on various and diverse protection goals in order to point out the bearing of their decisions to the beneficiaries of waste production.

Following international recommendations [G107][G175] the respective Swiss Guideline R-21 mentioned states that the "overall objective of disposal is ... that ... no undue burdens are imposed on future generations ..." [P133:2]. "The provisions for radioactive waste disposal are the responsibility of the present society which benefits from the waste-producing activities and shall not be passed on to future generations" [*ibid.*:3]. The individual person is to be protected, in protection objective 1 by means of an individual dose (0.1 mSv/yr), in protection objective 3 as a risk limit. The individual dose amounts to one tenth of the dose of 1 mSv admissible to members of the general public according to the Swiss legislation on radiological protection [P265], allowing additional sources of radioactivity in the impact area of a repository[54]. In accordance with a Swedish-Swiss advisory document [G246] doses have to be calculated until achieving their maximum value; contrary to that, the US American Environment Protection Agency (EPA) requires the demonstration of compliance based on cumulative releases of various radionuclides during the mandatory 10,000 years [G265].

Since we are concerned with uncertainties of several types over long periods of time it is advisable to approach demonstrating safety as a moving target along diverse pathways[55]:

- Conventional approach "individual radiation protection": An anthropocentric perspective is the paradigm of (traditional) radiation protection. It is assumed that human beings radiologically belong to the most sensitive organisms and their protection encompasses the protection of the entire ecosphere. In addition, the human as an individual is the protection goal whereas in biology "only" species or at least populations are to be protected. This is the line of reasoning of the regulatory bodies [P133:2] in agreement with the International Commission on Radiological Protection, ICRP [G127][G128][56].
- "Optimisation" approach: Indication of the total collective dose expected provides another decision-making instrument as one indicator of repository performance [G246:19] or to compare alternatives of disposition, like direct disposal *vs.* disposal after reprocessing [G251:3*passim*].
- "Activity" approach: Familiar from many fields of life and easy to see are restrictions of radioactivity releases as recommended in the "Nordic Flag Book" by the Scandinavian safety authorities [G224:434].
- "Biodiversity" and "sustainability" approaches: Following the concept of sustainability, the formulation of Article 1 in the Radiation Protection Act of 1991 has to be recognised whose purpose is "to protect ... the environment against dangers caused by ionising radiation" [P265]. SSI 1997 of Sweden recommends respective regulations [G251:5*passim*]. "Given the lack of a systematic and structured approach that has wide support", in 2000 ICRP set up a Task Group on Protection of the Environment with the purpose of developing a

---

[54] Allowing an individual dose of 0.01 mSv/yr "in a region" (but 0.1 mSv/yr to the critical, *i. e.*, the most highly exposed group), the Swedish National Institute of Radiation Protection (SSI) takes into account that the benefit from using nuclear energy is relatively short termed and that about 100 additional activity sources are possible to make full use of the dose limit for the general population [G251].

[55] Secondary criteria such as demands on waste and containments are not discussed at this point. They are integrated into the concept of robustness covered in Section 13.2.

[56] But see below and consult footnote 130.

protection policy and establishing a corresponding framework to be fed into ICRP's next recommendations [G130]. Since dose calculations for a very distant future are not conclusive Miller and colleagues propose to set release limits being a fraction of the natural geochemical material fluxes [G166][G167]. The underlying idea is that an eventual flux from a repository may not significantly modify the natural radiation background.

- "Toxicity" approach: Independent of a disposition design and a geological location, Kirchner 1985*passim* proposes a toxicity index by which the necessary isolation periods and retention factors may be calculated solely from waste properties [G150].

From 1994 the IAEA has accepted the challenge of expanding protection and has, *i. a.,* examined various safety indicators [G106][G109][G114] [G123][G126]. The issue is put forward within the 6th Framework Programme of the research in the European Union, namely under the titles of geological disposal and radiation protection [G57]. A concrete application is attempted in Section 13.2.

2.4 Empirical findings on risk

Some pertinent views and tendencies following the headings "general risk issues", "risk definition", "risk analysis" and "risk concern" are outlined below.

<u>Protection goals, risk targets</u>

Opinions differ as regards the fundamental question "how safe is safe enough?" [R23] between the "technical community" and the "lay public". It boils down to reducing an inevitably remaining "residual" risk by means of risk assessment or to requiring "absolute safety"[57]. The Bernese anti-nuclear movement "Aktion Mühleberg stillegen" (Phase out [the nuclear power plant of] Mühleberg Initiative) got to the heart of it in 1993 and maintained[58]: "Whoever ... demands upper damage bounds [as is done with the protection objective 1 in Guideline R-21] ... demonstrates in advance that one has to refrain from a safe and permanent containment of the radioactive elements from the environment. In the long run, an unknown number of people may

---

[57] See also context of *Figure 10-2*.
[58] This is made the notion absolute of a "smeared" risk in *Figure 1-5* and disregarding the fact that technical environmental protection in general takes damage mitigation as its basis, and not damage prevention.

be impaired over centuries. Hence, risk increases to an extent undreamt of. And, thereby, the requirements Nagra has to meet ... are lowered at the very start" [P98:6*passim*]. With regard to the potential site at Wellenberg, the anti-nuclear citizens' group "Stopp Wellenberg" follows up in 1995: "For a long-term storage with radioactive waste only the best possible solution is good enough" [P1].

Principle of waste management

The harmfulness of radiotoxic substances was detected early but the necessary actions were not taken for a long time. Even in 1963 the newly passed Radiation Protection Ordinance required a release of radioactive substances "to be washed down with plenty of water" [P266]. From 1969, altogether in the sense of this principle of dilution, low- and intermediate-level waste from Switzerland was dumped into the Northern Atlantic Ocean. Since man was the only protection goal, risk analyses, being at that time utterly anthropocentric, could defend this option over a long period of time [G106:11]. The requirement of confinement, already put forward at the UN's First International Conference on Peaceful Uses of Nuclear Energy in Geneva in 1955, was not adopted until the late 1960s; hence in 1969 Graf & Zünd demanded the "activity to be prevented from whatever dissemination into the environment during an indefinite period, as much as possible without surveillance" [P111:21].

Robustness/Uncertainty

Acknowledging that a strict mathematical proof of long-term safety for highly toxic waste is not possible the so-called "Co-ordinating Committee Radioactive Waste (KARA)" proposed even in 1976 to execute "experiments" in repositories [P154:14]. Buser & Wildi, in 1981, advocated "test facilities" [P47:114], Kasser *et al.* 1984 so-called "test final disposal facilities" [P155:4]. Corresponding actions were, however, if at all, not taken before the mid-1990s, *e. g.,* with the Belgian experiment named "Coralus" [G228]. The frame was given, even if not always explicitly and in practice, that a demonstration of safety could only take place in a phased, stepwise approach, with interim results. Nidecker of the Physicians for Social Responsibility (PSR), went a step further, in 1995, and gave, in the end, priority to the process element: "... proving the long-term safety of repositories is not a matter of the actual feasibility of a site but whether the proof submitted can be accepted by the institutions and the general public participating in the procedure" [P228]. Consistent with this, a joint position paper by the Swedish and Swiss safety authorities postulated: "The procedure of scenario development has to be well documented in order to allow for independent review" and "For public acceptance it is important that the site selection process is as transparent as possible" [G246:2*passim*].

Such traceability was not given for a long time and for two reasons: Factually not even fundamental issues were clarified (like the inventory), institutionally no external experts were consulted. Greenpeace wrote in 1993: "The 'safe final disposition' is out of the question.... Whether the regulations concerning radioactivity (HSK Guideline R-21) are complied with is not auditable because controllability of a finally sealed facility is not part of Nagra's concept" [P98]. Accordingly, and by itself consistently, the traditional concept of "final disposal" "without the intention of retrieval" has been criticised for long by the non-technical community. Hence, comprehensibility and reviewability materialise around the key issues of "controllability" and "retrievability" as well as participation of third parties ("independent"[59] experts and the public). This is discussed below.

As mentioned, the immanent problem of "non-demonstrability" of long-term safety has been tackled since the mid-1990s within the technical community [G106][G182]. The findings from the review debate display that the radiological protection goal as the only indicator has to be complemented also for transparency reasons because its calculation is based on complex modelling at a high level of aggregation. The search for so-called exclusion criteria is looked into below.

3. TYPE OF HAZARD

3.1 (Amount of) damage potential

Several methods have to be selected to substantiate the "robustness" of a safety demonstration because in risk perception the aspect of damage is the centre of attention yet the state of the facts is very complex and not easily accessible to non-experts.

Material flux perspective: strive for short fluxes

The nuclear industry advertises its case by pointing out the CC waste principle of concentration and confinement as opposed to the chimneystack policy of dilution by other industries. Nagra's bulletin of August 1997 reads as follows: "... wastes arising from energy production... are more or less predestined for disposal by confinement – because of the favourable benefit-

---

[59] "Independent" stands in inverted commas not to infer that there actually is somebody "independent". The term shall indicate the origin of knowledge or viewpoint that adds to multiperspectiveness and is used likewise from here on.

cost ratio (only a very small amount of waste is produced per kWh) and because the process of nuclear fission can be completely isolated from the biosphere. The resulting wastes thus arise from the very beginning under contained conditions" [P224, no. 30:12].

Save for the fact that a reactor emits radioactivity during normal operation via air and water (every energy converting technical system emits noxious substances) the official nuclear "disposition system Switzerland" is still based on reprocessing [P225:1996:14][60]. This technology is termed a "recycling path" in a 1994 position paper of the so-called Fuel Committee of the Swiss nuclear power plant owners: "With reprocessing, 96 per cent of the input materials are used for a second time [P37:2]. An investigation revealed that, to date, about 2.6 tonnes of plutonium have been separated out of the reprocessed spent fuel from the five nuclear reactors in Switzerland. By way of plutonium mixed oxide (MO_x) fuel elements 1.9 tonnes of that have been re-used, mostly in Beznau I and II, and since 1997 in Gösgen as well. Mühleberg and Leibstadt refrain from inserting MO_x so far [P91:84]. Plutonium accounts for about 1 per cent of a spent fuel element, uranium for 96 per cent [P37:1]. Only about one third of the plutonium in MO_x is potentially "burnt" [P91:54]. Not before 1997 did a Swiss operator, Gösgen-Däniken AG, decide to re-use reprocessed uranium.

Significant emissions are associated with currently practiced reprocessing. Reprocessing of Swiss fuel in the chemical factories of La Hague and Sellafield leads to an emission of radioactivity by a factor 1000 higher than in Switzerland [*ibid.*:24][61]. Generally, recycling is associated with lower, and not additional, pollution[62].

If the Swiss NPP operators rigorously pursued the CC approach they would have to give up reprocessing in the European facilities actually in operation – it prolongs the nuclear material flux and multiplies, in spite of efforts, the waste volume. During the separation process sludges, resins and deposits accumulate and increase more than sixfold [P91:36]. At least in the case of France, since 1991 by law [G227], these wastes have to be returned

---

[60] The option "direct disposal" has ever been "kept open" but never been thoroughly investigated [P180]. The initial March 2000 draft of the Nuclear Energy Act revision provided for a ban on reprocessing; during parliamentary debate, however, this was attenuated to a 10-year moratorium [P279].

[61] This is not process-inherent as the emission values of the Japanese facility of Rokkasho or the calculations for the German project Wackersdorf demonstrate [*ibid.*:22].

[62] NEA 1995 argues similarly: "In comparison with many chemicals the toxicity of radioactive substances is well understood. However, unlike some industrial chemical wastes, most of the radioactive inventory of nuclear wastes is the inevitable by-product of power generation by nuclear fission and, except in the sense of packaging into a small volume, is not very amenable to further reduction by recycling or process improvement" [G175:]. See also Flüeler & van Dorp 2000 [G78:14-15].

to the customer countries. Problems for the disposal operator could arise because he completely depends on the quality assurance of the reprocessing companies. Cogéma, operating La Hague, examines one or two out of 2,800 glass ingots, the number piling up during 40 years of operation of the five Swiss reactors [P91:55].

In 1998 Nagra was "restructured" in conjunction with a significant cut in staff and finances. It is determined to establish itself as a "competence centre" for radioactive waste. Also, with regard to the optimum design of the Central Interim Storage (ZWILAG) in Würenlingen, with the pilot plasma incinerator still under testing, the Co-operative could take the lead in managing the nuclear material flux in Switzerland. This proposal bears in mind that in their appraisal of radioactive waste management, the public not only evaluates a single project but a technology with its front and back ends [R96:169].

In the light of the hazard potential and the longevity of radioactive waste, the principle of concentration and confinement of waste was proposed in the 1950s and codified in the 1970s with the concept of disposal, internationally and in Switzerland. This was not upheld altogether, since from 1969 (until 1982) landlocked Switzerland participated in sea dumping as mentioned, with 10 per cent of the total activity [P140:18]. The discourse on concepts of disposition is given in Sections 9.6 and 13.2.

It is conceivable, however, that a massively developed reprocessing technology might lead, in combination with separation (partitioning) and transmutation, towards the "closed fuel cycle" aimed at. In spite of longstanding efforts, the technology will not be there in the foreseeable future, but in view of the long decision periods in radioactive waste management this strategy has to be considered particularly, since for the NEA it is "not justified" to favour direct disposal over reprocessing at present [G187:10*passim*]. In this perspective the decision pro or contra reprocessing depends on the direction and intensity of subsequent decisions, a dilemma highlighted in *Table 9-1*.

Risk comparisons

The approaches just presented do not directly relate potentially affected persons to the hazard but shall allow a review of the work done by the institutions involved. Direct comparisons with various risks are attractive, although this has to be handled with great care. If they are not good comparisons they meet with even greater refusals. In Flüeler 2002e some reflections are made regarding radon and radioactive waste [G72:78*passim*].

3.2 Emergence/onset of impacts, time dimensions

As regards time dimension one has to distinguish between the long-term effects of ionising radiation ("objective" time dimension) and the "institutional" time dimension meaning the period of the decision-making processes and control. We and our descendants have to deal with a chronic (non-punctual) and creeping (low-level) damage potential extending over a long period (*Figure 1-5*). Originally, in 1978, Nagra took "some thousand years" as a basis for isolation needed for high-level waste [P283:[21]]. For the project "Guarantee 1985" it noted "more than 10^4 years" as the retention potential of bentonite, an engineering waste barrier, and "the first radionuclides reach the biosphere after 1 million years" [P201:189]; the effectiveness of the technical barriers were specified as being "around 100,000 [years]" [P203:27]. The regulatory Guideline R-21 stipulates that the individual dose of 0.1 mSv be exceeded "at no time" [P133:4][63].

Concerning the "institutional" time dimension in "Guarantee 1985" it was until lately assumed that the spent fuel or the vitrified glass would have to be stored for 40 years after discharge from the reactor for the cooling down of the decay heat to a tolerable temperature for disposal. At that time, "the repositories of type C [for high-level waste] (and possibly for A [for low-level waste]) should be ready for operation at the raising of the respective waste categories"[64] [P201:20]. The IAEA points out the pronounced long-term character of the project "disposal" [G114].

3.3 Scientific uncertainties/controversies

We are confronted with a series of uncertainties in the context of radioactive waste. In addition to the uncertainties in the technical system, *i. e.*, in risk assessment, there are the political-societal uncertainties which play a role in the choice of concept[65]. Complex systems like human societies are evidently not stable as is demonstrated by the fall of the Berlin Wall only in 1989, although relevant intelligence reports dated a few years earlier did not give any indication of such a development. Besides there are more decisional uncertainties [R41][R1]: multiple goals (see Section 9.6) and

---

[63] It goes without saying that there are comprehensive modelling and scenario building between the indication of technical barriers and the compliance with protection goals.
[64] It is referred to reprocessing and decommissioning wastes.
[65] Evidently there is a variety of classifications, *e. g.*, integrating decision rules (assessment of risk, summation and financial measures as well as utility functions) [D20:7-23]. See also Section 9.8.

legal indeterminacies, *e. g.,* with respect to protection goals (see Section 6.2.3).

Even in the technical field there are manifold uncertainties. I distinguish among models/concepts, scenarios and data/variability of measurements (see Sections 6.2.2 and 9.8)[66]. One example shall suffice, *viz.,* the change of the geological view of the crystalline basement in Northern Switzerland which is traced back to a knowledge extension within a few years :

- 1983: The crystalline basement is described as undisturbed and gently dipping towards south [P199:73, Fig. 14].
- 1985: After seismical investigation and drilling of six deep boreholes the so-called Permo-Carbonic Trough is discovered and characterised as a simple tectonic rift or *graben* (ditch) structure [P202:Fig. 3.8].
- 1985: Integrating regional geology shows that marginal faults of the valley were activated to thrust faults, resulting in a supremely complex pattern [P184].

Technical people sometimes hold that (portraying of) complexity undermines the intentions: "Complexity kills confidence". I come to a different conclusion. As it is in the treatment of risk concern and protection goals an amplification of the spectrum of methods and a proactive discussion of uncertainties inevitably lead to an increase in redundancy and diversity – in the sense of a scientific "defence in depth" [G100][G117]. This is a means of achieving more reliable and secured knowledge leaving, in the end, more room for error tolerance [R61:83]. According to Sprecher & Turner 1991 "uncertainty is a factor that cannot be overstated" [G253:771], probably out of negative experience in the USA [MA27].

Convincing arguments will be necessary particularly when scrutinising long-term phenomena because, as was repeatedly mentioned, a strict "proof" of safety is not possible. Observing and measuring "hot" material during a sufficiently long period in underground rock laboratories of the "second generation", with test galleries, will render valuable services, as done in Belgium and planned in France [G138:307][G228:22*passim*][MA24:94*passim*].

An institutional method to improve the concept of risk analysis is reviewing, the re-examination of activities and institutions by external parties. By this, scientific deficiencies and differences may be openly addressed [R63:128], and the risk of suppressing dissenting opinions which might offer novel approaches is minimised [R67:65,200*passim*][R96:168], the chance of taking good, or better, decisions is increased [R96:169]. In Swiss radioactive waste management it was not until 1980, after the discussion with a group of distinguished geologists [P5:3*passim*], that the so-called Subgroup Geology

---

[66] See also Rowe 1994 [M78], Flüeler & van Dorp 2000 [G78], and Zio 2000 [G282].

of AGNEB was set up, having an industry interest-free second opinion [P8:X]. Its successor was, in 1989, the Commission on Radioactive Waste Disposal (KNE) as a sub-committee of the Federal Geologic Expert Commission [P11:17]. This decision also signified that a reviewing body was administratively removed from the Ministry of Energy, even though its members are still nominated by this minister. Both bodies considerably stimulated the process of site selection in Switzerland [P274][P165]. If the terms of duty and the activities are transparent and independency is guaranteed, also peer-reviews may be of use [D43][D81]. Today internationally reviewing is considered mandatory and meaningful [R18:192][R88:344][D10:1754*passim*][MA12][MA13][G279][G93].[67]

Often it is not until "concerned" groups provoke a change of thoughts by persistent questioning and pressure because other stakeholders are ponderous and manifold interconnected[68]. Only the disclosure and publicity of the toxic waste scandal around Love Canal in the US state of New York paved the way to change the official waste policy [R55:349]. Not before the chemical contamination of the Rhine at the Schweizerhalle incident of 1986 a corresponding Swiss ordinance was tackled [P57:3]. The accident of Three Mile Island (Harrisburg) in 1979 and the catastrophe of Chernobyl in 1986 forced the "human factor", including safety culture, to become the new centre of attention in reactor safety [D65][D66][D51][D67][G100].

Since the authorities have a leading role to play in radioactive waste governance they have to structure the discursive risk debate. Ravetz 1980 goes as far as to say that "the problem of risks is not so much one of decisions as one of regulation" [R68:52]. Evers & Nowotny 1987 conclude: "Hence, the controversy over the institutionalisation and regulation of the progress of technological knowledge is, at its heart, also a controversy over the possibility of democratic rules of living together; and it reveals the inadequacy and the crisis of those traditional types of democracy which were based on the circumstance that the issue on the direction and generation of progress was not yet publicly and politically disputed" [D15:247*passim*][69]. Such an open approach presents a challenge to the management and governance process [R61:8][G253], also formally, by having to duly

---

[67] One of the main reasons for Nirex's failure in their siting search was probably insufficient reviewing [MA24:50-82][G276].

[68] According to Peters *et al.* 1990 so-called critical scientists and citizens' groups challenge the "established" actors and are a "warning system" as well as a "counterbalance" to them [G220:132*passim*]. According to the US DOE "public participation is essential to identify problems, ... generate additional alternatives and solutions, ... and mediate differences among competing interests" [G53:1345].

[69] One of the main reasons for the setback experienced in the USA programme was the disregard of these considerations, see, *e. g.*, [MA25:164].

consider expert and dissenting opinions in time [R65][MA12:1741][G161]. Even the US DOE speaks of the necessity of "strategic planning ... in a dynamic environment" [G252:1559].

For the Swiss authorities the embedding of diverse opinions is important all the more "regulation of waste management in Switzerland is done by a very small number of persons within the Energy Department and the Nuclear Safety Authorities (less than 10!)" as Charles McCombie, then director of Nagra, emphasised in 1994 [P190:242]. "They are backed-up by a small team of consultants but they undertake almost no independent programmes of data collection" [ibid.]. This holds for 2004 still[70]. Thus, the administration is challenged with respect to novel attempts to solutions. For Canada Ballard & Kuhn 1996 [MA2:823] propose a "three-party liaison group" with an "impartial mediator committee" and two other committees each to explore the pros and cons of a potential site. All bodies would be composed locally and would dispose of an equal budget for expertise and public relations. Such a procedure would meet the requirement of evidentiary equity of the interested public and professional actors which has been claimed by social risk research for a long time [R38:183][R56][R57][G28]. In search of an interim storage facility in the USA, so-called "Strategic Principles and the Draft Mission Plan Amendment Workshops" were held with public participation, to serve the formation and formulation of a respective programme [G252:1563][G18:1918*passim*].

It is not meant to produce foolproof guidelines which could pave the way to an ideal solution of the radioactive waste issue. But the proactive and process-oriented engagement of the authorities is paramount because the preparation of a long-term responsible concept lies in their hands [R20][R37] [R62] (see Section 6.4.4). "Some of these interventions require abandoning customary practices, perhaps sacrificing some short-term efficiency for a chance at greater long-term efficacy", as Fischhoff 1990 wrote [R20:327] (see Section 9.10).

3.4 Experience with danger

In the long-term governance of radioactive waste but also of conventional toxic waste, it is particularly difficult to build on experience [G45]; this is true for the scientific community and all the more for the broad public. Risk assessment and "risk learning" are impaired if experience feedback is delayed (this takes place already due to the latency period of carcinogenic

---

[70] Five regulatory posts deal with radioactive waste at HSK [P132], about three at FOE; Nagra had 63 full-time posts in 2002 [P225, 2003:24]. See page 82.

substances [R19:91] or radiologically induced cancer types). Consequently the availability heuristics of laypeople fails[71].

Even if in such cases a transfer of insight, experience and knowledge from related or even different fields is problematic: Exactly then it takes place because the new situation has to be anchored somewhere. Indirect "experience" plays a role as well: the military and/or civil management of waste in the American plutonium facilities of Savannah River and Hanford [G160][G230][G137] or the nuclear weapon factory of Rocky Flats [R67:37-77], or the continuous environmental contamination in the Russian nuclear facilities of Mayak, Tomsk and Krasnoyarsk [G22] and the environmental problems due to bomb testing [G17].

The sea dumping of radioactive waste by Switzerland from 1978 to 1982 was brought dramatically to public attention with images of barrels washed up on the Breton coast in the 1960s [G81:62*passim*]: The dumping techniques had been the same but in the earlier case carried out by France. It was also from actions with Swiss participation where heavily damaged barrels with radioactivity leaking were raised [TA, 1991-1-21].

Contaminated sites are scrupulously registered by "publics typically having long memories and employ broader contexts" [R38:182]. This may be the case when the bulletin of the Swiss Association for Atomic Energy (SVA) says: "Two dumping sites having been erected almost 40 years ago in the nuclear research centre of Dounreay in Northern Scotland are now being remediated.... This implementation will last for about 25 years, cost GBP 214–355 millions and secure some hundred jobs. One dumping site is a silo with liquid radioactive waste, the other is a pit. This one is doubtlessly the most difficult nuclear waste issue of Great Britain. From now on, the partly undocumented wastes containing approximately 4 kg of plutonium, 100 kg of uranium-235, tools and further equipment shall be recovered" [P262, 9/98:13].

In spite of assurance to the contrary, questionable management of radioactive waste is still being practiced: "Deliveries of radioactive waste from the South African nuclear power plant of Koeberg ... to the state-owned ... waste treatment facilities of Vaalputs were resumed after issuance of a new licence by the Council for Nuclear Safety (CNS). In September 1996 CNS had revoked the operating licence from Vaalputs after deficiencies in documentation and operation had been detected. To save costs, Escom, the operator of Koeberg, had suspended transports three years ago because a decrease of waste was realised. Subsequently the ditches in Vaalputs were not filled up as planned and the containers were exposed to the weather for a longer period than expected" [P262, 2/98:23].

---

[71] See footnote 40.

Nagra does not give a credible reference in October 1996 when it lists Vaalputs but also Morsleben (Germany), Centre de la Manche (France), and Drigg (UK) in an article entitled "The L/ILW repositories worldwide – information compiled by the IAEA" adding the following remark: "The list is probably not complete but the number of facilities appearing on it came as a welcome surprise to both expert and layman. In times when it becomes increasingly rare to hear of a job 'well done'..., it is pleasing to note the efforts that are being made worldwide towards achieving 'environmentally friendly' disposal ..." [P224, no. 28:25].

To make matters worse for the industry and the proponents, the phenomenon of "signal value", meaning the potential impact on society, is to be taken into account: Accidents or incidents with comparatively low consequences but a high informative content on the system under examination – "as a harbinger of further and possibly catastrophic mishaps" [R77:284] – are valued as more severe than others [R81][R84]. A test by Slovic et al. 1980 revealed that the failure of a seemingly poorly designed steering system in a new model car resulting in three fatalities was perceived as more serious than a bus skidding on ice and running off the road with 27 people killed [ibid.]. According to Jungermann & Slovic 1993 "the signal characteristic of accidents implies that additional efforts and expenses might be required to reduce the possibility of accidents with high signal impacts. Accidents due to radiation and chemical substances apparently quickly display a potential for far reaching economic and societal repercussions. Thus decisions involving the control of such critical technology should take into account its possible 'overshooting' impacts" [R33:94].

Positive experience, however, also has its effects: After, in the 1970s, half of the LLW sites under the auspices of the US DOE had been shut down for environmental and safety reasons, several individual states co-operated. Illinois, e. g., constructed a site for a region of nine states altogether and lowered the federally admissible individual dose limit of 0.25 to 0.01 mSv. The success was largely determined by the increased participation of the public, e. g., citizens' advisory groups with considerable competences were established, in Illinois and other states [G27][G29].

3.5 Voluntariness and inevitability

In the wake of the crisis about the mad cow disease in the United Kingdom, people polled in 1996 were particularly afraid of dangers they were informed about but still exposed to [R53]. The disposition of radioactive waste prototypically is a technological constraint. Thus, in my view, it is not appropriate to make voluntary potential siting communities the centre of attention, as Ballard & Kuhn 1996 do [MA2:823]. To the contrary, Colglazier & Langum 1988 demand not to overrun the potentially concerned population (such as by appealing to volunteer) but to proceed with particular caution in such imposed cases [MA9:352]. In involuntary situations, additionally, no discounting is tenable, neither substantively (*e. g.,* in dose rate) nor financially [G175:17]. The aspect of voluntariness is controversial, esp. in respect of compensations, and dealt with in Sections 10.4 and 12.4.

3.6 Individual controllability, damage defence

Regarding controllability Alvin Weinberg 1972 got to the heart of the matter with his legendary yet disputable article named "Social institutions and nuclear energy" [G275]. In exchange for the virtually inexhaustible energy nuclear source procured by the "nuclear people", society would have to guarantee its own long-term stability and vigilance. "In exchange for atomic peace" these "nuclear people" previously had established a "military priesthood" [*ibid.*:27,33-34]. This reasoning, by itself, was a logical claim, today's wording would be under the sign of the principle of causality. For Weinberg, current light-water reactors were no "transitional technology" but one step towards closing the nuclear fuel cycle with reprocessing, fast breeders, the utilisation of as many isotopes as possible for industrial and medicinal purposes, transmutation, *etc.*

The insight that this control including damage defence would not be feasible for thousands of years has already been gained by the US National Academy of Sciences in the 1950s; that is why it proposed deep underground disposal as the preferred strategy in 1957 [G267][72]. As to the hazard potential, without taking into account foreseen barriers: Had all four L/ILW waste types been deposited at the Wellenberg site as planned in 1994, after 600 years half of the radiotoxicity would still remain in the "repository for short-lived low- and intermediate-level radioactive waste" [P215:29]. The initial potential in high-level and long-lived intermediate-level waste is 500 times greater accounting for ingestion [P213:60].

Since damage defence cannot really become effective (some waste remains there), protection, as the primary goal, has to have a high value. In consequence , the need for protection measures with radwaste is revealed to

---

[72] No systematic option analysis was carried out though. The original idea was to reprocess radioactive waste. It was not before the renunciation of this technology that geological disposal was properly addressed (see Carter 1987 for an extensive account [MA6:129*passim*]). In Germany reprocessing was viewed by law as an "innoxious reuse" of the waste [G16:177*passim*]. See also footnote 14.

be high in a European survey of the end-1990s: rank 2 right after nuclear weapons and above reactor operation [R69:17][73]. Society, and hence also individuals, may only exert control indirectly, via institutional paths [G236] [G256][G146][P45], via pressure on safety authorities and applicants. Colglazier & Langum 1988 cite the example of the "Environmental Evaluation Group" [MA9:353] which, as an oversight committee, accompanied the decision procedures with the US American project WIPP, which received the operating licence in May 1998.

3.7 Reversibility of actions

Via the term of freedom of action, reversibility of actions is embedded in the concept of sustainability, together with monitoring (see Section 9.6). Transferring to radioactive waste management it corresponds to retrievability as the *ultima ratio* based on the principle of minimum regret.

Nagra assumed, and still does, "a definitive removal without the intention of retrieval": Retrieval, "albeit always possible in principle, becomes more and more costly, in technical as well as economical terms, with the progressive implementation of the graded measures of confinement, and concomitant with an increasing radiological hazard for the operating personnel" [P201:15]. To address this aspect, however, risk control and retrievability must be integrated into the project design, since retrieval is only feasible and defensible if the location of the waste in the facility is known. Only in a well planned disposal site can the official Guideline R-21 be complied with requiring that "[a]ny measures which would facilitate surveillance and repair of a repository or retrieval of the waste shall not impair the functioning of the passive safety barriers" [P133:3]. For Thunberg 1999, the president of the consultative Swedish National Council for Nuclear Waste, KASAM, the inclusion of retrievability is an "extension, not substitution" of the original (final and sealed repository) concept. According to her, it is "a precarious act of balance" between two ethical principles: the admissible burden (we impose) on future generations *vs.* equal opportunities for them: "The concept of retrievability seems to me to function as a coordinating symbol of this shift" (between two phases with the two principles) [G259:130-133].

---

[73] On a scale of 0 to 7 Sweden ranks highest with 6.17 concerning "need for mitigation of risks" [*ibid.*]. Incidentally SKB chose an "overdesign" for their site of Forsmark; only operational waste is admitted, the radiological design values are by far not made use of, the capacity is relatively moderate. The Eurobarometer 56.2 of April 2002 did not touch upon this issue.

Pressure to retrieve would hedge the freedom of action of future generations, also by presuming a certain economical welfare and sufficient technical know-how. The need will therefore be inevitable to secure adequate financial back-up for eventual actions to come[74].

3.8 "Commonplace character", familiarity

The given evidence makes it clear that, at least for laypeople, no familiarity with the risk of radioactive waste may evolve. As Evers & Nowotny 1987 set out, the older, known and supposedly familiar technologies draw on some "background of safety" [D15]. In the case of radioactive waste, this inherent deficiency has to be compensated for as well as possible by the establishment of trust in the institutions involved (regulators, applicants, "independent" science). According to Vlek & Stallén 1981 the public has to be convinced of the "degree of organised safety" achieved by these institutions [R91]. This holds for the field under scrutiny as well as for related areas, such as the landfill community, whereby the diverse responsibilities make things difficult in implementation. It is of equal importance that fair procedures are ensured (see Section 6.4.2).

3.9 Perceivability to the senses

Radioactive waste as such is not perceivable. Accordingly, the public heavily depends on the proxy criteria just mentioned. The institutions in charge have to demonstrate their competence and sense of responsibilities, now, tomorrow and in the past.

3.10 Empirical findings on hazard

For a long time the CC approach (Confine and Concentrate) was not preferred to DD (Dilute and Disperse), by way of sea dumping and substandard

---

[74] Paradoxically, democratic society builds on reversibility; technology, however, and partly nature, does not. Single processes like the "waste management" of nature are closed (composting) thus in a certain sense reversible, whereas integral processes, like evolution, are not.

reprocessing[75]. It was not before 1992, formally not before 1998, that the Federal Council also refrained from dumping, after most signatory states of the so-called London Dumping Convention had refrained from such activities. The decision to abandon dumping had not been taken freely but, in the Government's own words, "due to the strong opposition … as well as the increased public understanding of the global interrelations" [P14:4/7]. The IAEA, time and again, admonished Member States in vain to apply the CC principle. According to experts of the Swiss Agency for the Environment (SAEFL), this never was the intention of the Swiss position (TA, 1991-1-29).

The demand for a "complete control of radioactive material" of the "entire material flux" was raised in 1988 by the Swiss Energy Foundation, a major anti-nuclear NGO [P254]. Against the draft of the new Nuclear Energy Act of 2000, the Federal Parliament did not enshrine a waiver of the rights for reprocessing but imposed by statute a 10-year-moratorium in 2003 [P84]. Having constructed the so-called Central Storage Facility, ZWILAG, in Würenlingen, the NPP operators apparently do not hold reprocessing to be strategically indispensable any more; as opposed to the shortage in the 1970s there is now sufficient storage capacity for the years to come. At any rate, interim storage and reprocessing, as well as sea dumping, did set the frame for the disposal concept in Switzerland. In 1968 the Minister of Energy, Ritschard, said: "In the mid-1960s the Department of the Interior planned a building for storing waste. Shortly after, though, the dumping actions began" [P72:443] – this seemed to mean that interim storage was, for the moment, unnecessary. Until 1975 it was officially assumed that highly radioactive waste, i. e., reprocessing waste, would remain abroad (in France and UK), thus no national programme for high-level waste would have to be set up. In the same year Federal Councillor Ritschard announced that only "in the distant future" the waste would have to be taken back, i. e., in "10 to 15 years" [P120:879].

All time dimensions – objective, institutional and project-related[76] – have been, until recently, dealt with in a very heterogeneous and unrealistic manner. For instance, Nagra assumed in 1976 to start operating a repository (unclear of what type) "within 5 to 10 years", i. e., by 1985 [P197:14]. At that time, feasibility of construction engineering was prioritised, not long-

---

[75] It has to be emphasised that DD does not always have to be "worse", e. g., in the hazard analysis of individual radionuclides (incineration of waste containing C-14 vs. emplacement into deep disposal).

[76] In a rigorous sense, only the "objective" time dimension should be subsumed in the main category of "hazard" but deadline and project aspects are portents for the priority the issue was given (or not). Abundant evidence is given in Tables 23 to 25 in Flüeler 2002e [G72:282-314].

term safety. Scattered voices conceded that this issue had been underestimated [P286][P185].

To make matters worse: Political considerations were at the core of many time indications by stakeholders. The model case of this was the fixing of the legendary deadline of 1985 for the Project "Guarantee" because, by that, the Federal Decree of 1978 on the Atomic Energy Act had to be satisfied, requiring "the permanent and safe final disposition and disposal of the ... radioactive wastes". This bill was a so-called "counter proposal" to fight the (first) popular initiative against nuclear power of 1979, a common tactic against unwanted political undertakings. Still, in 1991, political considerations framed Nagra's time table: "Nagra wants to utilise the 10-year moratorium on nuclear power to deliver the lacking proof of the achievable disposal of high-level and long-lived wastes in Switzerland until the year 2000" [NZZ, 1991-4-10].

The time targets were set according to the political background and situation. In 1990 the headquarters of the Swiss Association for Atomic Energy (SVA) demanded an "implementation without delay of those disposition facilities which are certainly needed in Switzerland: storage facilities, repositories for low-level and intermediate-level wastes" [P117]. Three years later, Nagra wanted to go ahead with the submission of the general-licence application for Wellenberg, the start of construction would be at the "end of 1990s" [TA, 1993-6-30]. The President of Nagra, Hans Issler, declared: "Low-level and intermediate-level wastes are there in types suitable for disposal even today. We think that the technology for the construction and operation of such disposal facilities are nowadays developed" [P147:13]. After the vote of June 1995, when the Wellenberg application was turned down by the electorate in Nidwalden, the President of the Association of Swiss Electricity Companies (VSE), Kurt Küffer, said, unexpectedly: "We need a [LLW] repository when the power plants will be dismantled after their period of operation, because then great quantities of material will arise. This will be the case from 2025 onwards" (BT, 1995-9-9). There has never been the slightest hurry in the case of high-level waste as is explained in detail in Section 10.3. As the Director of the Federal Office of Energy, Eduard Kiener, said: "... with high-level wastes there is no hurry necessary.... In my judgement in this field an international solution is even imperative" [P162].

Mainly because the issue of radioactive waste is badly accessible to the public and very complex, the "hazard" aspect focuses on two sides: institutional aspects, above all the external reviewing and, in general, the involvement of so-called "third parties", as well as key areas like controllability and retrievability. It particularly was pressure which made the "official" stakeholders[77] such as the authorities and the applicants recognise unwanted issues – pressure exerted by external experts or by political opposition. Even the NEA located needs in 1999 in their "Strategic areas in radioactive waste management": "... it should be helpful to examine how far the present concept of deep geological disposal would need to be modified to ensure retrievability/reversibility at several time scales" [G184:14].

Controllability

The insight that radioactive waste management is associated with a long-term hazard potential led, in the course of the late 1970s, the international community to demand an isolation of radiotoxic substances from the biosphere, *i. e.,* "final disposal" [G172][78]. This philosophy of deep geological repositories explains, at least partly, the applicants' resistance against conceptual modifications.

Nevertheless, there were, even in those days, repeated demands for controllability. The reason for this is manifold: Disposal was only debated in a narrow circle – in Switzerland in the so-called Co-ordination Committee Radioactive Waste (KARA), where authorities and the nuclear industry alone were represented; only little experience with long-term issues existed altogether; the geosciences were not consulted until a late stage, though they were, and are, the expert field for the geosphere as the main barrier. In the legendary "geologists' discussion" of 1979, when representatives of the Swiss geological community were confronted with the disposal concept for the first time, their tenor was sceptical, especially as statements on the future behaviour of geological systems were an entirely novel perspective for this field [P5]. Already in 1984 the potential siting canton of Uri, in central Switzerland, made the claim for "controllability" of the waste, following a recommendation of experts who had prepared a study for the siting communities [P155].

Since the conceptual issue of control was not thoroughly and broadly discussed, the ideas to modify disposal remained vague and undifferentiated: Physical accessibility of the waste packages stood and still stands in the forefront of most critics, all the way to storage "on the site of production" (at the NPPs) as demanded by Greenpeace from 1992 on [P112][P113][P114].

---

[77] This term was used in the RISCOM 2 project, there also including environmental NGOs [G6:5].

[78] It has to be pointed out that Switzerland was a forerunner in terms of the, at the time, modern concept of "final disposal". It adopted this philosophy in 1975 [P185:51][P154] whereas the NEA 1977 [G171] was still in doubt – in fact, a NEA group of experts ascribed the "urgency" for disposal facilities to "pressing ... public opinion and sometimes political authorities" [*ibid.*:67].

Even Nagra found an "open access shaft" to be appropriate for control as they presented the so-called "adapted disposal concept" after the lost vote in 1995 on Wellenberg [P219].

The discourse was unstructured and dominated by political standpoints. Depending on the situation and the location of the statement, surveillance in the eyes of Nagra was either "contained ... in the concept" [P171], "no part of the safety dispositive" [P172:103] or even "yet technically possible but at the expense of long-term safety" [P192:23]. As a rule, neither party differentiated among near field or far field monitoring or environmental surveillance (on the surface); the conservation of documents was often mentioned. But there was no discussion of surveillance targets, time periods, methods or effort. After the defeat of 1995 the Federal Government established a "Technical Working Group Wellenberg" expressly for that purpose, it sanctioned Nagra's mentioned supposed compromise of 1998, the "adapted disposal concept", as corresponding to the "current international state of the art" [P271]. The scope of interpretation has, until recently, not been used, as given by the regulatory guideline R-21 in its revised version of [P133].

The aspect of control is an impressive example how in a complex factual field the dimensions are often debated in reverse order, especially pertaining to a technological constraint like waste: first the technical and commercial aspects prevailed, then political and economical issues dominated, afterwards the social dimension crept in and, last and least, ethical aspects were mentioned. Theoretically, it should be the other way around: First, one should have a broad debate and decision on political principles over ethical guidelines, this should in turn lead to the selection of the corresponding optimum technical variant, in consideration of ecology, economy and society. A catalogue of criteria for decision making in radioactive waste governance is proposed in Section 12.6.

The actually chosen procedure is by no means a Swiss peculiarity. Twice it was Sweden, *i. e.,* the advisory committee KASAM, to take the lead: KASAM put forth a concept in 1988 which "makes controls and corrective measures unnecessary, while at the same time not making controls and corrective measures impossible" [G139:15]. In 1993 it followed up with the recommendation to the Swedish proponent SKB to construct a "demonstration facility" with 5 to 10 per cent of the high-level waste quantity before implementation of the definite repository [G140:15]. It was not before EKRA 2000, though, that a so-called "pilot facility" for the demonstration of long-term safety was proposed (see *Figure 13-2*) [P60:51-55][G145:91-92]. The international community still takes up a wait-and-see attitude towards a potential extension of the final disposal conception [G123][G191] though gradually the notion is being reported [G271][G33:164*passim*][79].

Retrievability

At the outset, internationally, as also in Switzerland, retrievability was conceived as a matter of course in the mining industry until final disposal was established. After a close examination of long-term safety the international scientific community came to the conclusion to favour the concept of final disposal; officially the term "disposal" is defined as a disposition "without the intention of retrieval" [G99:4][80]. In the sequel, retrieval was mostly negatively evaluated though never totally disregarded.

Like controllability and along the known (political) arguments, the issue of retrievability was, across the countries, not really probed [G191][G83]. In 1993 the Swiss national environmental organisations demanded "to produce concepts considering retrievability" [P227] whereas two years later Nagra still did see "no need" for a close examination [P33]. This is peculiar, particularly as Nagra had authored an internal report on the subject 15 years before [P198]. Also, the Subgroup Geology of the Administration's strategic committee AGNEB had stated in 1986 that waste would be retrievable "without a major additional effort" if commensurate provisions were made, *i. e.*, if retrievability conceptually was envisaged in planning.

Counter to the more complex issue of control, retrievability was put forth as being much easier to communicate. Kreuzer of the anti-nuclear "Forum for Responsible Science" said in a 1993 seminar with Nagra: "… retrievability is … the most urgent demand before starting to bury [waste]" [P98]. In parallel, Greenpeace concluded in the same year that "surface storage in the site of production (the atomic power plant) is the least bad option: Monitoring and retrievability are thereby guaranteed at any time" [P113]. In 1999, the environmental organisations emphasised retrievability as "the central theme of our concept idea": "Retrievability may not be reconciled with the final disposal concept. Our conceptual idea wants to secure a permanent access of control to the facility surroundings for the generations to come so that a possible event of damage can be recognised early and prevented or limited, respectively" [P257].

One has to be alert if reversibility, as a further step of retrievability, is postulated by various stakeholders with diverging backgrounds. Along these lines, the French Government took the following decision in 1998: "It is

---

[79] Discussions in several countries have started and do explore the various dimensions, *e. g.*, in UK [G31][G233:23][G234], Canada [G211], but also at an international scale [G13].

[80] Interestingly enough, there is no talk about "final disposal" or "final repository" in international circles.

decisive that future generations are not tied down by decisions previously taken and that they may change the strategy, in view of the technical and societal developments having occurred in the meantime." Departing from the causality principle, the government merely deduced its duty to research and not to implementation: "The duty of today's decision makers" rests with the "preparation of all possible paths of research" [G85, transl. suppl.].

Consequently, a target conflict opens up, if retrievability serves reclamation of resources, becoming an instrument of a non-safety-oriented target-setting (*Figure 6-1* overleaf).

"Retrievability" is the result of a cascade of considerations and "hidden agendas" (see Section 9.6) which have to be uncovered in reasoning:

Criterion safety (passive safety against damage)

Retrievability in itself is not of top-priority. The primary goal is long-term **passive safety** (**bold** arrow in *Figure 6-1*). In this approach, retrievability is the last logical step necessary if control via "living" safety case and surveillance (see, *e. g.,* EKRA concept [P60]) shows that the system has failed and one should proceed with remediation[81]. So there is a sequence of preconditions for a special case of retrievability, *viz.,* **recovery**: controllability (the ability to control at all, *i. e.*, retrievability and, therefore, controllability had to be and *were* foreseen in the disposal design) – institutional control (including knowledge, technology and funding) – effective (overall) engineering control – monitoring (to detect abnormal and non-tolerable situations) – retrievability (concept effectively incorporated into the design) – recovery. This reason for retrievability would include all waste types.

Criterion resource facility

The material, esp. plutonium and uranium, may not be considered waste but a **resource.** This has to be a secondary goal because, otherwise, the substances should be kept in an (interim, above-ground) storage. As mentioned, so far international consensus (in the technical community) favours "disposal" going along with the non-intention of retrieval. Opposed to this is not only the main goal of (passive) safety but also the (political?) goal of security implemented with safeguard measures. The act to obtain the material would be **retrieval.**

---

[81] There might be the following alternative line of reasoning: Since passive safety is the primary goal and compromised by any intervention, *all* active measures are jeopardising and consequently banned; passive safety has to be "demonstrated" via an *ex ante* – and exclusively *ex ante* – safety case. This is a matter of debate. I believe that some validating *in situ* measures should be adopted, at least thoroughly explored, as proposed by EKRA [P60]. See Chapter 13, especially Section 13.2.

Risk perception in radioactive waste issues

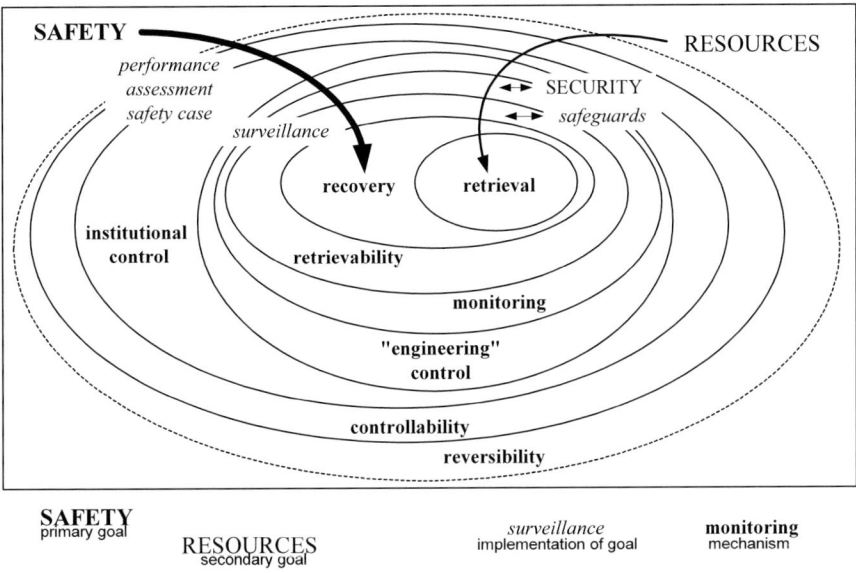

Figure 6-1. Retrievability with goal conflicts. Retrievability is, on one side, a sub-goal of safety, as an ultimate step of monitoring and control in the case of grave system failure (remediation/recovery). On the other side, retrievability may be an expression of a resource policy but being opposed to security, *i. e.,* measures against the abuse of fission products. For further consideration refer to the text. The onion ring metaphor is to make clear that individual goals or strategies are subsets of others or assume others.

Criterion reversibility

Somewhat apart from the above is the notion of "reversibility". This approach extends retrievability to all or most system properties, measures and corresponding decisions. In my view it is illusionary since even with storage one has to condition substances in a way which is not fully revocable (see footnote 74).

As the goal discussion below (Section 9.6) demonstrates, controllability should be ascertained and demanded in the first place, to check against the main goal of long-term safety, prior to retrievability. Retrievability and retrieval would, in this sense, be control of second order if and when the controlling evidence necessitated such an action. The EKRA proposal of a monitorable pilot facility [P60:51-55] makes it clear that the crucial point of future research is the issue of controllability in the post-closure phase, and not retrievability as such. It was in 1998 the consultative committee KNE stated: "Lately the argument has come up that one may not impair the scope of action of future generations by irreversible steps. This claim is met in the current disposal concepts foreseeing stepwise construction, operation and closure in several phases. The waste principally remains retrievable over a long period of time" [P167][82]. The perspective of prevalence of controllability over retrievability has later on been endorsed by others [G33:159-173].

Internationally, a sound of retreat from disposal philosophy can be detected (especially in USA, UK, and France), where the motives are manifold. They stretch from an anti-nuclear attitude via an anticipated increase of nuclear acceptance to the strategy of resource storage on behalf of subsequent reprocessing (see *Table 12-2*). If there is no goal discussion or their results are open, every stakeholder group may put forward their motives to legitimise their particular strategy (see the debate in the Netherlands [G9][G272][G280]). In this sense it is understandable that the safety experts are temporising, for whom "retrievability should never be used as an excuse to make any compromise with respect to the level of scientific and technical soundness" [P287].

4. SOCIETAL CONTEXT

4.1 Benefit and responsibility: Possible strategies of main stakeholders

Some direct "benefit" cannot really be attached to radioactive wastes as such because they arise from a preceding benefit, the electricity production in nuclear reactors, medicinal radiotherapy, *etc.*[83]. Waste is "material ... for which no use is foreseen", a loose definition adopted in the nuclear community, and only for "legal and regulatory purposes" [G107:20]. There is the danger that the issue is instrumentalised. This indeed is the case: Opponents of nuclear energy production associate the waste debate with phasing out of "atomic energy"[84]. They contest the benefit of nuclear energy. In an implementer's view, the claim that the waste issue is 'unsolved' "has now reached a stage...where it has an aura of certainty about it and it is no longer challenged in some quarters" [P173:3]. This fits the long-standing–pro-nuclear–conjec-

---

[82] Concerning the development of concepts, footnote 81 makes it clear that this is not my view.
[83] In a completely "integral" approach, the "back end" of the nuclear fuel "cycle", *i. e.,* the waste, should be considered in a comprehensive life-cycle analysis as part of the overall nuclear system. Due to the long-term dimension this poses a severe methodological problem [G80].
[84] There has never been a struggle to phase out industrial, medicinal or research applications. Today's problems were created by yesterday's decisions. Insofar it is "rational", and legitimate, to link the waste with the operation of reactors (see Section 7.3).

ture that "the problem is a political one but not technical" [P94][85], and was denied by the implementer-expert McCombie in 1997 [P193:33]. Similarly, the nuclear supporters would deploy the "solution of the disposal issue" as a free ticket for an expansion of the technology. As a consequence, both political factions, both internationally and in Switzerland, put the issue high on their agendas [G261][G216][P238].

The "dispute over the use of nuclear power is directed at the wrong target" [P173:3]. Before the relevant time frame, *i. e.*, the isolation period of radioactive substances, it is our current society which is benefiting from the produced nuclear energy. This includes, whether we like it or not, the nuclear opponents; it is as "unsolvable" like the "permanent and safe final disposition and disposal" of the waste, as stipulated by Swiss legislation. Both statements are inescapably in the same basket and remain as a paradox in the conflict about nuclear power.

Is this the end of the story? By no means. Nobody can be released from their responsibility. Beneath I propose some possible strategies for the main stakeholders to follow:

The implementer

They entirely adopt their "environmental duty" and, consequently, distance themselves from nuclear energy. They would have to become independent, not being any more "an arms length operator of the industry" as was said of Nirex in a 2001 workshop in Manchester [G31]. David Wild of Nirex admitted in 2001: "Nirex was always mistrusted because it was 100 per cent owned by the waste producers" [G276]. Nagra still is fully owned and controlled by the Swiss NPP operators.

The energy industry

Their main purpose is not the production of nuclear electricity, they even evolve from mere producers to service companies (at least for Switzerland a mid-term phase-out has been shown to be economically, societally and ecologically feasible [P59]). This strategy opens up new perspectives, technically (creation of value, diversification) as well as societally (away with the scapegoat image). It has to be emphasised that the industry does not, in the case of Switzerland, consist of four reactor operators only. These reactors are to be safely operated, without increasing risks, *e. g.*, by power increase, and later on they are to be responsibly decommissioned. Following the

---

[85] Tables 28 through 32 in Flüeler 2002e [G72:Vol. 2] document the extensive use of the pairs political *vs.* technical and solved/solvable *vs.* unsolved/unsolvable. Their reciprocal utilisation by antagonist stakeholders confirms Hypothesis 1 in Section 1.2.

causality principle, Nagra – or their successor – are to be fostered not to cut down as in the late 1990s. This is associated with costs but all history shows that legacies are more expensive. Cost wise it has to be mentioned that the "front end" has received way more money than the "back end"; R&D was predominantly paid by the state [P136][P137][P138]. In a comparative study of seven western industrialised nations Kemp 1992 comes to the following conclusion: "Radioactive waste management has been the Cinderella of the nuclear power industry, with a disproportionate amount of investment being placed on reactor development as opposed to waste management" [MA24:5]. Sweden – and possibly France – is the exception, having some prospect to write a success story (Section 11.3).

The state

Firstly, the safety authority HSK complies with their "trustee function regarding the government, the parliament and the population", as HSK's former director Serge Prêtre and top management declared in 1998, in front of the domestic nuclear industry [P237]. "Cost reduction by a cutback in safety would not be compatible with Art. 11 of the Nuclear Safety Convention, which Switzerland signed in the end of 1996" [*ibid.*:1.4-3/1]. According to the HSK, in international comparison the Swiss nuclear authority's stock of personnel is "in the lower bound of the staffing requirements" as recommended by an IAEA code [G98]. The benchmark refers to a country with few reactors of the same type (four out of five Swiss units are different). This is aggravated by HSK's terms of reference stipulating oversight in waste matters, a field not covered by the code. A comparison of nine regulatory bodies identified some lack of consulting continuity [G277]. The Finnish colleagues of STUK are in the same situation, but get by with intensive international collaboration [*ibid.*:830]. As to the building up of trust (see below), Hunt *et al.* 1999 identify as most advantageous three institutional characteristics: independence, high level of technical expertise and a specific dedication to the interests of the general public [R31:179].

Secondly, the state fulfils the obligations in the conventional toxic waste area and keeps the cantons at their duties [G78].

Thirdly, the general goals in energy policy, like stabilising and reducing the consumption, are attained and specified regarding sustainability.

Finally, the state realises that – in the long run – it has to take up liability for the waste, even if the polluters have done their best. Only public institutions have some chance of very long-term continuity.

After all, the politicians, as the superiors of the public servants, are called in. They, referring to the laypeople's supposed NIMBY syndrome: " Not In My Back Yard" [R52], do not relapse into the NIMTO(O) effect: " Not

In My Term Of Office", *i. e.,* they take contested and unpopular decisions in their terms (see also page 119).

The environmentalists, citizens' groups, the public

The controversy is led over the "right target", *i. e.,* the right object[86], and at the right time [P173:3]. The mere demand to do away with the implementer, Nagra, is not productive. Benefiting from nuclear power, today's society has to assume responsibility and ownership of the problem. But: There seems to be a need for pressure on the traditional stakeholders. As stated on page 66, according to Evers & Nowotny 1987 we have to deal with a "controversy over the institutionalisation and regulation of the progress of technological knowledge" and, thus, with the "controversy over the possibility of democratic rules of living together". Nowotny & Eisikovic 1990 soberly write: "Risks do not arise as quasi-natural consequences of technical progress. Rather, they are to be understood as consequences of decisions under conditions of uncertainty, of social action and diverse situations of interest. Conflicts thus sparked off, which appear to be factors of disturbance in a scientific-technical view, turn out, from a social science perspective, to be normal processes of negotiation and evolution in democracy which may be integrated through institutionalised forms" [R59:13]. In this sense "dialogue", "involvement" and "participation" cannot merely signify achieving acceptance, at any rate, radioactive waste at the most can best be tolerated, and energy policy is a debate of the whole of society.

4.2 Spatial and temporal risk distribution, concern, participation in procedure

"What concerns all has to be solved by all." The Swiss poet's, Friedrich Dürrenmatt's, requirement cannot be met regarding radioactive waste governance. The potentially concerned parties may participate today and in a specific location but this is not the case with future generations. In any case, an "export of risk" into the future takes place.

Spatial risk distribution, procedural equity

Factual based and transparent, broad siting and an inclusive involvement of the stakeholders in the decision-making process may fulfil intragenerational and, to some extent, procedural equity standards. In a rigorous sense,

---

[86] In reversing the argument that "the dispute over the use of nuclear power is directed at the wrong target" [P173:3].

procedures will never be fair because most potentially affected parties are not born yet. That is why Kasperson *et al.* 1983 call for a "public defender of the future" [G147:366] and Posner 1990 advocates a "Council of the Future" [G223]. Many states violate geographical or intragenerational equity by siting waste facilities around existing nuclear locations [P233:41]. A spatial distribution of risks may only be obtained if there is substantial need for more than one facility in a given area (see the US American debate in the 1980s).

Cutting means of legal redress seems to be counter-productive in view of optimising options and involvement of the public in the technology debate. The experience in Germany shows that opposition increases with a highly centralised decision-making process [*ibid.*:24*passim*]. Having a similar purpose shifted the revision of the Swiss Atomic Energy Act in the 1990s to a debate over federalism; this involved the danger of a broad political destabilisation because of which the bill was withdrawn [P32].

If procedural and geographical fairness are grossly disregarded, as it happened in the USA, some stigmatisation may take place [G249][G52]. The bad image may entail negative economical consequences [G152][G164].

Temporal risk distribution

Potentially affected parties mostly, future affected parties are never identical with the decision makers and the people responsible for the risk [R86:287]. The people polled by Svenson & Karlsson 1989 were aware of this dilemma which we live in (*Table 6-2* overleaf): In taking decisions we can only be responsible for three to barely six future generations; in radioactive waste governance, though, the (Swedish) parliament ought to be responsible for a planning horizon of around 50 to 600 – the lowest value stems from experts of the Swedish implementer SKB. The radioactive hazard potential, however, is reported to last for more than 58,000 years [*ibid.*:392]. In the survey test persons expect Sweden to be non-existent as a state after 400 years from now at the latest.

Consequently, people questioned voiced their support for, in principle, holding today's experts and politicians responsible for future incidents during this period. In other words, only a moderate discounting took place; a quarter of the test persons opposed any dismissal from responsibility [*ibid.*:395-396]. Results found by Baird 1985 are in accordance: The risk profiles (in this case with NPPs) were constant from 20 up to 5000 years [R3:242]. The conclusion, indeed, has to be drawn that a phenomenon bears a long-term character if it persists for some human generations. Thus, a distinction between 500 and 500,000 years is of minor importance in human terms (even though not in terms of the physical hazard potential of radionuclides).

Table 6-2. Time notions for planning horizons and possible events in radioactive waste management (spent nuclear fuel). RWG radioactive waste governance. Source: Svenson & Karlsson 1987 [R86:392-393].

Question	SKB experts	Students in engineering	High-school students	Retired subjects
	Planning horizon (*future events) in years (*geometric means)			
Responsibility in general	5.4 Generations	3.4 Generations	5.9 Generations	4.2 Generations
Responsibility of Parliament in RWG	49	588	375	121
Societal planning in RWG	110	458	376	100
Spent fuel not dangerous any more	31,623*	58,076*	5,781*	1,675*
Sweden ends as nation	266*	392*	248*	248*

Against this background, the "NIMBY" effect is shown in a different light (see Section 8.3). A survey of Kunreuther *et al.* 1987 in Nevada revealed that people viewing risks to future generations as "serious" were by at least 30 per cent less in favour of a HLW facility at Yucca Mountain than others for whom future risks were irrelevant (both at a given level of risk to self) [MA15:106-107]. "NIMBY" – or the less derogatory term of "Locally Unwanted Land-Use (LULU)" [R25] – becomes "NIABY": "Not In Anyone's Back Yard". Not just selfish interests but general value conflicts matter [MA24:3], integrating future generations. One cannot speak of "NIMBY" either, in the cases of Central Swiss and Zurich Weinland respondents perceiving the total risks of a local potential repository to be greater for future generations than for themselves [P234:81]. See also page 119.

Recognising that future generations only bear potential detriments from radioactive disposal without benefiting, the NEA postulates an equal level of protection for them as for us today, this in accordance with IAEA (principle 4, [G175][G107]. The US American National Academy of Sciences calls in additional protection, so to speak, a negative discount: "Moral intuition tells us that our descendants deserve a world that we have tried to make better" [G269:16]. Silini 1992 even goes a step further: First he wants to equalise exposed professionals with the general public, *i. e.*, to make operational radiation protection much more rigorous; second he proposes to integrate the non-human environment into the radiation protection standards [G241][87].

---

[87] In the meanwhile, in 2002, the ICRP issued some corresponding recommendations "to fill a conceptual gap in radiological protection" albeit it "does not intend to set regulatory standards" [G130].

Financial risk distribution

A comprehensive risk distribution debate entails securing financing for planning, construction, operation, post-closure activities including competent control for a review of the long-term safety demonstration and possible correction measures. In the light of the opening up of the energy market the issue obtains the more relevance the more provisions for decommissioning and disposal are made by carrying the NPPs under operation as assets [P58:141]. Stranded investments might be imputed on generations to come, particularly as the current economic decision models assess future damages only partially [D79]. For the issue of "Postponing decisions?" refer to Section 9.6.

4.3 Degree of information, understanding, knowledge

As illustrated above, also experts have some bias in recognising long-term phenomena. With laypeople, increased knowledge correlates with a negative attitude towards radioactive waste management (Lee 1989 in [MA24:19]). We are faced with an interactive complex system, tightly coupled (see Section 7.0), aggravating learning [D85][D86]: The processes are incompletely known and restricted as regards to time, they are unidirectionally defined and there is little room for manoeuvre [D67][D51:192]. The connotations made by laypeople with radioactive waste management are correspondingly negative [G250][G247][G249].

Additionally, in risk management, it is of key importance that "evidentiary equity" at least is striven for between concerned parties and institutions responsible [R38]; the proof for evidence lies with the applicant and not with the potential risk bearers (Colglazier 1991 after [*ibid.*:183]). The quality of siting and risk analyses has to be transparent and traceable [G168]; an approach is suggested via explicit feasibility and exclusion criteria (see Section 6.2.2).

4.4 Credibility, trust

It is especially the issue of trust which demonstrates that the risk concept does not and cannot cover the issue of acceptability. Trust (and confidence) are an indirect yardstick for dealing with risks, in radioactive waste management they have become a central theme, *e. g.*, [MA27][R26][R38][G161] [G149][G18][MA32][P191]. Slovic and colleagues even raised them to some sort of recurrent theme, *e. g.*, [R77][R78], although the seemingly encompassing explanatory power of the various concepts of trust is in discussion [R24:136-137].

This is not the place to discuss the, heterogeneous, concepts of trust and confidence[88]. One can assert, though, that in issues byzantine for laypeople, even more so in authoritarian procedures, both are pivotal aspects and both are – in the latter case – even the "only source of legitimacy" [G27:718] if there is little active public involvement. The field is further explored in Section 9.6.

Kasperson *et al.* 1992 point out that controversies around waste facilities are not as risk-oriented as risk analysts might believe they are: "Controversies over hazardous-facility siting ... are less about risk than they are about institutions" [R38:175]. Pijawka & Mushkatel 1991 reason in a similar way with regard to the US American repository project Yucca Mountain when they list independent sources which located "serious and critical credibility and trust problems" [G221:191]: the General Accounting Office, the Office of Technological Assessment [G213][G214], the National Academy of Sciences and the majority of the potential "host" State of Nevada [G221:-191]. The same held true for the British Government and their behaviour during the mad cow disease crisis [R53]. Accordingly only 11 per cent of the respondents trusted (the administration) of the European Union as conveyed by the Eurobarometer 56.2 in autumn 2001. This is just one per cent more than the value of the nuclear industry. As opposed to that, almost a third of the 16,000 people polled in all 15 EU Member States at the time did trust the "independent scientists" and "non-governmental organisations" [G133:4].

It is probably exaggerated to speak of "confidence crises" in Switzerland. The share of "discontented" people rose from 20 to 44 per cent from 1986 to 1997, though, and the percentage of test persons trusting "political actors and institutions" fell by one third between 1989 and 1996. In a national barometer on "content clients" the value for the "public administration" at all levels was more or less stable at fifty per cent [P105][P251].

According to Oberholzer *et al.* 1995 between 54 and 69 per cent of their interviewees ("affected" with LLW, "non-affected" with HLW) believed

---

[88] See, *e. g.*, [R49][R72], or [R31] and [R75] and [R78][R79:31-33] as well as [R60:353].

that in case of a future incident "mishaps would be suppressed and the danger would be extenuated" [P234:160]. Oberholzer 1998, nevertheless, holds it to be trust if 43 per cent of the respondents in the potential "host" regions in Central Switzerland believe that Nagra would "inform objectively" in case of an "accident" [P233:88-89]. As to this author's comparison with the US American Department of Energy, DOE, this might be so [*ibid.*:59]. One, however, should not overestimate the issue of trust as the US American research perception makes believe: There has not been much progress either in siting low-level and intermediate-level wastes in the USA even though neither DOE is in charge nor are these wastes associated with some "bomb stigma" [G15:314].

All statements in regard to "trust" have to be checked for credibility, authenticity, consistence of argumentation and the (historical) proof of evidence [R66]:

1. It is not *credible* if the Forum vera, a lobby organisation paid by Nagra, postulates "safety before energy political tactics" and, at the same time, accepts an indefinite postponement of closure, maintaining that this satisfies the request that "the room for manoeuvre must remain open" [P96:2]. Such an attitude is tactics and jeopardises the deep geological disposal concept whose safety is based on closure. The necessary surveillance has to be carried out on behalf of the demonstration of long-term safety, like sealing[89] as specified and implemented to a great extent by the waste producers; both cannot "be left to our successors" [*ibid.*].

2. Credibility is a necessary but not sufficient requisite to create trust: The quality of an actor on stage depends on his/her interpretation of the character who may not have to do anything with the actor. The argumentation put forth has to be *consistent, coherent and as free of contradictions as possible.* The "Technical Working Group Wellenberg" (TAG) on the one side, denied nuclear guardianship "which has, to date, in no country been seriously taken into account"; on the other side it was in favour of "surface final disposal facilities" *(sic!)* which "continuously have to be monitored and if necessary maintained" [P271:7*passim*]. Who favours final disposal has to reject both nuclear guardianship and "the internationally accepted disposition method ... surface final disposal facilities". The redefinition and weakening of the project "Guarantee 1985" on "disposition and disposal" is no example of coherent argumentation either [P43]. This project was used as a powerful political lever in the run-up to the plebiscite on the first antinuclear initiative in 1979, esp. its deadline of 1985, by which final disposal in Switzerland should have to be proven. The passage that this deadline could be "adequately" prolonged reached the public only in the summer of

---

[89] Sealing, as it is defined here, denotes final closure of a facility after backfilling.

1979, after the vote, and by way of the media [*ibid*.:70,161]. Even if the principle of asymmetry is true (trust is easily lost but more difficult to achieve) [R78:677], gaining trust may be achieved by confessing mistakes or recognising deficiencies.

3. Trustworthy statements have to be *authentic,* genuine and honest. If we, rightly, have to adopt responsibility and "the issue must not be passed abroad" [P95:3] the search for siting a domestic HLW repository has to be intensified. In case of failure, Nagra has to be expanded to examine the "option abroad"; contradictory action may not, as it happened, be exerted. Sea dumping, and partially reprocessing, was passing the buck abroad. According to surveys people are opposed to such a strategy. More than two thirds of Oberholzer's sample of directly concerned stakeholders would "rather take the waste at their place" than to export it [P233:95*passim*]. According to a comprehensive poll in Switzerland of November 1992, commissioned by Nagra, only one out of ten favours disposal abroad [P56:12][90].

Colglazier & Langum 1988 identify two options in difficult siting decisions: "overpower a weak constituency by sheer political force, or set up a more objective selection process that might be perceived as scientifically credible. In the latter case, the host area, even though opposed, might be willing to accept its fate" [MA9:352]. This would falsify Wynne's 1983 assertion that the social organisation and psychology of some modern technologies are "to be not only *internally* incoherent, unable to find a viable level between fantasy and demoralization but externally conflictual too, because their structural need for high morale and commitment tends to generate authoritarian and dogmatic attitudes" [R95:25]. The international nuclear community has finally recognised that involvement of stakeholders is a crucial determinant, and requisite, of a possible success. See Section 11.3.

A clear determinant of trust – and trustfulness – is the proof of evidence, a good record in the eyes of the respective body where trust is sought. A strict separation of powers is indicated: Promotion and oversight/protection may not be reconciled; this is valid for applicants as well as authorities. Even though the industry took advantage of it, the US and UK promotion of nuclear technology was, in the long run, counterproductive [G27: 714][R95:22*passim*]; nuclear issues were, rightly or not, more and more perceived as some sort of "collusion" between authorities and the industry [R16:83]. Only the establishment of the Nuclear Regulatory Commission in 1975 brought about the separation of the oversight body in the USA

---

[90] This attitude is consistent with other surveys, of 1990 [P101:31], 1997 [P105] and 2001 [P263:8]. Polls in Germany suggest some other direction; 55.6 per cent of a sample asked in 2001 imagine that a site "might be found on an international level, especially somewhere within the EU" [G4:15]. See Section 6.4.5.

[*ibid.*:85]. Switzerland even awaited the Nuclear Safety Convention of 1994 [G110] until steps were taken to formally disentangle the Nuclear Safety Inspectorate, HSK, from the Office of Energy. In this sense the degree of "organised safety" also applies to the responsible institutions [R91].

After years of "failure" a so-called "restructuring" of Nagra, *i. e.,* cutting down, is understandable from the investors' point of view. With regard to the general aim to responsibly implement the legal mandate it is disastrous. Up until a short while ago the appeal that "time is running out" sounded from the nuclear industry, but today's risks rather lie with cutting down financial and human resources of Nagra (see, *e. g.,* Sections 6.4.1 and 9.4.1). On the contrary: Expansion would signify good performance with regard to the mission of "permanent and safe final disposition and disposal".

Trust-building good performance on Nagra's side would be *not* to offer compensation to potential siting regions. If envisaged at all, unlike hydropower taxes and licences in Switzerland indemnifying today's siting communities, the affected population should be recompensed in the periods of maximum risk of release from the facility[91]. If the risk is considerable no financial discounting is allowed, the fund for a "risk premium" should be enormous; if, however, the risk is negligible, nothing and nobody has to be compensated at all. See Sections 10.4 and 13.4.

4.5 Empirical findings on social context

The aspects of "perceived benefit and responsibility" are difficult to explore because in the course of time goals, activities and areas of responsibility were persistently redefined[92]. Radioactive waste is a constraint resulting from politically desired and officially promoted activity (see Section 9.2). The diffuse discharge of duties among the Confederation, the waste producers and the waste organisations entailed an indeterminate leadership of the programme. It is symptomatic that the conclusion AGNEB (federally responsible) reached in 1980, when it had to clarify the duties and an eventual "federal solution", was: The Confederation should act "in a subsidiary way" in case the NPP operators and their waste organisation Nagra digressed from the polluter-pays principle "in a serious manner" [P6]. This sentence was not specified, even when mandated again in 1988 by the Federal Council "if this option should be needed" [P76]. Such a corporate identity makes it under-

---

[91] Nagra's reference case scenarios indicate 1,000 years for the LLW site of Wellenberg [P215:166], 300,000 years for the HLIW site in the crystalline [P214:294] and about 1 mio years for the corresponding site in Opalinus clay [P222:330].

[92] As documented in detail in tables 19 and 20 of Flüeler 2002e [G72:244-261].

standable that the separation of the state's promotional and supervisory functions was only initiated upon the signature of the Nuclear Safety Convention in 1994 even though a parliamentarian committee had demanded it way back in 1980 [P100]. The independent advisory body EKRA called for taking care of this central issue in their second and last report on institutional structures in 2001 [P61:14].

In such an ambiguous situation it is obvious that further difficulties and inconsistencies arose. Under the heading "supervision" some prominent examples are cited below. I distinguish between waste political responsibility, political and strategic leadership and technical supervision.

Disposition philosophy

Even though there was a national plebiscite on the Federal Disposal Decree, in 1979, no broad debate took place on the disposition concept in Switzerland in the 1970s. Until the mid-1990s all demands on modification of the (final) disposal philosophy were officially ignored.

Political responsibility

The Federal Council recurrently heralded the implementation of final disposal sites to be "a national task", *e. g.,* 1977 [P24], 1989 [P246], 1994 [P80]. The Waste Convention of 1997, signed by Switzerland, states in its preamble (xi) that the Contracting Parties are "[c]onvinced that radioactive waste should, as far as is compatible with the safety of the management of such material, be disposed of in the State in which it was generated" [G110]. Nevertheless, the government kept the disposal concept for HLLW "open" in their draft for the revision of the Atomic Energy Act, it did not "at this time" want to determine "when the duty of disposal is discharged" [P277:19]. Even at a time when disposal was not contested by the Swiss majority (1980), the then acting Energy Minister, Leon Schlumpf, conceived it a "practicable path" to "domestically controlled and retrievable interim storage facilities as well as multinational disposal facilities" [P223, 3/80:12].

The letter of the law and the official statements differed. There was a powerful waste organisation (Nagra) as well as a regulatory body but the main recommendation of IAEA of 1995 with regard to "Establishing a national system for radioactive waste management" has not been met: "Member States shall ensure continuity of responsibilities" [G108:7].

"Backlog"

A "need to catch up" with other nations in nuclear research was officially criticised in the 1950s and 1960s. Authorities and Nagra complained that such a need existed in the 1970s and 1980s with regard to disposal as such (frequently referring to the model example Sweden[93]), during the 1990s there still remained the "need to catch up" in the construction of a LLW site. In recent years, though, the external committees have found fault with the "insufficient" progress of the disposal programme, particularly in the HLW field, and the cutback in Nagra's resources [P167][P178]. In accordance with the "continuity of responsibilities", as internationally recommended, the advisory group KNE proposed in 1998 "that phase-related programme flow chart and scheduling are drawn up ... where goal agreements and estimations of work load of the separate phases are laid down" [P167:5]. Interestingly enough, KNE also referred to Sweden whereas from Nagra and the authorities the Nordic country was usually cited in the context of a "timely" realisation of facilities (see above).

Regulatory supervision

It is inconsistent with a "national task" and the causality principle to have a minimum standard oversight authority. The former director of HSK, Roland Naegelin, diagnosed in 1996: "In spite of staff increase we continue to abide by the Swiss principle that in this country fewer people are brought in for a certain problem than somewhere else" [P195]. Substantive expansions of personnel were only allowed due to massive external pressure, after the accident of Three Mile Island or the catastrophe of Chernobyl. Equally low were the resources allotted to external reviewing; such instruments were only established upon pressure and late in the process (in 1980 when the Subgroup Geology of AGNEB was appointed). Since 1987 the supervisory authority HSK has been employing just two geoscientists and one safety performance specialist in the disposal field (see pages 67 and 82). A "National Geological Institute" as demanded by Buser over 25 years ago, is not on the agenda anymore [P42]. Until the end of the 1990s there was no official review of Nagra's inventories and data bases (which they actually produced).

State leadership, role of AGNEB

The lack of governance by the Confederation culminated in self-fulfilling arguments, such as in 1981 when the Federal Government supported Nagra: "The unconventional geological research programme of NAGRA is largely determined by time pressure" [P194:910], knowing that it was the administration which had defined the deadline of "Guarantee 1985" in 1976, follow-

---

[93] The deadline of 1985, set in 1977, for the project "Guarantee" was based on the plans in Sweden to achieve a demonstration of feasibility within two years [P5:35]. Other references, *e. g.,* are for 1980 [P146][P242], 1982 [P223, 82/1:3], 1984 [P25][P223, 84/8:1], 1987 [P204], 1990 [P83:120], 1994 [P13:3], 1996 [TA, 6-27], 1997 [P192], 1999 [TA, 4-28].

ing a proposal by the energy industry [P154:14]. Relevant issues, such as reprocessing, were and are subject to private law and thus classified "private", being in the industry's scope of decision [P53]. A similar attitude was taken regarding waste export, *e. g.,* 1989 in the context of a Chinese option: "The operators of the Swiss nuclear power plants have informed the Federal Council that the negotiations ... have been discontinued" [P241]. Mandates to draw up concepts were commissioned time and again (to a federal committee in 1957 [P82], to the energy industry in 1969 [P263, 16/69:3], or to AGNEB in 1988) but only partially, if at all, handled.

The role of the Interdepartmental Working AGNEB is characteristic of the loose institutional federal oversight of the disposal programme. As per order of their terms of duty of 1978 they would have initially been in charge of "producing a project to guarantee a safe disposition of nuclear energy [*sic!*] in Switzerland" [P4:Annex no. 1]. A year later they were assigned the task to do the conceptual work and to accompany the activities of the directly responsible bodies[94]. Apart of the ambivalently worded mandate, the Group has not come up to expectations due to organisational reasons: It is a body of merely public administrators having neither competences nor adequate resources. Only in the initial programming phase it released fundamental documents, such as on the definition of "Guarantee" [P7] and on a potential state solution, *i. e.,* the option that the Confederation would have to implement disposal instead of Nagra or another private institution [P6]. Later on, until 2000 [P16], they confined themselves to summarising the activities of others involved or giving them room for self-portrayal. An analysis of their documents reveals that AGNEB mostly took the industry's view or even placed less stringent demands (as in the case of putting into service the HLW repository which, in 1992, it rated as "no urgent task" [P12]).

It has to be esteemed that in 2000 the Group required Nagra to set up a "strategic plan" for HLLW by 2002 [P17:3]. By 2003 it planned to create a "schedule" for these waste types itself [P18:2]. In 2002, this was specified in so far as to "draw up a schedule ... on the basis of the technical conditions" [P20:5]. It appointed a Subgroup Inventory and initiated two closed-door meetings among the federal administration and its advisory bodies to discuss a concerted course of action [*ibid.*:4-5]. For a further discussion see Sections 13.3 and 13.4.

---

[94] In its 1979 revision the document runs: "The Working Group ... produces the decision base.... By that the prerequisites shall be created on the part of the Confederation so that the nuclear disposition can be implemented safely and in due time" [P4:Annex no. 4].

Self-image and external image of Nagra

The role of Nagra may be qualified in terms of the diffuse strategic discharge of duties. The "solution" of a "national task" by a private institution is associated with several areas of tension, opposite expectations, behaviour and conditions (see *Figure 9-4*):

- *State:* political and administrative pressure on a "solution" (deadline of 1985 with the project "Guarantee", pressure to succeed), increased control compared with deregulated approaches but pressure on a unanimous public appearance (since the waste institution complies with a public mandate);
- *NPP operators:* financial and substantive provisions (cost reduction, influence on programming and parallel search abroad), energy political pressure (link of disposal with nuclear path);
- *Public:* divergent, non-constant demands and attitudes (domestic mplementation, cost reduction, "NIMBY" effect, pressure to perform; criticism of technology and/or atomic energy, preservation of the basis of life; federalism/devolutionary trends);
- *Opposition, critics:* divergent expectations, need of cohesion in view of the – real or perceived – "coalition" of Confederation/operators/waste institution; criticism of technology and/or atomic energy;
- *Media:* reception and reproduction of opposite points of view (black and white), personalisation of a complex discourse on technology and environment (expert controversies, dichotomy of experts *vs.* laypeople, technocrats *vs.* ecofreaks, urban *vs.* rural contrasts, profiteers *vs.* affected parties), misuse of topicalities, misuse of popular patterns (David against Goliath, failure story, waste of money, *etc.*);
- *Waste institutions (Nagra, former GNW):* "... the political monopoly of the para-public institutions weakens ... their capability of learning and adapting in a society placing high participatory demands", Nagra "is forced to [include divergent interests in planning] only by increased external pressure" as is stated by Wälti 1993 [P284: 216-217].

As soon as Nagra had to expect or even deplore failures they backtracked to "scientific/technical" issues or demanded Federal decisions. When it became clear that the deadline of 1985 could not be met, it stated (in 1982): "We understand our mission solely as a scientific/technical one.... The Federal authorities will ... provide the base for political decisions. These decisions clearly rest with the Federal Council" [TA, 1982-2-2]. When the officially required additional proposed site for the LLW programme was not in place, Nagra announced in 1986: This issue "shall be kept open for the time being.... Claims for a low relief [*i. e.,* flat topography, tf], however, are not met. It is now up to the Federal authorities to decide on this issue.... At any rate, Nagra is going to search for a fifth site for exploration. It would probably be situated in the Swiss Midlands" [P223, 10/86] (this promise was not kept). After the negative vote at Wellenberg Nagra declared in 1995 (basically as in 2002): "For that it is now up to the Federal Council and the political authorities to change the further points and, eventually, to set out for a fundamental re-orientation" [TA, 1995-10-14].

A logical consequence of such behaviour of the stakeholders is the formation of political blocks. According to Buser & Wildi 1984 "the Federal authorities behaved as coalition partners of the energy industry and of Nagra" [P48:26]. By way of the diffuse role play of strategic and operational leadership (the Confederation and Nagra, respectively?) it was an easy job for politicians to intervene, such as the representative Franz Jaeger in 1990 did: "It really is comprehensible if today the Federal Council has to, for instance, admit that they have no overview where the waste is stored and how large their quantity is. Data bases do not exist" [P83:109]. In 1998 an IAEA expert team in a review of HSK recommended a definition of "low-level", "intermediate-level" and "high-level radioactive waste" [P141]. In 2002 KFW made recommendations regarding LLW [P159], and from the same year a respective Subgroup of AGNEB is taking care of the issue [P20:4].

The formation of blocks was not conducive to the mentioned capability of learning and adapting. With increased time and cost pressure, opposition, and criticism in general, was interpreted as an obstruction and preclusion. Escalation on one side, appeasement on the other side, was on the agenda. Adaptations were made only after a defeat, all the more ill-prepared, such as the "adapted disposal concept" of 1998 [P219]. What Nagra launched as a "specifying presentation of controllability and retrievability" and a supposed adaptation to criticism in the aftermath of the Wellenberg bargaining of 1995, was in actual fact a loss on safety [P167]. To leave access shafts open for an indefinite period of time is associated with oxidations, a hydraulic sink, and the prevention of saturation. Such are all phenomena which should be impeded by a deep underground (geological) repository.

The mixture of tasks and positions led to demands for disbanding Nagra and transferring the mission to the Federal state, at times with the bizarre argument that the co-operative is "too expensive ... and their mission to seek a repository for radioactive waste has become inopportune. An arrangement has to be made at their place which brings about more sensible and more economical solutions of elimination (by way of a retrievable and monitorable storage) and of atomic waste avoidance" [P285]. Lately pressure is exerted on the co-operative by the operators trying to reduce the production costs of nuclear energy on the liberalised market. Flüeler 2000a commented as follows on such a constellation: "The management of radioactive waste is at a turning point .... Pressure on the radwaste proponents and implementers is not only executed by large parts of the public but also by their owners themselves.... As a potential consequence, an alliance could build up, *viz.,* between pro-nuclear parties (shareholders with a wait-and-strike-later strategy) and anti-nuclear groups (nuclear guardianship instead of final repositories). This might result in indefinite intermediate storage" [P89:304-305].

Options abroad

The shown behaviour, inconsistent but forced by condition, comes to the fore when the debate around so-called "solutions abroad" is analysed. An export strategy was chosen in the end of the 1960s, when reprocessing and sea dumping were applied. H. R. Lutz of NPP Mühleberg was of the opinion in 1970 that "the waste problem does not belong to the immediate purview of the nuclear power plants. The spent fuel elements are yielded to reprocessing plants" [P186] (the first contracts did not encompass clauses of repatriation). Although the Radiological Protection Act of 1991 and the revised Nuclear Energy Act of 2003 state that disposal has to take place "in principle" in Switzerland, in all political camps and at all times the option abroad has been either admitted or expressly demanded. All Swiss cantons, except Basel-City, Fribourg, Nidwalden, Schwyz, and Zug (Appenzell Innerrhoden refrained from commenting), favoured international options as a must [P29:9]. In accordance with multiple statements by the Federal Office of Energy, the technical supervisory board, HSK, wrote in 1997 regarding a HLW facility: "A common multinational project may as well have benefits in terms of technical safety.... Thus, it is not advisable to already now take steps towards the realisation of such a repository in Switzerland" [P124:8]. Nagra, on the one side, kept the option abroad open and temporarily was involved in the company Pangea, on the other side, it had to dissociate itself from activities of their owners abroad (*e. g.,* in Russia or on Pacific Atolls). The attitude of these main stakeholders is in stark contrast to the revealed public opinion being in favour of a Swiss or domestic solution in their vast majority [P56]. This is true for the European Union: In a poll in autumn 2001 [G133] there were still 63 per cent who held the waste producing countries responsible for the disposal (in 1998 it had been three quarters). It is not surprising that Dutch respondents showed the best support for a "regional solution"; the Netherlands do not follow an independent programme and uphold continued retrievability (Section 11.3 and 12.3). It is consistent that interviewees apparently are more concerned about the situations in other countries than their own [*ibid.*:8].

Involvement of third parties

Given the constellation laid down above, it is congruent that third parties were not broadly involved in forming disposal concepts and programmes: External experts, NGOs, and the general public have only gradually entered the scene. In 1980 it was, upon pressure by the geoscientific community, up to the just established Subgroup Geology of AGNEB to demand "a categorically necessary public involvement and publication" [P8:Annex V,1]. It is equally striking, though, that the issue of stakeholder involvement has much gained in attention and importance (Chapter 8, Sections 11.3 and 13.3). One is tempted to call it a change of paradigms in the international arena. The then director of the Swedish safety authorities SKi, Sören Nörrby, put it in 1997 as follows: "There has been a steady trend to move from activities solely focussed on information and education of the general public to a new attitude that is more focussed on dialogue and involvement of the public and local politicians" [G206:15]. The milestones of the involvement of third parties are displayed in Section 13.3., particularly in *Table 13-1*.

PART III

PERSPECTIVE "FROM ABOVE": DECISION PROCESSES

Chapter 7

INSIGHTS FROM DECISION RESEARCH

1. PRELIMINARY REMARKS

An analysis [G64] of the risk perception aspects – for the Swiss Federal Department of Energy – revealed that this approach is, indeed, relevant and necessary for adequately dealing with the radioactive waste issue, but it is not sufficient. Thereby, the following points play a role:

- Because evidently there is the real constraint of radioactive waste, ultimately a decision for action is indispensable if a responsible management of the material is strived for.
- Risk perception issues ordinarily relate to individuals whereas the decision needed is taken by collectives.
- Even risk perception research, after a 20-year individualistic phase, demands a "politicization" of itself [R79][R80]; individual risk perception does not explain non-acceptance sufficiently [G243].
- The "decision anomalies" mentioned initially refer not only to persons but groups and organisations, which, in turn, presupposes an institutional perspective.
- On a theoretical basis, one has to agree with Nowotny & Eisikovic 1990, "that it is not 'technical progress' resulting in changes but decisions, it is the consequences of interests and certain actions, irrespective of whether the intended consequences coincide with the factual ones or not" [P230:10].

- In order to tackle the issue in an integrated way, systems theoretical insight [M72][M71][M34][M88][M35] is used together with decision science.

Below findings from normative and descriptive decision research are utilised for the formation of categories of the content analysis[95]. The development of risk analysis as a method serves to interweave rationality/risk, risk perception and decision issues. The content analytical findings based on that are the starting point for an optimisation of decision problems (see Chapters 12 and 13).

2. DECISION, PROBLEM, INFORMATION, AND UNCERTAINTY

Via decisions, possibilities to act or alternatives become either active or passive actions. "Deciding" consequently is, on the one side, a selection of alternatives during a mental phase, and, on the other side, an elicitation and implementation of will during a phase of realisation [D56:2-3][96]. Principally, decisional problems are informational problems[97]; complete information on an issue would make a debate on deciding superfluous since there would be no deviation from the initial/factual state to the final/target state – there would not be any problem to solve (see *Figure 7-1* overleaf). But if the information is incomplete, even variable, the question is not just what to do or not to do but whether additional information should be obtained or not. Aside of the issue of need for decision on actions there is the need for decision on information [*ibid.*:V-VI]. Information is purpose-oriented knowledge in a decisional situation aimed at the future; it serves to reduce the decider's uncertainty on what will actually happen in the future.

Deciding is not just the preference of an option [D44:17]; in decision making one has to deal with the following questions:

- How is sufficient knowledge collected? (*Figure 13-7*).
- How to judge in the presence of uncertainty? (Section 9.8)
- How to integrate individual values? (discourse on dimensions in Chapter 12)

---

[95] The detailed analysis is done in Flüeler 2002e [G72:126*passim*].

[96] It is assumed that "judgement" is followed by "choice", a proposition debated in social psychology [D1:235]. Decision science usually assumes that decisions actually are taken.

[97] This is probably the reason why information-processing and cognitive sciences are so productive for, and the current underlying conceptual basis of, problem solving psychology [D39:39*passim*].

Insights from decision research

- How to assess the potential implications or side effects? (*Table 12-2*)
- How are the options perceived? (PART II)

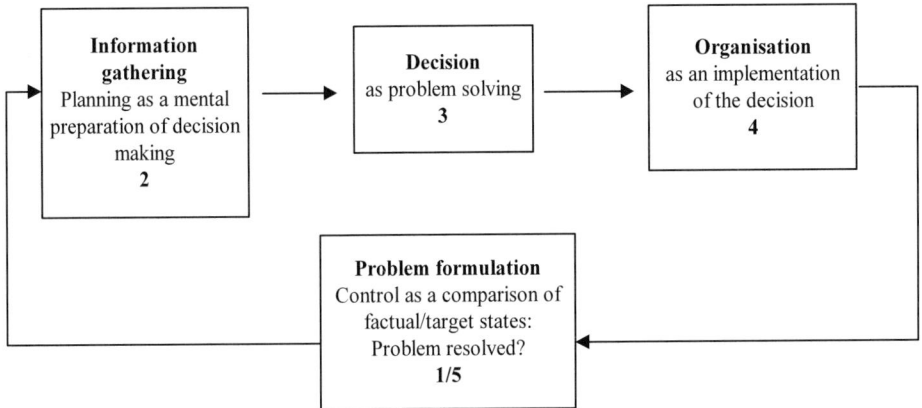

Figure 7-1. Decisions in the feedback model of an institution (modified after Mag 1990 [D56:3]). It is important to link the iterative and cyclic decision process with the loop of Problem formulation (relation of factual/target states, **1**), Information gathering (**2**), Decision ("first" problem solving, **3**), implementation ("Organisation", **4**), and Control (eventually "second" problem solving, **5**).

Decision making, thus, includes the process of deciding, the judgement made, the choice taken and, ideally, the decision implemented.

Complex issues like the present one mostly require phased collective decisions which form a chain of, ideally iterative, partial decisions over a long period of time. Mintzberg *et al.* 1976 proposed the following structure of the decision-making process (*Figure 7-2* overleaf).

The starting point of an adequate problem solving strategy is a thorough and thoughtful analysis of the situation concomitant with suitable system modelling. The phase of problem identification is accompanied by the formulation of goals, because "optimum decisions are ... always goal-oriented decisions" [D56:28]. Against this background it is amazing that 21 out of 25 decision-making processes investigated by Mintzberg *et al.* 1976 were dominated by the problem development phase, *i. e.*, phase of elaborating options for solution [D62:255]. The authors rated it as "rather curious" that decision research was heavily focussed on the evaluation – choice routine [*ibid.*:257]. Abelson & Levi 1985 detected a scarcity of research in problem recognition [D1:271]. According to Janis & Mann 1977 good decisions usually are characterised by a careful processing of several alternatives [D42]. The search routine fits the requirement of decision problems less than working out alternatives along the design routine [D1:273].

Kahneman and Tversky distinguished two decision phases: first, an editing phase where options are organised and reformulated, then, a second evaluation and choice phase [D45][D46]. It is precisely in the editing phase where the evidence has to be carefully formulated, this bearing a great influence on possible – divergent – appraisals (the key word here is "framing" [D82]). Thus unwanted so-called context effects may be avoided [D38:40]. Often a decision is prepared already in the development phase when solutions are sought. The pivotal role of the process and, thus, the procedure is dealt with later on.

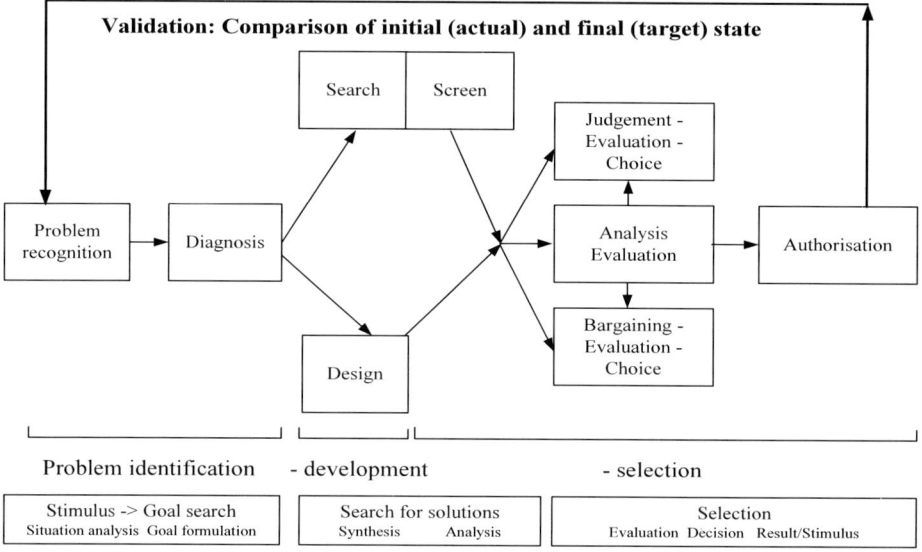

Figure 7-2. General model of the strategic decision process (modified after Mintzberg et al. 1976 [D62:266]). The indications below refer to the "problem solving cycle" in systems engineering [D32:47*passim*]. "Stimulus" denotes closing the feedback cycle, "Synthesis" means the elaboration of options, "Analysis" means the reviewing thereof. With ill-defined problems the phases mostly are not sharply detachable [D1:274]. See text below.

To reach sustained decisions among individuals, groups and organisations—as in radioactive waste governance – there is a need for "informed consent" which, in turn, requires an explicit elaboration of many possible ways and consequences of courses of actions [D21].

3. ILL-DEFINED PROBLEMS

As mentioned, problems are defined by the perception of the difference between a final state (sought after) and an actual state (unwanted) [D1:270] [D62:253]. Decision problems are well-structured if the decider is familiar with their initial state and the goal state as well as a defined set of transitions [D55]. Additionally, according to Simon 1973 any proposed solution for such problems has a definite test criterion and only "practical amounts of computation" are required, meaning that the information needed is "effectively available" [D77:183]. Related to problem structuring is its complexity (see Section 9.6). Funke 1991 defined complex problems as being non-transparent, having multiple goals (called polytely), situational complexity, and time-delayed effects [D27].

Environmental problems usually are complex and ill-structured or ill-defined [D70][M80:26*passim*]. In such situations decision research does not offer a dominant paradigm but resorts to concepts and methods put forth by many scientific fields, like sociology, administrative sciences, political sciences, or psychology [D1:269]. Cognitive strategies of participants, be they individuals or groups, may greatly differ [D17]. "The optimum solution cannot be unambiguously determined. Only the relatively best of the solutions found can be detected" [D64:128].

"Good" decisions are "good" in relation to the goals envisaged. Thus, the problem recognition and – indirectly – the goal discussion are important (Sections 9.4 and 9.6).

In addition to the difficulty of problem definition it has to be acknowledged with radioactive waste that it poses – in terms of the theory of decisions – a so-called "implicit problem", *i. e.,* it was caused by a preceding activity or decision (to utilise radioactive substances) and now constitutes a (factual) constraint. To this extent it is "rational" to link the issue of radioactive waste with the operation of nuclear power reactors. The uneasy situation, however, also has to be accepted that research in this area – in whatever direction it goes – is "supportive" research, this term coined by SKB in their R&D endeavour to implement final disposal, mildly criticised as "supporting research" by KASAM 1995 [G141:59-60]. This underlying factual constraint determines the debate; Sundquist 2002, in an analysis of the Swedish case of radioactive waste governance, calls it the "technological imperative" [MA43:222].

4. CRITERIA OF DECISION SCIENCE

A review of the literature (References, 2.2) results in the following compilation of relevant criteria of how to qualify a decision or a decision-making process (*Table 7-1*). The approach is mainly based on system theoretical insights. For the issue of recognition of perception and communication problems it is referred to Chapter 6.

Table 7-1. Criteria to assess decisions and decision-making processes (Cont. to page 109).

Decision criterion	Elucidatory remarks[98]
System understanding	Systems consist of components and relations. The type of interrelationship of the components defines the structure and the qualities of the system. Complex systems are characterised by the interrelated connections of their components/subsystems, their manifold dynamics and or incomplete information on them. These features have consequences regarding objective, modelling and information gathering, forecast and extrapolation, action planning, decisions, implementation and assessment of the strategies of action. Socio-technical systems consist of a technical and a societal part; in contrast to technical systems, they may not only change their state but also their system structure, *e. g.,* by way of a change of the users' requirements. Problem solving has to be targeted, the goals have to be efficient yet flexible. Complex systems need a variety of analysis tools to attain system understanding regarding their ways of functioning, the strength and weakness and corresponding causes, the relevant influence factors, the system environment, and the future chances and risks of the systems. Each problem identification is part of the whole; the problem solving process, thus, always has to be aligned to the exterior, to the system and problem environment.
Avoidance of logical fallacies and biases	Because complex systems, let alone socio-technical ones, exhibit no clearly defined structure and, therefore, are not predeterminable, it is pertinent to avoid logical fallacies and biases. Environmental issues are too multifaceted to tackle them solely single-mindedly. Gomez & Probst [2]1997 ask to consider the following: Problems are never given "objectively", they are defined by somebody; a joint learning process is the prerequisite of communication among the participants [D30:241-242]. Unexpected side effects, thresholds and positive feedback often are beyond the capacity of linear causal reasoning. Chains of effects are riddled with dynamic components of diverse degrees of activity. In dynamic systems analyses suddenly may not reflect the evidence any longer. Short-term interventions may lead to sys-

---
[98] Extensive references to every statement are given in Flüeler 2002e [G72:115*passim*].

Decision criterion	Elucidatory remarks
	tem <u>destabilisation.</u> The information base on complex systems is too limited to serve <u>forecasting.</u> Certain aspects may not be <u>steerable</u> in spite of great efforts. <u>Manageability</u> is a myth, there are no <u>rigid solutions</u> for complex systems. One should not act *against* the system, *i. e.,* the system mechanisms are to be considered. – <u>Formal criteria</u> for formulating possible courses of actions alone are not purposeful with socio-technical systems (such as mutual exclusion, exhaustive enumeration)
Consideration and adaptation of problem structures	Problem decisions are triggered by multiple stimuli. Their power, *i. e.,* cumulated amplitude, depends on a multitude of factors, *i. a.,* the influence of their origin, the interest of the decider, the perceived benefit for action, the related uncertainty and the perceived probability of a successful termination of the decision. Ill-defined problems often occur in the context of groups and organisations. Criticism may be suppressed by "groupthink" and pressure to conform. Feedbacks are particularly relevant in long-term dynamic systems; they provide information on interrelations of effects, the system environment, variability and eventual side effects. Solutions may virtually ripen and blossom out.
Decomposition into subsystems and re-integration	Systems theory may provide a framework for organised, "systematic" problem solving. After rough structuring, the resulting subsystems facilitate the handling of partial problems (technology, resources, policy, *etc.*). These always have to be resolved as part of the total system; the partial solutions have to be integrated into a total solution. The definition of components and systems depends on the point of view, the criteria for investigation and the powers of perception.
Goal relations investigation, complex goals	Talking of complex problems the goals have to be formulated in a comprehensive manner to check, specify and revise blurred notions. This does not signify diffuse complex goals – the precise definition of goal relations is useful: target equity, target complementarity, target neutrality, target conflict and target contradiction. Weighing the targets facilitates a more effective trade-off of alternatives and reveals the preferences of the participants. "Must" goals define the minimum requirements for a system or a solution; "shall" goals may stabilise the solution as additional requirements. Thus, results may be appraised with regard to type/content, level/dimension, temporal and locational relations, as well as safety. It is helpful to use the investigation concept of the policy programmes: programme goals (clarity, range, period of validity), instruments and provisions (financial, regulatory, persuasive instruments, usefulness).
Adequate treatment of diverse levels	Controversies over technical systems, *e. g.,* waste facilities, refer not only to risks but are "are less about risk than they

Decision criterion	Elucidatory remarks
	are about institutions" [R38:175]. The technical dimension is added by the societal one. Values may have a leading function in complex and non-transparent situations. Procedural issues become important – whereas proponents focus on technical benefits (*e. g.*, good leakage behaviour), concerned and affected parties may prioritise "process utilities" [D78:297], *viz.*, their involvement in the procedure. The so-called policy style comes in: type of the problem solving procedure (openness, transparency), type of stakeholder contact (communication behaviour, type of negotiation), and problem solving behaviour (target consideration, problem relevance, degree of activity, planning perspective).
Decisions under uncertainty	Decisions are future-oriented and are, thus, always taken with incomplete information. If probabilities can be assigned to the relevant results, we speak of decision under risk, if not we are faced with such under uncertainty[99]. As a type of privileged alternative in a decision under uncertainty there is the "*status quo* bias" as Samuelson & Zeckhauser 1988 term it, "to follow customary company policy, to elect an incumbent to still another term in office, to purchase the same product brands, or to stay in the same job" [D74:8]. Hanson 1991 concludes that the preferred strategies under uncertainty are the *status quo* and adaptive alternatives such as the keep open option and the deferral of decisions [D35:49-50].
Co-operation problem	Environmental action is a problem of co-operation of the prisoner's dilemma type[100]: Individual members of society would rather not co-operate and take advantage of the environmental contributions by other parties; paradoxically, this individually rational strategy leads to a collectively inefficient balance. Co-operation would prove worthwhile for all, but each individual leaves the environmental share to the others and takes advantage of the common good "environment" as a free rider. Franzen 1998 proposes five variants

---

[99] There are more specific classifications, *e. g.*, safety is associated with deterministic knowledge, risk with complete probabilistic knowledge, uncertainty with partially probabilistic knowledge, and ignorance prevails when neither deterministic nor probabilistic knowledge exist [D53:13][D2:265]. Also refer to *Figure 8-1*.

[100] According to Frey & Bohnet 1996 this analogy is not valid with many environmental issues such as the greenhouse gas effect [D26]: If an individual's environmental contribution is close to zero, a society with many members loses the incentive to deviate from the co-operation strategy as soon as the difference between environmental benefit and cost is greater than or equal to zero. The egocentric, "rational" individual, though, has no incentive for co-operation; therefore, the prisoner's dilemma is useful for illustration [D23:25]. I will come back to the asymmetry in radioactive waste management – due to the absence of the future generations as "players" – in Section 9.9.

Decision criterion	Elucidatory remarks
	to solve the co-operation problem of which four may be relevant for environmental action [D23]: Individuals do collaborate more frequently if they know that their own game history is put on notice to future players or if they are threatened by retaliation. Additionally, suitable external interventions are conducive, in a limited manner, to moral appeals by which the dissonance between attitude and behaviour may be reduced.
Utilisation of latency periods as an opportunity for learning	Ill-defined problems need a thorough analysis of the situation and of effectiveness. "While initially an abridged procedure may seem to increase efficiency, it does not pay, however, in the long run. For if chances and problems are analysed insufficiently or too late, in the end constraints determine action" [D64:66]. With solving procedures of ill-structured problems Ninck *et al.* 1997 refer to "incubation phases" whereby deciders disengage themselves from the problems for a while after an intensive adsorption and before solutions variants are specified and elaborated [*ibid.*:-129-130]. In political processes, Freiburghaus & Zimmermann 1985 detected a "phase of latency" when, *e. g.*, the state is confronted with a new task or an old problem arises in a new fashion, in which the political arena is newly set, where the stakeholders have not yet taken up position (or abandoned it) and where the rules of the game are not fixed yet [D24:88-89].

Kleindorfer *et al.* 1993 [D48:388*passim*] propose the following procedure[101] to improve the quality of decision making and, supposedly, to finally reach a "good" decision [D12]:

Element 1: Problem finding

- Define the problem (problem identification).
- Relate the problem to goals, values, and needs.
- Frame the issue and acknowledge biases.
- Investigate at the appropriate level (level of analysis, scale).

Element 2: Institutional arrangements

- Identify the primary and secondary stakeholders.
- Measure their goals, objectives, views, constraints, and agendas.
- Assess similarities to and differences from your concerns.
- How do they interact with each other?

---

[101] The approach in principle corresponds to the problem solving cycle of systems engineering [D32:47*passim*]. See *Figure 7-2*.

- How are the social levels interrelated? (individuals, groups, institutions, society)

Element 3: Information gathering

- What information do you need regarding facts, assumptions, stakeholders' values to develop a systematic approach?
- Identify the biases – how do you address them?
- What are the cost and benefits of collecting additional information?
- Relate specific solution procedures to the types of information.

Element 4: Choice process

- Consider the choice approaches in addressing the problem.
- Will the choice process involve others?
- Appraise techniques.
- Make a trade-off between effort (cost) and accuracy (benefits).
- Formulate the decision criteria.
- Evaluate the implications of so-called "nestedness" (interrelations), complexity and legitimation on the selected solving procedure.

Element 5: Implementation

- Consider feedback, control and accountability in implementing.
- Formulate legitimacy criteria (results as plausible solutions for the relevant problem).
- Periodically review the decision-making process.

Such building blocks of a roadmap for decision-making processes are supported by five guiding principles and concepts:

1. Context dependence: analysis of the reference points and (re)framing of the pertinent questions;
2. Process matters: "Good decisions result from [a] sound process, in much the same way that great golf or tennis shots result from great swings" [*ibid.*:392], need for long-term process observation, analysis of sources of error, cost-effective strategies to process improvement;
3. "Nestedness": examination of complex issues from multiple perspectives and divulgence to outside parties, long-term perspective (over several negotiation periods);
4. Bounded rationality: need for trade-offs due to limited information-processing capabilities, decomposition and structuring of complex decisions into more manageable subsystems;

Insights from decision research

5. Legitimisation: joint formulation of criteria relevant to key stakeholders because a "decision can only be successful if it is perceived as legitimate by all the relevant parties" [*ibid.*:397], broad-based participatory process, development of trust among the stakeholders.

There are no "simple" decision rules, patterns, or strategies how to proceed in complex situations such as radioactive waste governance, especially as today's deciders cannot assess the quality of the outcome of their actions taken (and as they cannot be called to account). And yet, according to Jungermann *et al.* 1998 some requisites or positive features of a task may be formulated [D44:271*passim*]:

- With a rising number of options one's decisions are rather attribute- than option-oriented – no information is looked for regarding *one* option on *all* attributes (of this particular option), but information regarding *all* options on the most relevant attribute, then on the second-most relevant one, *etc.*
- An increase of attributes enhances the confidence of the deciders in their judgements and choice.
- Time pressure raises the error rate.
- Concrete information is preferred by the deciders to implicit information.
- Clear information facilitates deciding and is utilised more readily.
- The completeness of options influences the decision behaviour; if an option comes off well on one attribute, it is inferred that it comes off correspondingly on an attribute with less information.
- The format of presentation comes in, *i. e.,* the manner how options are presented ("framing").

Even if decision science does not offer a unified model all invoked theoretical constructs share the ability of being system- and process-oriented (also refer to the life phase model in systems engineering [D32:47*passim*]). In Chapters 12 and 13 attempts are made to consider the insights from decision research.

Chapter 8

DEVELOPMENT OF DECISION MAKING IN TECHNICAL SYSTEMS

Regarding radioactive waste approaches and issues in decision making, risk methodology, and risk perception have to be interlinked. Below the historical change of decision concepts, perspectives of "rationality"[102] and risk perception including the involvement of stakeholders, are outlined on the basis of the development of risk analysis[103]. The aim is to propose to amplify the assumption of "decision anomalies" starting out from the notion of single valid purpose-oriented rationality [D40][D50][D66][D67][D69][R96]. It is suggested that the discourse of rationalities leads to an adequate description of the decision issue under scrutiny, a logical step which has not been taken frequently yet [R19:95]. Insights from risk perception research were displayed in *Table 5-1*, respective criteria were crystallised in *Table 6-1*. Criteria with regard to decision making were given in *Table 7-1*, and conclusions regarding the decision-making process in radioactive waste governance will be drawn in *Table 9-1*. A proposal leading to an integral risk analysis will be made in Chapter 13.

A note has to be made that the respective courses and formats are sketched out and diagrammed in a simplified manner (as an example, there are neither just "the public" nor the "opposition" nor "the authorities" nor "the engineers").

---

[102] Rationality is "a consistent, adequate, and meaningful behaviour, based on insight, with regard to a given situation" [R8:87]. Von Foerster even defined it as "acting while keeping open possibilities to the greatest extent", according to [R50:185]. The terms of "rationality" and "risk" are used in parallel in the present context. See *Figure 1-3*.

[103] This presentation is given after and in extension of Otway & Thomas 1982 [R64] as well as Fischhoff 1995 [R21]. Also refer to Covello & Mumpower 1985 [M16] for technical and Nowotny & Eisikovic 1990 [R59:25*passim*] for sociological aspects.

1. EARLY DAYS (UNTIL 1960s)

Process:

A proponent/implementer proposes a project (decision step **1**) (see *Figure 8-1*) – the safety authorities review and decide (steps **2** and **3**, according to the motto "Decide–Announce–Defend, DAD" [MA5:24]). Analysis, assessment, evaluation, and management go hand in hand.

Mental model:
- "Absolute" rationality: "objective", in principle entirely quantifiable, mostly not yet though according to the formula Risk (R) = Probability of occurrence of the event (P) x Consequence (C).

Techniques:
- Technical and deterministic safety analyses (allow sufficient safety margin).

Actors/stakeholders:
- Engineers and scientists (with proponent and authorities);
- public: left outside.

Aim:
- Definition and determination of quantitative risk targets.

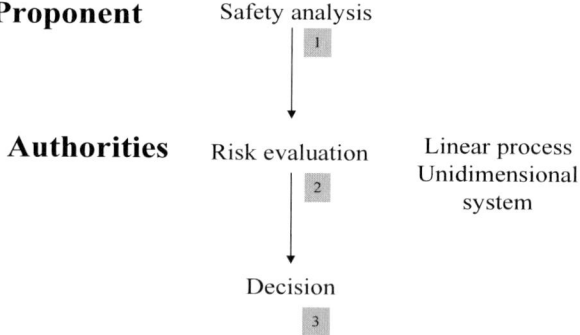

Figure 8-1. Linear decision making according to a simplified 3 steps-2 actors model: The authorities, more or less directly, base their evaluation and decision on the analysis of the proponent.

In the described decision-making model it is assumed that risk perception shall be governed by "objective" data and it can be changed by rational arguments – the "technical" rationality of engineers and scientists [R58:126]. This type of rationality is unidimensional in the sense that it reduces risk to a quantifiable entity [G62:123][104].

Institutionally it has to be said that the issue of radioactive waste is not tackled within a discipline of its own or an expert community, but directly by the waste producing organisations and the evolved companies. This, doubtlessly, occurred and occurs in collaboration with universities and research labs, but since the beginning of waste generation, issues and problem-solving patterns have been defined and elaborated by the producers. Risk knowledge, the monopoly on understanding, and the resources have been with the producers. This, in turn, has weakened the position of the other stakeholders, i.a., the oversight bodies, and limited the reviewing activities to the nuclear waste community[105].

The management of radioactive waste in Switzerland in the 1970s corresponds to the two models mentioned, the rationality model and the institutional model. Going along with this, the first official working group on the issue of disposition of radioactive waste, the Co-ordinating Committee Radioactive Waste, KARA, which created the so-called "national concept" of final disposal, consisted of representatives of the Confederation, the NPPs, Nagra, the Federal Institute for Reactor Research (now PSI), and Motor Columbus, a related engineering consultant. The institutional structure was minimal, "in concert", in 1975, "it was refrained from appointing a Federal Expert Commission" and, with the Committee, "a flexible, less formal solution was found" [P153:3]. One single (newly employed) scientific assistant within ASK, the predecessor of HSK, was in charge of radioactive waste.

---

[104] It has to be clearly noted that the promoters of quantitative risk analysis – like Kaplan & Garrick – receded in an early phase (1981) from a simple product as the valid risk definition; they warned of simplifications [M50:97].

[105] It is symptomatic of the radioactive waste management that, although models and safety analyses have always been debated and harmonised within the international waste community, it is only in the past few years that they have appeard in scientific peer-reviewed journals [G225][G229]. KASAM 1999 had to repeat their 1995 demand to intensify publishing in scientific journals [G144:107][G141:8]. The state of largely unpublished information was criticised in Switzerland by the Sub-Group of AGNEB in 1980 [P8:Appendix 5], by USGS in 1999 in view of Yucca Mountain [G266:17], incidentally a year before a special issue on the project in Reliability Engineering and System Safety [G225].

2. EMERGENCE OF (PROBABILISTIC) RISK ANALYSIS (1970s)

In addition to its function as a tool to analyse weaknesses, the quantitative, entirely or partially, probabilistic risk analysis (PSA) was established to substantiate the acceptability of low risks and to calculate cost-benefit analyses[106]. The setting of this was the growing public opposition against technological undertakings, which had to be faced.

Process:
A proponent/implementer proposes a project (decision step **1**) (see *Figure 8-2* overleaf) – the safety authorities examine the deliverables in a review process (step **2**) and decide (step **4**); the public "appears", often as opponents, and is informed (step **3**).

Mental models:
- "Absolute" rationality for risk analyses ($R = P \times C$), systematic approach by the (official) experts;
- "Bounded" rationality of the public, heuristical approach by laypeople, leading to so-called "cognitive biases" [D76].

Techniques:
- Deterministic safety and probabilistic risk analyses (with eventually high consequences but low probabilities), followed by
- "Risk communication" (education) of the public to gain acceptance of the decision.

Actors/stakeholders:
- Engineers and scientists (for the analysis of "objective" risks);
- Social scientists (for public perception studies of "cognitive biases");
- Communication experts (for educating the public, at least mitigating their "fears");
- Public (as pupils of the others in need of education).

Aim:
- Definition and determination of quantitative risk targets in consideration of non-quantitative risk perception aspects ("risk acceptance").

---

[106] It has to be emphasised that probabilistics are used only hesitantly and selectively in the field of radioactive waste performance assessment. See footnote 52.

Development of decision making in technical systems 117

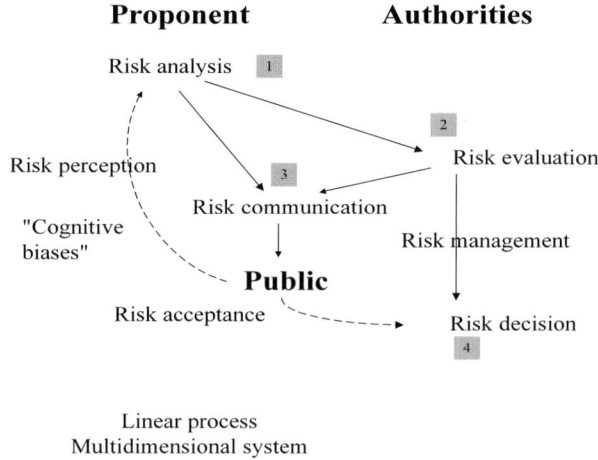

Figure 8-2. Sophisticated linear decision making according to a 4 steps-2+1 actors model: The model in *Figure 8-1* is broadened by risk communication, the means to instruct/inform the public as a "passive" actor. The process still is linear but the public as a disruptive factor makes the system multidimensional.

The management of radioactive waste in Switzerland way into the 1990s resembles the models in *Figure 8-1* and *Figure 8-2*: It essentially was based on the "Top-down" paradigm according to the terminology of TRUSTNET [M18:11-12][G88]. The involvement of third parties, besides proponents and authorities, *i. e.,* external experts, NGOs and the public, initially was minimal. As mentioned, in 1980 the just appointed Subgroup Geology of AGNEB demanded "a categorically necessary public involvement and publication" [P8:Annex V,1]. Six years later, the administrative lawyer Seiler observed: "In … the field of disposal there has never at all been a procedural involvement of third parties in so far as this entire topic was ignored in the issuance of operation licences … nor when fixing the deadline of [Guarantee] 1985 were third parties given a stake" [P253:31]. One-way communication dominated, it was risk communication from the experts "onto" the public. The authors of the so-called RISCOM Transparency Model [G56][M85] phrased it as follows: "Traditionally transparency has meant explaining technical solutions to the stakeholders and the public" [MA1:7]. The official stakeholders behaved in a rather paternalistic way, according to the "deficit model", and this twofold: First with respect to communication, they wished to convey information, so to speak, *onto* others [G281:38]; second, there was some sort of "democratic deficit" [MA1:15], following the former deficit, not showing up, though, in a lack of public participation but in a lack of legitimacy of governance.

Accordingly, Nagra replied to objections lodged by interveners in the procedure of application for HLW test drilling in Siblingen: "Nagra is willing to listen to concerned persons with genuine worries and in dire straits.... It fulfils ... a mission of environmental protection. This cannot be done with cheap political propaganda. The psychological impacts of drillings, to a large extent, depend on the concerned people themselves .... It is up to anyone to inform themselves in an objective manner and, thus, to encounter irrational anxiety" (after [P44:39]). Stakeholder involvement has gained much attention since (Sections 11.3 and 13.3), although there are relapses such as the statement in an NEA document of 1999: "The common perception amongst the public that there is a strong body of technical opinion challenging the feasibility of safe disposal does not reflect the realities of the debate" [G181:10]. A reflexive position, which is insinuated here by a title like "Progress towards geologic disposal of radioactive waste: Where do we stand?", should at least be aware of the fact that there are diverse "realities" (see *Figure 1-3*) and that the technical experts might not have addressed the questions raised by the public (see PART II).

3. SOCIAL RATIONALITY

The escalation of the risk issue takes place at two levels: On the one side, technologies and concurrent risks have become more complex, on the other side, the decision procedures have become more complicated by an increase of the gap between the risk analyst (who calculates the risk), the decider (who seeks or allows for it), and the affected person or risk bearer (who, in the case of long-term risk, eventually does not even exist yet). In addition, several levels of conflict have to be distinguished. According to von Winterfeldt & Edwards 1984 [R92:56] one may differentiate among "factual" and "value" conflicts about

- Data and statistics;
- Estimates and probabilities;
- Assumptions and definitions;
- Risk-cost-benefit trade-offs;
- Distribution of risks, costs, and benefits:
- Basic social values.

The dispute on nuclear energy, and hence on radioactive waste, ranges at all levels, whereby counterparts may simultaneously be at different levels and, thus, be talking at cross purposes (see *Figure 1-1* through *Figure 1-3*).

It is precisely the indications of diverse types of uncertainty (see Sections 6.3.3 and 9.8) which makes it clear that "social rationality" [D67:321*passim*] linking various aspects may embrace the characteristics of highly centralised technology and "evolutionary risks", even though (or because) it is diffuse in the eyes of the expert. Risk attributes not dealt with in risk studies play a vital role in the public dispute [R28][R96], sometimes not prone to be dealt with by experts, due to "overcomplexity". In nuclear technology this may be the connection of civil and military use or proliferation[107], the "normality" of disasters with system immanent failure [D67][D65] or the longevity and irreversibility of potential impacts. According to Wynne 1983 the public appraises a technology as a whole including its institutions [R96].

The imposed character of high technology makes the "NIMBY" notion to appear in a different light, even the role of the opposition to nuclear energy[108]. According to Mazmanian & Morell 1990 "NIMBY" arises from more than self-interest but from a broad technology-critical attitude [R54]. Kemp 1992 could show how local opponent groups became providers of information and trustees for a community which, otherwise, would have found itself in a passive position in decisions taken by others [MA24]. This was also beneficial to fairness of debate [G15]. Krohn & Krücken 1993 hold that NIMBY "is a veto principle ('not at our place') which proves its worth as a social motor for the production of future-oriented solutions" [R43:30]. Zillessen *et al.* 1993 go even farther and maintain that the laypersons' risk awareness is, in principle, more relevant than the experts' awareness [D94]. At any rate, as Jungermann & Slovic ^{2}1997 state, "laypersons fit in with needs and values which are seldom considered in conventional risk analyses" [R34:202].

Morone & Woodhouse 1986 view "[t]he actions taken by concerned groups and individuals" as "an important component of the catastrophe-aversion system". Their reason? "To relax the vigilance of those who monitor errors and seek their corrections would be to change [this] system. Quick reaction, sometimes even overreaction, is a key ingredient of that part of re-

---

[107] In this context it has to be noted that the IAEA may systematically only control facilities in Member States of the Non-Proliferation Treaty; military sites of nuclear states are not [G32:17], reprocessing facilities only partially under international supervision. Hug 1987/ 1998 proves that Switzerland kept the military option open, even though it was under secrecy [P136][P139]. This was confirmed by the official historian of the Federal Department of Defense [P261]. There is empirical evidence for the perceived respective connection [G55:107].

[108] Logically one might define as "NIMBY" to have adopted a nuclear energy programme without a conception of what to do with the waste. And yet, the utilisation of "NIMBY" is not productive in one way or the other; it leads to branding and blaming without the potential to recognise patterns [P233:205*passim*]. This is underpinned by evidence: In Slovic *et al.* 1991 only 273 out of 100,000 textual or image associations could be ascribed to "NIMBY" [G247:1605]. This was confirmed in other studies [R25][G243].

gulating risky technologies that relies on trial and error" [D63:174]. Luhmann 1991 makes the following statements: "... protesting reflection accomplishes something that is not accomplished elsewhere. It pursues issues that none of the functional systems would recognise as their own, neither politics nor business, religion nor the education system, science nor the legal system. It compensates obvious deficiencies of reflection of society – not by doing it better but differently. The rapidly gained attentiveness for ecological issues is due to them like the increasing challenge of trust in technology .... After all ... the sensitivity thus attained for impacts of structural decisions in modern society ... is an asset which one does not have to just value negatively" [R51:153-154]. In concrete words, Bullard 1992 praised in the siting issue of LLW in the USA "[i]ndeed, the environmental and public interest groups have been in the vanguard of the public participation process, and have contributed most of the ideas overcoming the problems associated with [low-level radioactive waste] technology's complexity and reach" [G27:719]. With respect to WIPP, King *et al.* 1991 of the US Office of Civilian Radioactive Waste Management recognised that "classic NIMBY does not exist in those communities affected by WIPP.... More, a reading of the efforts of other programs could suggest for viewing NIMBY as a rational response" [G149:475]. That not just minor modifications will do was shown by a task force report for the US DOE on trust and confidence of 1993: "Efforts to restore and sustain public trust and confidence cannot simply be appended to ongoing activities. There must be a recognition among senior policy-makers and managers that most choices have consequences for institutional trustworthiness" [G237:39]. The factual and substantive contribution by "citizen experts" is backed up by studies in other fields [R87].

According to studies done by the Research Center Jülich, the German public did not behave "emotionally" in the aftermath of the Chernobyl accident: "Most people were aware of the contradictions and uncertainties in the available information. Thus, subjective uncertainty about the level of risk caused by the Chernobyl disaster can be regarded as an objectively rational and appropriate response to a situation that has a risk potential which cannot be defined with absolute certainty" [G220:132]. This fits in with Renn's 1990 observation that the whole-body doses measured in 15 countries highly correlate with the increase in public opposition [G226:158]. Accordingly, high public pressure was the main factor in issuing dose reduction measures by the authorities (and thus resulting dose savings) [*ibid.*:165]. It was the so-called new social movements and citizens' initiatives which forced society to become sensitive [R43:29]. This is honoured by society's high esteem of "alternative" experts, they have a comparably credible status as "established" research institutes [G220:132-133].

Lidskog & Elander 1992 [R47] view the NIMBY effect as a matter of the choice of perspectives. Whereas conventionally the central authorities claim to represent the "public" and "national" interest with respect to repository siting, motives for protest might go deeper than "merely provincial selfishness". In terms of democracy theory, it is also a matter of who has the power over the local territory; environmentally, the local protest raises the issue of the explicit problem, the generation of waste by nuclear power. The conventional interpretation might be looked at as a logical fallacy (see Section 9.3).

Within the framework of, according to Beck 1988, "large-scale hazards of late industrialism" [R6:77], nuclear technology stakes out a forerunner-type position. It includes radioactive waste, even though its hazard potential is by far smaller than the one of a nuclear reactor. This contradiction goes along with the "rationality" of the following perspectives: the claim not to separate waste and production (the anti-nuke strategy) as well as Nagra's frustration to be liable for disposal due to their corporate identity as "active environmentalists" [P93:7] that they are the dog in the anti-strategy of "kicking the dog and meaning the master" [P65].

Due to the enlarged scope of "risk" and associated aspects, risk assessment, evaluation and management cannot be maintained separately [R2], unlike the official stakeholders could as long as they were able to enforce the linear decision model. Based on the criteria verified in Chapters 6 and 9 risk analysis is expanded, the approach of "robustness" and the disposition concept is developed to "extended" final disposal and, in the end, the decision model is augmented by a comprehensive stakeholder involvement (see Sections 11.3, 13.3 and 13.4).

Chapter 9

DECISIONS IN RADIOACTIVE WASTE GOVERNANCE

1. TRANSFER OF CRITERIA TO RADIOACTIVE WASTE ISSUES

The decision criteria investigated in Chapter 7 are applied to the radioactive waste field in *Table 9-1*. This transfer is elucidated and developed in Sections 9.2 to 9.10.

Table 9-1. Criteria for decision making in the radioactive waste field. Possible approaches for future management are outlined. Explanations follow in Sections 9.2 to 9.10 (Cont. overleaf).

Criterion for decisions on complex problems	Transfer to the radioactive waste field	Perspectives for radioactive waste governance
System understanding	Complex socio-technical system with floating system boundaries and an active environment; the societal part consists of today's and future generations.	Design a careful system modelling (technical and societal) with the aim of an integrative issue understanding; pursue a system-, cause-, and solution-oriented approach.
Avoidance of logical fallacies and biases	Predictability of the system: bad with partial system "technology", impossible with "society"; long-term safety demonstration possible only with post-closure data; consideration of individual and institutional decision anomalies.	Demonstrate the robustness of the solution chosen; establish a monitoring and controlling programme in the post-closure phase; intensify review process.
Consideration and	Radioactive waste as an "implicit	Strengthen supervisory autho-

Criterion for decisions on complex problems	Transfer to the radioactive waste field	Perspectives for radioactive waste governance
adaptation of problem structures	problem" (created by the use of nuclear energy, which in turn, addressed the energy (shortage) problem); system environment (groupthink).	rity and intensify review process; discuss contradictions and inconsistencies proactively, consider time dimensions (facility construction and impact, dynamics and variability).
Decomposition into subsystems and re-integration	Long-term character of project management and construction, long-term impacts of the disposal system.	Verify the consistency of the partial systems.
Goal relations investigation, complex goals	Goal analysis: system performance (sustainability); goal-resource relations (investment of resources for goal achievement); involvement in procedures ("process utilities"), trust issue.	Primary goal as system performance: Formulate as stability (protection against release of radioactivity); complementary goal is flexibility (intervention): Specify monitoring programme, quality assurance (integrated management), intensify reviewing.
Adequate management of diverse levels	Problem is never detached from environment (surroundings); comprehensive appraisal with multiperspectiveness only; broad discourse on values with complex issues.	Launch transdisciplinary discourse; consider diverse perspectives; bring in the issue of radioactive waste production.
Decisions under uncertainty	Final disposal is a decision under uncertainty (concomitant with the knowledge situation of partially probabilistic information [D35:16]).	Confront types of uncertainty with each other and make a trade-off (reduction of uncertainty by deferral of decision?).
Co-operation problem	Intragenerational and intergenerational equity principle.	Analyse compatibility: international solution/intragenerational issue; active control/intergenerational issue.
Utilisation of latency periods as opportunities	Phases of latency in times of change (concept discussion, cost pressure).	Set off a broad discourse, check arguments against consistency, take up suggestions.

In the Sections to follow, the respective issues, classified according to *Table 9-1*, are dealt with in the radioactive waste field and introduced and specified. At the respective end of the Sections 9.2 to 9.10, the propositions are empirically backed up and illustrated by insights from the content analysis done for Switzerland [G72:Vol. II].

2. SYSTEM UNDERSTANDING

Radioactive waste management is characterised by all features of a complex situation (of action and decision) [D9:58-66]: One has to consider many single attributes – from the waste characteristics via logistics, suitable interim storage and long-term disposal up to surveillance and quality assurance,

- namely often simultaneously and interconnected, *i. e.*, with side and long-term effects (of a technical but also institutional and political type);
- not as a static but a dynamic and extremely long-term issue;
- the internal dynamics of the system with its subsystems (technical, institutional, political) has to be assessed in its possible evolution;
- the situation is non-transparent for the stakeholders – they do not possess complete information, they do not even know in which situation they are at present;
- uncertainties about the state of the facility once built increase with time, all the more about the social environment [G231];
- the assumptions and the knowledge of the stakeholders on the system structure, their "model of reality", are probably incomplete and largely incorrect.

Already, the waste definition as such is "ill"-defined: What "wastes" are is not absolutely determinable and depends on their prevailing user or owner. According to the Federal Radiation Protection Act of 1991 [P265] they are "radioactive substances or radioactively contaminated materials for which there is no further use" (Art. 25, para 1). As mentioned in Section 6.4.1 the international definition is no better. The boundary between valuable substances, products and wastes is floating and dependent on various framing conditions. It is contentious, for example, whether plutonium or uranium from spent fuel are wastes (for disposal) or valuable material to be recycled in reprocessing. Waste – and this goes to the core of the issue – does exist but is a socio-technical construct.

The type of the disposal concept determines the system characteristics to a large extent. The hazard potential is associated with the toxicity of partly highly concentrated long-lived substances. If the choice is made in favour of a repository, *i. e.*, a disposal in the deep geological underground "with no intention of retrieval", no catastrophic events have to be expected – due to the chemical and physical waste properties (no explosives, no criticality) and the absence of large driving forces. The main mechanism is a low-level, but long-term, chronic release of radioactivity into the environment (see *Figure*

1-5). Such a possible impact is hard to detect with respect to location and time (an exception may be certain scenarios of human intrusion [G78]).

In an actively controlled long-term storage, though, non-chronic low-level release scenarios might dominate, be this by way of an intentional intrusion into a surface or subsurface facility, be it by an inadvertent emission of radioactivity. From a deep disposal facility of this type there exists a potential short-circuit with the biosphere via the necessary control channels. The complex system modelling requiring a high effort *per se* is complicated by the fact that human activity – be it unplanned intrusion or necessary controlling – cannot be sufficiently assessed in a safety analysis for (hundred) thousands of years (see Sections 9.8 and 9.5).

A thoughtful system modelling with technical and societal subsystems aims at an integrative problem understanding. Waste facilities are primarily designed to provide protection against detrimental emissions of radioactivity; this demonstration has to be transparent, for today's deciding generations and our potentially affected successors. Transparency must be institutionalised in the factual and political decision-making process, as illustrated in its dynamics and complexity in *Figure 9-1*.

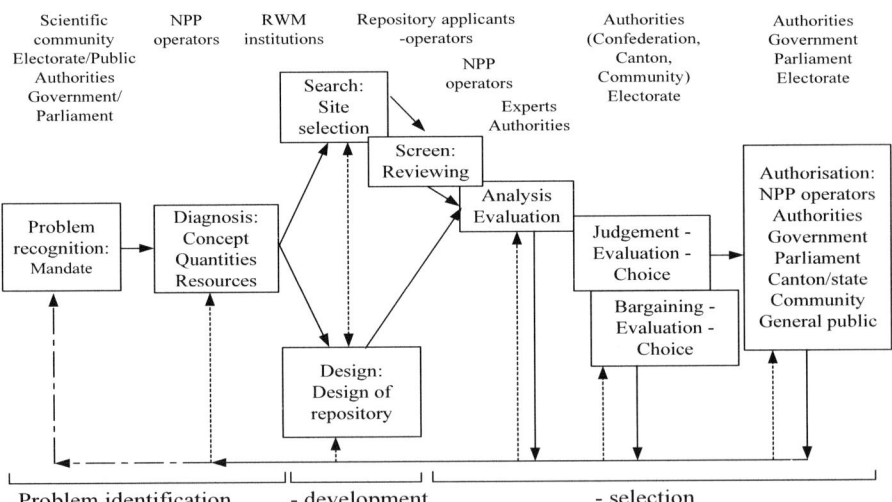

Figure 9-1. Phases of the decision-making process in Swiss radioactive waste management, modified after Mintzberg *et al.* 1976 [D62:266] and Hanson 1991 [D35:10]. The process of each project (interim storage, repository facilities) consists of several stages: from general licence to closure licence and sealing. The planned feedbacks (incl. criticism) are marked with dotted lines. Unplanned feedbacks, such as the rejection in a referendum or financial cutbacks of waste organisations, or even new claims for concept change, are shown as broken lines (bottom left). At the top, the main stakeholders are indicated. For a concrete suggestion refer to *Figure 13-9*.

The complex nature of the issue underscores potential fertility of using a systems approach. The OECD study of 2003 on "Emerging risks in the 21st century" gets to the heart of it when it concludes that "emerging systemic risks demand a systemic approach" [M66:257].

2.1 Empirical findings on system understanding

Integral system understanding requires a wide comprehension of the issue, which does not exclude the designation of gaps. Particularly in the initial phases, the radioactive waste issue, however, has been misused as a political vehicle by both opponents and proponents of nuclear energy: "insolvability" as an "argument" for phasing out *versus* "solution" as a "proof" of the legitimacy for a prolonged use of nuclear power. As a recurrent theme even, the seeming juxtaposition of the "technical" (technically solved) and the "political" (*i. e.,* politically problematic) pervades all time periods. Implementers and authorities, on the one hand, have maintained that all technical issues are under control and the "reason for delay" is merely "political"; on the other hand, the actors mainly in the NGO-oriented sphere have asserted that not even the "technical" basics have been solved to date. From 1957 to mid-2002 I found over 150 such examples [G72:Vol. II]. This situation is reflected in Nagra's corporate identity. Their director Issler, now President, declared in 1982: "We understand our business exclusively as a scientific-technical one" [TA, 1982-2-2]. As for the demonstration of "Guarantee 1985" the former President Rometsch said: "As a result, one is looking for technical solutions for a psychological issue, which of course is impossible" [P223, 8/82:3].

This phenomenon is not unique to nuclear issues though; in another context the Swiss constitutional lawyer Max Imboden in 1964 coined the following expression: "The essence of politics is based on trading off technical options" [P143:26]. The political value of the radioactive waste in the debate on nuclear energy, however, has often been so high that factual aspects were put last. Accordingly, the Federal Commission on the Surveillance of Radioactivity (KUeR) received the mandate, back in 1956, "to take care of the issue" of radioactive waste [P119:613]. Apparently, this order has never been executed; this is understandable (and demonstrates the importance of top management commitment) since even the commissioning authority, the Federal Councillor in charge Max Petitpierre, valued the issue as "not very urgent" [*ibid.*]. In 1979 AGNEB commented on the disposal concept of the power industry that "essential decisions ... had not been brought up", such as retrievability and options "with deferred reprocessing" and "without reprocessing", or interim storage; in addition, an independent review was viewed as of importance [P4, Annex 3:3-4]. With the exception of this last point, the issues have, until very lately, not been dealt with. In 1983, the only independent expert body up to then, the Subgroup Geology of AGNEB stated: "In the long run it has to be strived for a full-time, competent, and neutral body of which the Confederation can dispose in issues of national interest and which is qualified to 'review geological experts' reports'. Such issues do not only bear on the disposition of radioactive wastes" [P273:1]. This demand vanished, since the positions were staked out, such as by the Federal Council in 1989: "To further ensure the aim of a sufficient, broad-based, and secure energy supply, as well as for ecological reasons, the nuclear option has to be kept open.... The yet unresolved issues of nuclear disposal have to be settled" [P77:53-54]. The counterpart had also taken their stand, such as Greenpeace in 1993: "The 'safe final disposition' is out of the question" [P113].

Senator Piller, in 1990, put the dilemma in a nutshell: There is "[t]his certitude expressing that the issue will be solved – for the first time I read that way back in the first half of the 70s, in a cloth paper pamphlet: 'The problems around final disposal are solved' –, this repetition slowly gets a magic spell.... Evidently the Federal Council is of the opinion that this disposition has to be possible. Exactly such a circumstance nourishes doubts about the [displayed] scientific objectivity. One would also have to be able to imagine that the disposition is not feasible in Switzerland and that something else has to be found" [P258:26]. The situation of the "implicit problem" has vast implications, up to wording such as "supportive" or "supporting research" as used in Sweden [G141:59-60].

The centre of attention was what was politically enforceable not what was really necessary. Obviously, the problem definition was floating, and this inevitably led to misunderstanding and complications.

Integral system understanding includes the societal domain, including present as well as future generations. This requirement has, until lately at least, been just another factor of disturbance given the primacy of energy policy and politics, although even from the perspective of a proponent this need not be the case (see Section 9.10).

3. AVOIDANCE OF LOGICAL FALLACIES AND BIASES

In PART II, Chapters 5 and 6, reasons for so-called individual decision anomalies (biases) were explored. In accord with the decision theoretical literature it is asserted that not only such "individual" anomalies (of concerned citizens) exist, but that there is also an institutional bounded rationality [D58] leading to institutional anomalies or biases [D49].

From the multitude of potential "fallacies", rather: constraint in their own mindset, one may emphasise the difficulties in system predictability. The partial systems "technology" and "geology" are difficult to foresee, the partial system "society" virtually impossible to foresee. Thus, a conception based on essential control is a bad solution from a systemic view. To assist in validating the traditional safety analysis, done long before waste emplacement, the demonstration of long-term safety requires data from the post-closure phase of a repository. Such a demonstration is more plausible if robust scenarios (see Chapter 13) are chosen for a basis; transparency is better achieved if the proponent's work is accompanied by comprehensive and broad reviewing.

3.1 Empirical findings on logical fallacies and biases

The inconsistencies mentioned in Section 9.2.1 are examples of institutional anomalies. Practically, this is manifested in the search for "quick" solutions and the incessant call of nuclear supporters for a reduction of the means of legal redress. Upon the governmental decision on "Guarantee 1985" in 1988, time pressure on Nagra, in principle, ceased because the Federal Council abstained from setting further deadlines. Factual pressure also decreased because the issue in principle (of feasibility) did not have to be answered any more. The co-operative, nevertheless, pushed on with the selection procedure for HLLW sites in sedimentary rocks, in HSK's and KNE's views too quickly. KNE criticised Nagra's procedure in the following words of 1990: "The selection of various formations and various siting regions must ... be based on a comprehensive, traceable exploration strategy, not to repeat similar mistakes [as in "Guarantee 1985" where localities for test boring were selected prior to seismics].... This issue also pertains to ... the choice and weighting of the parameters ... having led to the exclusion of the entire western Molasse basin" [P164:7*passim*]. Furthermore, the committee gives advice with regard to possible opposition, although not taken up by the receiver: "Experience shows that the selection of siting regions does not only have to be clear and transparent but comprehensive. Accordingly, the frequent question is asked: 'Why precisely here with us?' Comprehensive signifies that all possible formations and localities should be checked against the preset criteria and investigated. Thereby one may proceed iteratively stepwise so that in each phase stricter and more rigorous requirements are made. Such a procedure would mean that already the concept of the sediment study is based on a comprehensive exploration strategy" [*ibid.*]. The background of this dissent probably was that Nagra still preferred the crystalline basement at that time and viewed the sediment option as lines of extra work (for details see Section 10.3).

In the policy domain the Federal Council made another attempt in 1991 to speed up the proceedings in nuclear legislation – against harsh opposition among the cantons fearing some reduction of federalism. Accusing the cantons of having taken liberties, the Government threatened: "The modifications made in the legislation of Nidwalden would have the consequence that the Nidwalden Popular Assembly could, at least for some time, block the construction of a repository.... There is the risk that the Wellenberg project is deferred" [P80:1365]. It was, indeed, the electorate of Nidwalden who rejected GNW's application for a general licence in 1995; it was, however, also the Canton of Nidwalden (not the Confederation) which provided new impetus to the programme of "national" relevance with a phased approach and recommendations issued by a cantonal expert group (see Section 10.2).

Logical fallacies regarding system predictability of disposition concepts (Section 9.3) as well as uncertainties (see Section 9.8) were stated by the environmental organisations SES, Greenpeace, and MNA when they did their "Core Statements" on behalf of EKRA in 1999 [P256]: "The deficiencies detected in the final disposal concept [ignorance or absence of knowledge, need for information transfer, no intervention mechanism, too great risks] make it indispensable to look for an alternative. In contrast to it, the controllable and retrievable long-term storage provides the necessary increase in safety." This topic is dealt with more deeply in Chapter 12.

Since we are dealing with a complex socio-technical issue it is not advisable to revert to technical aspects even if "objective" evidence might call for it. An example for this is Parker's comment in a status report of 2000: "We should, however, keep the problem in perspective. This paper is about radioactive waste management and not radiation risks to humans, though that is what is of most concern to people" [G218:2].

4. CONSIDERATION AND ADAPTATION OF PROBLEM STRUCTURES

As mentioned, radioactive waste is an "implicit problem"; it was created by the use of nuclear energy, including medicine, industry and research, which in turn, addressed the energy (shortage) problem. The main theme of critics was to "kick the dog and mean the master" [P65] (see page 121). Time and again, the general energy political debate has an effect on the radioactive waste issue. This also explains the frustration suffered by the actors: Nagra, as the waste implementing company in charge, have been wearing themselves out in "providing scientific answers to political questions" as the former President, Rometsch, pertinently observed two decades ago (LNN, 1985-6-21). The opponent groups, however, are faced with a "technocratic" waste community (Nagra, authorities, experts) which do not even consider the "real" energy political problem, *viz.,* the continuation or phase-out of nuclear energy. The fathers of the "explicit" problem, the NPP operators, do not even sit at the table. The supervisory authority, too, merely regards the issue as a "technical" one; they hold themselves not responsible for energy political aspects. Such a constellation – as it showed up in the mediation exercises around the KORA group, "Conflict-solving Group Radioactive Waste" [P62][P63][P168][P169] – intensifies groupthink, by which the stakeholders isolate themselves, be it intentional or not. This, in return, may lead to not admitting (internal) criticism and to the impasse that no common solution range may be discovered.

Consequently, the problem range, including the resulting problem pressure, arises differently, according to the corresponding perspective. *Figure 9-2* depicts a general viewgraph of how a problem situation may be defined.

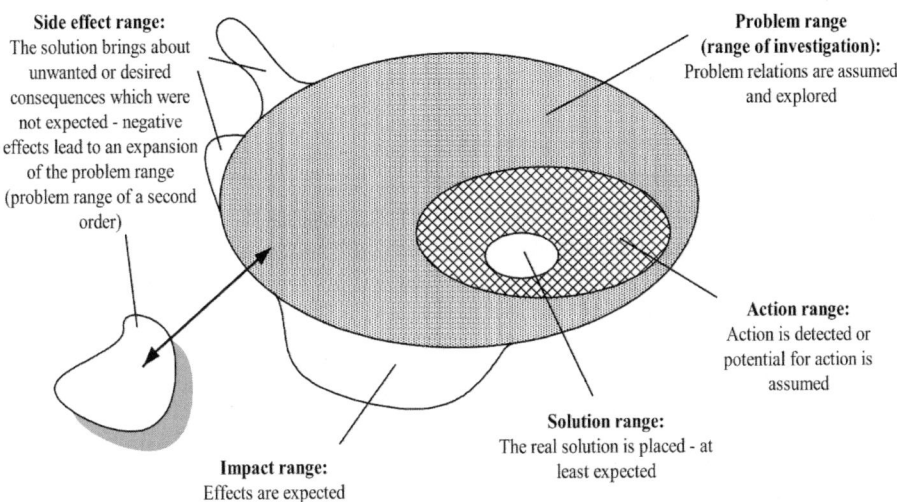

Figure 9-2. Divergent system units (after Ninck *et al.* 1997, considerably expanded). If an issue/problem is adequately tackled, impact range and problem range coincide. Depending on the resources used the action range is large or small. "Side" effects may be so grave that they lead to an expansion of the problem range (such as a problem range of the second order). A decision is well supported if it integrates relevant parts of both the problem and the solution ranges of the main stakeholders.

Table 9-2 below specifies the entirely different problem situation of the individual stakeholders:

Table 9-2. The main stakeholders define (at least: used to define) the system ranges and problem ranges (according to *Figure 9-2*) in a distinctly different manner. Consequently, it is evident that they are talking at cross purposes.

Stakeholder groups / System range	Radioactive Waste	(Continuous) operation of nuclear power plants
Implementer (Nagra)	Problem range and solution range, action range (technical)	Non-problem range (becomes problem range of 2nd order)
Potential siting state (canton)	Problem range as well as intervention and solution range (political and technical)	Problem range (political)
Opposition	Action range (political and technical)	Problem range and solution range (political)
Safety authorities	Problem range and solution range (technical)	Non-problem range (technical)
Department of Energy	Non-problem range (trad.)	Problem range (political)

The specific problem structure of radioactive waste is accentuated by the large time dimension involved, *i. e.,* the "objective" and the "institutional" time dimension, respectively (see Section 6.3.2). These aspects lead to the requirement that eventual contradictions and inconsistencies have to be proactively discussed, time dimensions duly considered with respect to the construction of disposal facilities and system impacts, the supervisory authorities strengthened, and the review process intensified (increased control by "third" parties).

4.1 Empirical findings on problem structures

How problem structures are dealt with is particularly well discernible if one analyses project management including work schedules. In sum, it may be asserted that management throughout was unrealistic in its assessment of the task complexity. As for the long-term character of a disposal project refer to Section 9.6.

Different stakeholders define different fields of problems, and thus exhibit a suitable proximity to respective other stakeholders, as *Figure 9-3* to *Figure 9-6* illustrate. This "proximity" is, *i. a.,* expressed by how the arguments of those are better understood – or worse (in case of greater distance). By this, they come up with common – or dissenting – conceptions. Such a mechanism impedes an exchange of resources, let alone a "novel coupling of policies" (see *Figure 12-3*). If the major problem is that "approximately every four years a new big electricity generator is needed" [P188] as the Federal Council believed in 1979, the radioactive waste issue becomes even more of a marginal interest. At that time, energy scarcity definitely scored higher than disposal on the political agenda. Accordingly, even in 1980, the continuation of the 1974 Federal Decree on the Restriction of Energy Consumption (in periods of crisis) was passed [P79][P73]. Waste disposal served to determine the course in energy policy and politics. All official stakeholders in radioactive waste management were at the service of avoiding "shortages in energy supply" [P223, 3/80:12].

Some rough sketches of possible perspectives of major actors in Swiss radioactive waste management are given below. Out of intrinsic logic for the shareholders of an electric power company, disposal is just an obstacle to their primary goal ("Focus"), *viz.*, the distribution of dividends (*Figure 9-3*). In times of opening energy markets cutting costs is of central importance. Corresponding signals were uttered, such as by Kurt Küffer, formerly with the Association of Swiss Electricity Companies (VSE), then with ZWILAG, in 1998, that operating NPPs should only be "as safe as sufficient" instead of "as safe as technically possible" [P182]. One of the consequences was a cutback in Nagra; the waste co-operative is not felt to be so close to the shareholders as a strong position of the power company on the market.

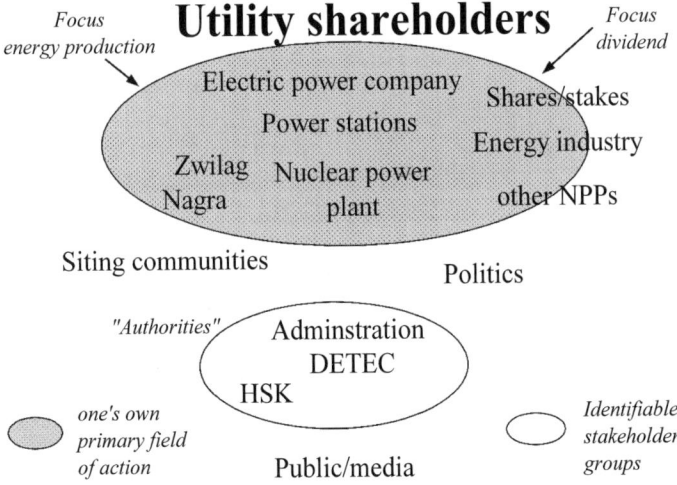

Figure 9-3. Perspectives of stakeholder groups (degree of identification or proximity): View of the shareholders of electric power utilities. "Focus" means the central target or field of action. Note the various positions the nuclear safety authority HSK has. See List of abbreviations.

Nagra, therefore, is not only under "traditional" pressure (of "green" and "left-winged" origin) but has had to defend its mission, the disposal of waste, against "internal" opposition. Accordingly, the energy industry is depicted on the verge of Nagra's field of action in *Figure 9-4*. In a bit of predicament between its mission and multiple pressures, the Co-operative is faced with the need for arguments when it has to link the imposed retrenchment with an increase of activities abroad (to acquire third party funds but with reduced staff) and the self-image of a "Competence Centre" (with additional tasks). It does not add to credibility if President Issler said in 1999, when approached with the loss of trust in Nagra searching lean solutions abroad: "The view across the boundaries is no question of trust but the consequence of opening markets and globalisation.... For countries such as Switzerland, with a small nuclear programme, a multinational solution is attractive" [TA, 1999-3-1].

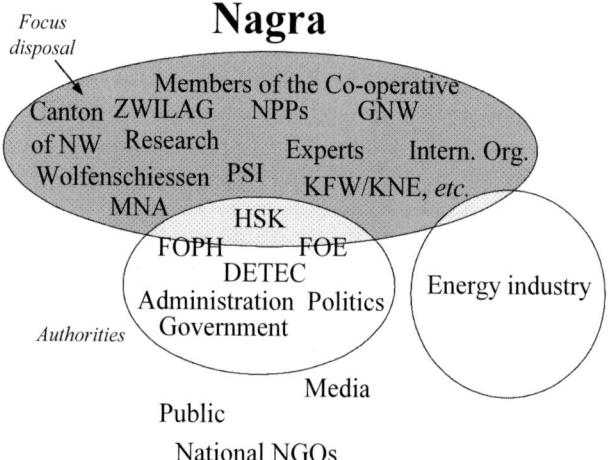

Figure 9-4. Perspectives of stakeholder groups: View of the proponent or waste implementer. Refer to *Figure 9-3*. See List of abbreviations.

Also the safety authority, HSK, is confronted with massive pressure against the background of a liberalised market and a concurrent call for an "efficiency-oriented" – cheaper – administration (*Figure 9-5* overleaf). Re-organisation and restructuring place great demands on constancy and stability of their focus, given the relatively low staff number.

With major stakeholders it is decisive to clearly define and sustainably orient their activities to their central and long-term problem (and thus solution) ranges, eventually regarding the formation of new coalitions (see Sections 6.4.1, 9.9, and 13.3).

Decisions in radioactive waste governance 135

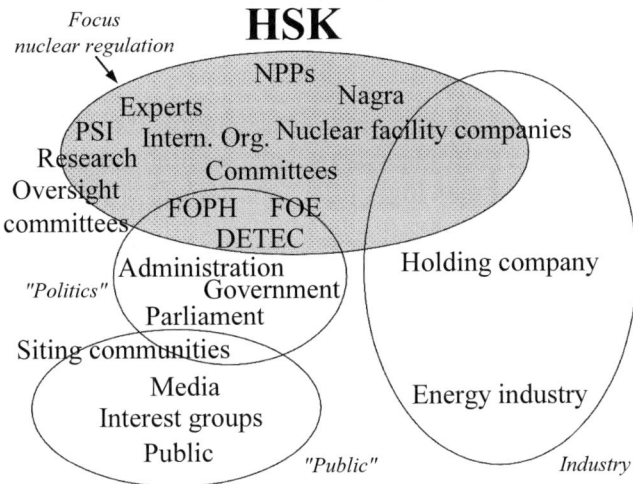

Figure 9-5. Perspectives of stakeholder groups: View of the safety authority (HSK, oversight and regulation, together with FOE). Refer to *Figure 9-3*. See List of abbreviations.

The general public is without a particular perspective (*Figure 9-6*) because they have no project- and task-oriented focus:

Figure 9-6. Perspectives of stakeholder groups: View of the general public. Contrary to the other stakeholder groups the public has no real focus. See List of abbreviations.

Surveys revealed that the public views the waste issue in an ambivalent way: It is a "national task" [P56] which is "not solved" [P104][P101] whereby one is not prepared to collaborate [P102]. Statements and actions by the main stakeholders are thus pivotal.

5. DECOMPOSITION INTO SUBSYSTEMS AND RE-INTEGRATION

The systemic approach also serves "systematic" problem solving. After a broad structuring, the resulting subsystems facilitates the treatment of partial systems (technology, financing, policy, *etc.*). Their solutions always have to be appraised within the scope of the total system – the partial solutions must be integrated into the total solution. A comprehensive safety performance assessment of a site (including validation) is, on the one side, staked on corresponding political provisions (the goal of a "permanent and safe final disposition and disposal" according to the Federal Decree or a continuously controlled long-term storage); on the other side, it has to embedded in an adequate implementation strategy and funding. Part of it is a consistent goal-means relation: Performance as well as the capacity for development must be intact. The disposition strategy has to be coherent, also when opened to its environment, *i. e.,* when it integrates external aspects and requirements. Nevertheless, such an opening does not have to be contrary to the programme coherence because it may add to an internal programme stabilisation [D6:274]. Refer, *e. g.,* to Section 9.6.

What makes stabilising (neither encrusting nor abandoning) the programme even more difficult is the pronouncedly long-term character of the disposal programme going along with, or interfered by, related programmes such as the continuation of phase-out of nuclear, interim storage, eventual reprocessing, *etc.* (see *Figure 9-7* overleaf). This factor has lately been conceded by the IAEA 1999a [G114:245-246].

Decisions in radioactive waste governance 137

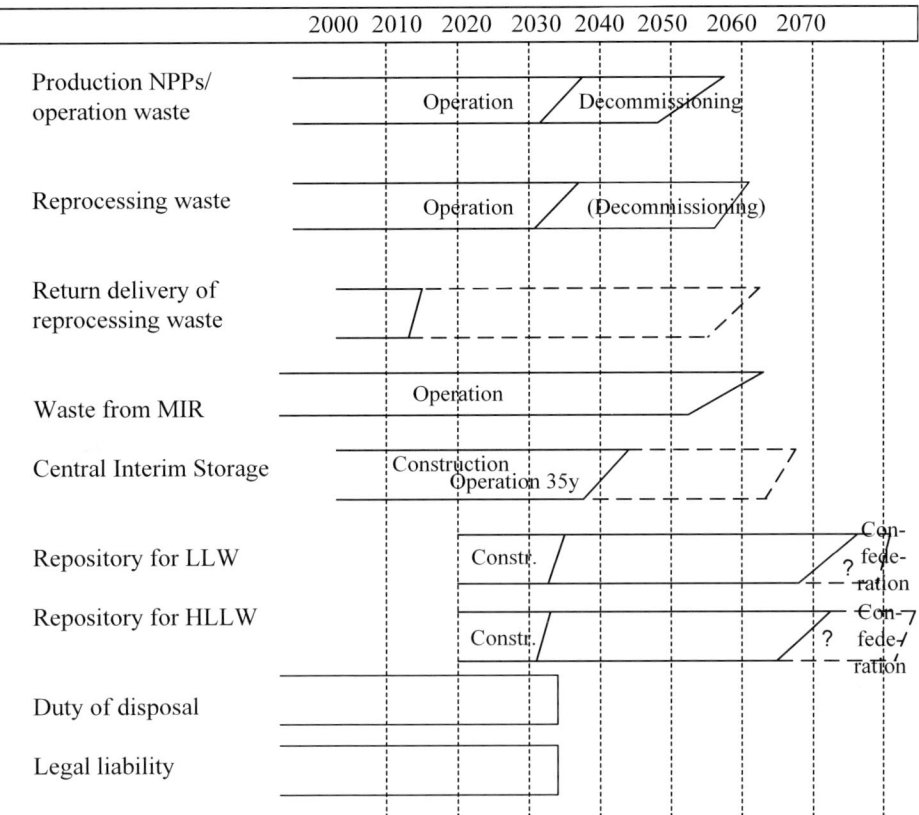

Figure 9-7. Time dimensions for duties of disposal and liability in Switzerland. For abbreviations refer to the List of abbreviations.

5.1 Empirical findings on decomposition and re-integration

Project longevity renders a consistent management and an appropriate allocation of resources arduous (*Figure 9-7*) even if one declares sealing in the foreseeable future as a "definite" termination of the project. Initially, when the first programme for disposal was conceived ("Guarantee 1985", around 1976/1977) people were not aware of, or did not want to admit, the complexity of the issue. Against this background one has to judge statements as the following, by the former President of Nagra, Rometsch: "We already know today that the project is feasible, it's just the demonstration of safety which is still pending" ([P243] in 1982), or: "Technical feasibility is not problematic…. The sensitive point is long-term safety …" ([P244] in 1984).

It was also in official policy where the goal-means relation was not adequate. In spite of the declaration that disposal was a "national task" it was left almost exclusively to the NPP owners' co-operative, Nagra, after a narrow interpretation of the causality principle. To exemplify this, in 1979 AGNEB noted that "the Working Group's job to elaborate a federally owned disposal project could be abandoned" due to the producers' responsibility [P4:6]. The insufficient provision of the supervisory authority has been mentioned several times, but also safety research in the disposal field has been almost exclusively left to Nagra or to the Paul Scherrer Institute (PSI), whose technical Waste Disposal Laboratory is predominantly remunerated by Nagra. In 1999, the responsible Federal Energy Research Commission (CORE) militated for cutting the resource of 2000 to 2003 into half: "Independent of the search and assessment of concrete sites for repositories (which is not the duty of energy research), the investigations of the last years have answered the essential questions pertaining to decisions sufficiently so that it is permissible to downsize the public means reserved in this field by half in the years to come (from today's approximately 8 million to 4 million francs)" [P50:38-39]. In the subsequent concept, for the years 2004 to 2007, the Commission concedes that in "some domains a sufficient basis of knowledge is still lacking for future decisions" [P51:47]. Nevertheless, "for these studies public funds have to be reduced. Until 2007 to 3 million francs per year". The "Priorities" are confined to geochemistry, radionuclide transportation and leaching behaviour of glass matrices [*ibid.*:49]. The need for integrative and anticipatory research in the disposal field is identified in Sections 13.5 and 13.6.

6. GOAL RELATIONS INVESTIGATION, COMPLEX GOALS

After Mag 1990 "optimum decisions are ... always merely optimum with respect to a certain goal, with respect to other goals often they are not" [D56:28]. But: Complex problems such as the present issue *are* defined by multiple goals, so-called polytely (see Section 7.3). The often quoted key word of "trust", however, is a "complex goal" according to Dörner 1989 [D9:81*passim*]. Such types of goals are difficult to analyse because they are either hollow phrases or they cannot be achieved and assessed at all. For example: Some states disarm whilst others re-arm – both do it for the sake of our planet's peace. In analogy, trust is to be broken down into partial goals, after Dörner, to be decomposed (see also [G135] and Section 6.4.4).

In an adequate goal analysis, the system performance strived for has to be examined as well as the aforementioned goal-means-relations (the deployment of resources to reach the goals) and the participation in procedures ("process utilities"). It is still the aim in Swiss radioactive waste governance that "the permanent and safe final disposition and disposal of the ... radioactive wastes has to be guaranteed". This was, in 1979, laid down in a national vote on the Federal Decree on the Atomic Energy Act [P39] and confirmed in the revised Nuclear Energy Act, in force from 2005 [P264][P30].

After all, the storage place for radioactive waste, being a technical facility, will have to be built according to technical criteria. An examination of variants is necessary thereto entailing conceptual decisions including goal issues. In the years past a comparison of option has been voiced internationally [G140][G182]. The NEA suggested in 1999: "Overall confidence must be developed in a much wider audience if a decision to implement disposal is to be acceptable", "the waste disposal community must be ready to discuss the merits of other waste management strategies" and "a phased repository development process keeps options open for very long times into the future [G184:8,10,22]. A year before, the IAEA had already stated: "In view of the current questioning attitude of many people to the established view of experts ... the possible alternatives: long term surface storage and disposal with the provision for retrieval, should be critically examined by independent international groups convened by the IAEA" [G113].

In complex issues it is very well feasible that conflicting goals exist. The magic spell of "sustainability", a complex goal as well, encompasses protection of, and leeway for, future generations (see page 13). In the case of a safe disposition of radioactive waste, both passive safety and "active" control or surveillance need due care and attention in parallel. Respective decisional situations were given consideration in Flüeler 1998 [D22]. In this context, some authors insinuate hidden goals, after Keeney & von Winterfeldt 1986 so-called "hidden agendas", i. e., opposition against an individual project be just used as a pretext to achieve wider goals [R41:420][109]. The concept of polytely does not have to resort to such constructions.

The conflict in the debate may be illustrated by means of a goal comparison (*Figure 9-8*):

---

[109] With malicious intent one might impute a hidden motive of the NEA if they want to involve all stakeholders in the process as put forth with the idea of a "Forum on Stakeholder Confidence" [G186], for "there is consensus that the management, storage and disposal... is a prerequisite to general acceptance of the continued and future use of nuclear power and should therefore be put in a special, top-priority category" [G180:43]. Accordingly, the 1956 Statute of the now more open-minded IAEA declares as the paramount objective: "The Agency shall seek to accelerate and enlarge the contribution of atomic energy to peace, health and prosperity throughout the world" (Art. II). It is only under Art. III A.6 that we find the task to "establish or adopt ... standards of safety for protection of health and minimization of danger to life and property" [G97]. It has not been amended by, e. g., the protection of environment.

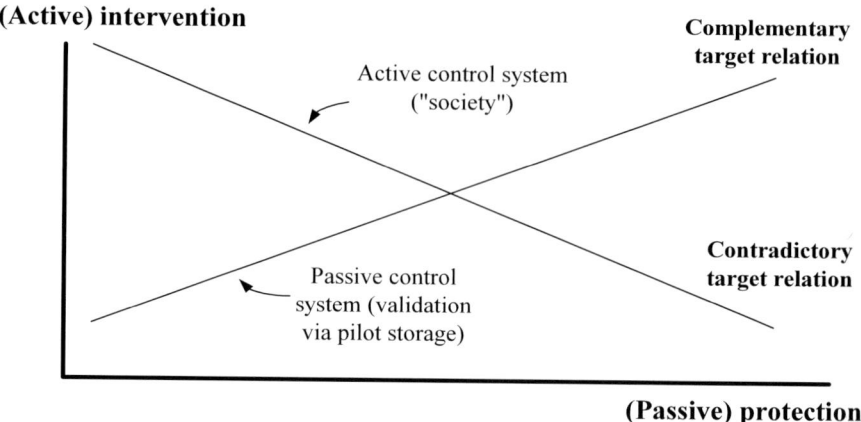

Figure 9-8. The principle of sustainability is based on two pillars: on protection and intervention of present and future generations. Depending on the disposal concept, one or the other target is valued higher: intervention with the controlled long-term storage, protection with the final disposal concept. With an active control system, intervention and protection are contradictory with respect to goal relation whereas a "passive control system" allows target compatibility.

The primary goal in radioactive waste management and governance is the stability of the disposition system as it performs: The protection against harmful emission of radioactivity is to be continuously ensured. The discourse on goals and criteria is developed in Chapter 12.

The input of resources to reach the goals (goal-means relation) is decisive in the achievement of goals. Repositories or stores require a sufficient support by proponents and operators to characterise sites, verify models, and validate safety cases. Controlled long-term storage requires an adequate allocation of resources during the period of control (see Section 9.5).

6.1 Empirical findings on goal relations

Different goals of different stakeholders need not even be "hidden agendas". Public authorities insist on enforcing decisions once taken, be they taken in a participatory manner or not (see Section 6.4.1). Proponents elaborate partial solutions (such as Nagra with technical performance assessments), and still have to keep an eye on the waste producers, for institutional and financial reasons (see Section 9.4.1). Citizens' initiatives and NGOs want to exert pressure and depend on donations and fund-raising campaigns.

The symbolic power and the effective instrumentalisation of radioactive waste in the energy policy debate are such that this eventually blocks the view on factual and protection issues. It happened when in recent years the question of "final disposal or long-term storage" was debated again[110] – in very different perspectives, such as the anti-nuclear protagonists, Macy 1989 internationally, and Kreuzer 1990 domestically, with the notion of a "nuclear guardianship" [G159][P175], or a former top manager in the Swiss nuclear industry, Lutz 1997, in analogy to the Egyptian pyramids [P187], or Flüeler 1991/1993 with the proposal of an "extended final disposal" [P85][P86].

If a problem, here the long-term governance of radioactive waste, is defined differently, it is reasonable that the goals are different as well. Except for the divergent problem definition the dispute may be traced back to a contradictory interpretation (and weighting) of sustainability.

Since the concept of sustainability has, up to date, not been specified in the context of radioactive waste (see the attempt in Chapter 12), its interpretation is left to each stakeholder according to their wish. Forum vera sees solely the aspect of protection in it: "The commandment of sustainability demands to impose no burdens on our children" [P97]. Damveld & van den Berg 1998, lately called in by the Swiss NGOs[111], give priority to retrievability: "The present generation places great demands on the disposition of nuclear waste, which should also be true for future generations to avoid negative effects. Permanent retrievability can comply with that. Thus, with permanent retrievability [each generation to come] gets the possibility of controlling the waste and taking commensurate measures" [P54:7]. The laudable idea of the Federal Department for Energy (DETEC) to base its ministerial strategy on sustainability and "to thereby disclose goal conflicts and substantiate the value judgements made" gets blurred where various "sustainabilities" are mentioned [P278:18]. The explanations are kept general: "Sustainability in energy matters means in detail: Ecological sustainability… The safe disposition of nuclear waste … economical sustainability… The internalisation of external costs …" [*ibid.*:3-4,18*passim*]. Neither helpful is the debate initiated by the Federal Office of Environment on sustainability and sustainability research, respectively [P144][P145][P252]. The Netherlands did officially base their radioactive waste governance on long-term storage, though without specification [G49][G9][G273][G238][G272].

---

[110] After very open options in the early days of industrial exploitation of nuclear energy the debate was revived in the 1970s: with the concept of an "underground retrievable storage" [P118][G87][P6:Annex IV] which, to a certain extent, was endorsed by some geoscientists [P5].

[111] The Swiss Energy Foundation (SES) demanded in the preparation of the EKRA discussions in 1999: "The authors of the Dutch report 'Nuclear Waste and Nuclear Ethics' (Herman Damveld and Robert Jan van den Berg) for the attention of the Dutch Commission for Radioactive Waste Disposal (CORA) are to be consulted by the Expert Group", *i. e.*, EKRA [P255].

Content analysis shows that goal discourses have been initiated time and again in Switzerland, *e. g.,* in [P5][P118][P155][P29]. The formulation in the official consultation report on the 1994 revision of Atomic Energy Act got to the heart of it: "It is throughout criticised that the elimination of radioactive wastes (Art. 87) shall be carried out in such a way that no safety and surveillance measures will be necessary after emplacement. It is pled that the biased formulation admits only the final disposal philosophy which is [according to the interveners] today no longer tenable ..." [P81:21].

A discussion on the management concept was repeatedly promised but – until EKRA – never carried out. AGNEB already in 1980 demanded: "Planning ... and construction of interim storages and repositories ... are to be pursued as long as comparable alternatives have come to hand for all radioactive wastes and their elimination" [P8:Annex IV]. To the contrary, demands or guidelines were blocked and attenuated. Even after the discussion in the framework of "Energy Dialogue" (see below) GNW "holds conceptual debates on repositories or controlled long-term stores to be superfluous" [NZZ, 1998-11-24].

In the meantime, even the "nuclear establishment" has come to see sense in having a broad discussion, in principle, if a sustained solution is to be found (see Section 12.7).

7. ADEQUATE MANAGEMENT OF DIVERSE LEVELS

The indication by Kasperson *et al.* 1992 that waste controversies "are less about risk than they are about institutions" [R38:175] makes it clear that the issue is never detached from its context. A comprehensive examination is only feasible if multiple perspectives are considered [G219:265][R30]. Increasing complexity enhances the relevance of a debate on values [D38:10][D18].

7.1 Empirical findings on adequate-level management

Whoever is interested in a comprehensive perspective and in sustained and resilient solutions should set off a discourse on the production of radioactive waste. It is correct that, according to Swiss constitutional understanding, the plebiscite is the epitome of citizens' involvement; consequently, further engagement is considered superfluous under the motto that "the

Popular Assembly is the consensus conference" [P174][112]. But evidently the issue of radioactive waste is too complex to cover it with a voters' booklet (which is sent to every household for information on each vote) or even with an intensive voting campaign (with heterogeneous resourcing of the parties). It was only after the national moratorium initiative had been adopted by the voting majority in 1990 that non-official experts were invited to talks in the nuclear field (except for the legendary Gösgen hearings in the 1970s in the licensing debate around the respective NPP):

- 1991/1992 "Action programme 'Energy 2000'": "Conflict solving groups" on reprocessing and LLW siting, with the participation of NGOs [P85];
- 1998 Publiforum "Electricity and Society": first broad consensus conference in Switzerland, *i. a.*, on the radioactive waste issue (the only experts were Roland Naegelin, KSA, formerly with HSK, and Piet Zuidema, Nagra), participation of volunteer laypeople [P152];
- 1998 Appointment of the working group "Energy Dialogue on Disposal" by the Energy Minister, Moritz Leuenberger (chaired by Prof. Hans Ruh, ethicist): examination of the concepts "controlled and retrievable long-term storage", comparison with final disposal and "adapted disposal concept by Nagra", with the participation of NGOs [P275];
- 1999 (– 2002) Establishment of EKRA by the Energy Minister, with the occasional participation of NGOs;
- 2000 (– 2002) Establishment of the Cantonal Expert Group Wellenberg (KFW), by the Cantonal Government of Nidwalden, with the occasional participation of NGOs;

The universitarian institute IDHEAP listed prerequisites for mediations and other alternative dispute resolving techniques [P168]:

- Voluntariness;
- Modified selection of participants;
- Consent on definition of the issue, mandate, rules of the game, time schedule (in extended negotiations);
- Transparency;
- Equal resources (financing, collaborators, access to information, political potential);
- Conflict of interest and not of values;
- Preparedness for compromise on all sides;

---

[112] For the ample and mostly unreflected use of "consensus" see Section 12.4. See also Wynne's 1975 "The rhetoric of consensus politics ..." serving as a warning [D91].

– Professional facilitation.

The attempts of the Federal Administration and of authorities with such "conflict solving groups" (*sic!*) [P169] did not fulfil these standards [M75] [M19][M20]. It is characteristic that after the defeat of Nagra/GNW at Wellenberg, a so-called Technical Working Group (TAG) was appointed by the Ministry of Energy in 1997. The opposition of Nidwalden was invited, on, however, the premise "matter of course that all participants are prepared to unconditional collaboration" [P69]. It has to be pointed out that it was the opposition that won the June 1995 vote, and the former Federal Councillor, Adolf Ogi, had promised after the fiasco of the official programme, "to investigate all options" [P281]. After all, the national anti-nuclear opposition was allowed to participate in the working group "Energy Dialogue" [P275], where, by the way, the financial and time resources were very limited as well.

8. DECISIONS UNDER UNCERTAINTY

Definitive disposal of radioactive waste is no decision under risk but one under uncertainty because the associated probabilities are not completely known [D35:16] and, therefore, a "risk" (=damage x probability) cannot be calculated. Whereas risk decisions are, in essence, dominated by the expected utility, in dealing with uncertainty, diverse models are competing (for a further discussion see Hanson 1991 [*ibid.*:41-50]).

Which types of uncertainties are pre-eminent is determined by the act of choice concept (see Section 6.3.3). As pointed out above, with deep geological repositories used as final disposals, the main relevant release scenario is not associated with an acutely induced dramatic failure but, if at all, with a slow system degradation. Long-term storage, however, is, by definition, based on controls by present and future generations. For our purpose[113], uncertainty (or uncertainties) may be divided into two main types[114] (see *Figure 9-9* overleaf). With "vagueness" knowledge is theoretically possible:

– *Stochastic and statistical uncertainty* depend on the efforts in gathering (parametric) data information, and *model uncertainty* may be

---

[113] Of course, there are many more ways to approach the notion of uncertainty, *e. g.*, all the way to decision uncertainty [D53][D2][D20][D35][M78].

[114] "Vagueness" approximately encompasses the principal causes of uncertainty #1 through 3, as proposed by the US Board on Radioactive Waste Management 1990, "insecurity" depicts the principal cause # 4 [G269].

reduced with increased model refinement, *i. e.,* systems knowledge is, at a given time, insufficient but, in principle, extendable.

With "insecurity", however, more or less:

- Plausible *scenario* assumptions are (and have to be) made with an accompanying *temporal and structural uncertainty* with regard to future developments and human behaviour.

In *Figure 9-9* overleaf the field of probabilistic application decreases from left to right. An eventual slow system degradation is not associated with a sharply defined event (except for meteorite impact or boring). Consequently, it is not possible to erect fault trees and to finally formulate direct probabilistic statements as simple risk figures. Nevertheless, to approach parametric and model uncertainties, probabilistic performance assessments (PPA) are useful in analysing the complex repository system where we have little if no long-term experience [G78:7-18].

One also has to recognise though that internationally relatively few "full scope" PPAs have been executed to date, and that the methodology is not PPA-dominated [G179]. International peer reviews, *e. g.,* [G176][G177] [G178][G200][G195] pointed out that types of uncertainty, variability and probability of scenarios were convoluted instead of discussing the distribution of the corresponding results, not all parameters were probabilistically dealt with, and not always clearly documented, identification of sequences of events and processes was difficult, the mixture of conservative, mean and probabilistic calculations and results was non-transparent. It was criticised that by focussing on cumulative radionuclide release, information on the parameter behaviour and on results of representative deterministic calculations was lacking. At any rate, "[i]t must be demonstrated that the simplifications introduced [by having to simulate reduced versions of detailed research models] do not lead to non-conservative representations of reality" [G102:88]. Hence, if concepts remain based on geological, rather than geotechnical systems, to be the dominant long-term safety barriers, PPAs will be useful (and necessary) in uncertainty and variability analyses but will supposedly not replace current mixed approaches. This comment is not to be interpreted as "anti-probabilistic" but as "pro-performance"-oriented: After all, the goal of safety assessments is to demonstrate that the chosen disposal system is sufficiently robust to guarantee long-term safety.

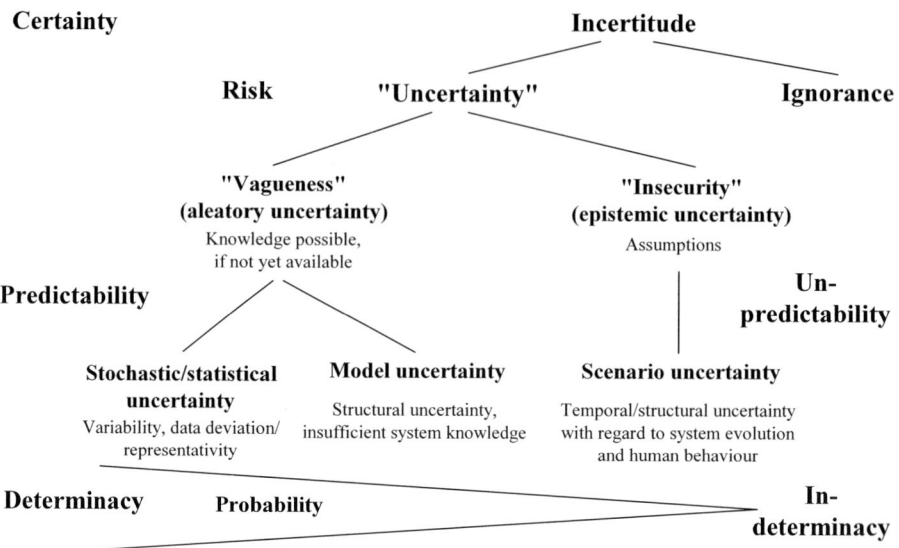

Figure 9-9. Attempt to classify the undifferentiated but key notion of "uncertainty" into two main types (acc. to Flüeler and van Dorp, 2000 [G78:18*passim*]). The three classes of uncertainty in the third hierarchical order are in line with NEA 1999a [G182:56-59] and [D37]. "Vagueness" corresponds to "aleatory" and "epistemic uncertainty", respectively, by Helton & Burmaster 1996 [D37] (see also footnote 114 for that matter). "Probability" stands for the degree of possible utilisation of probabilistic methods; high values would allow for a good treatment with probabilistics, low values would mean a crucial impediment in risk assessment. The notions of "certainty" and "ignorance" are added to indicate the continuum from determinism/determinacy (far left) to indeterminacy (far right) or (in this context) predictability and unpredictability. "Uncertainty" and "ignorance" may be subordinated to the umbrella term "incertitude".

As opposed to this, controlled long-term storage is dominated by noncalculable "insecurity", because neither values nor probabilities may be attributed to the effectivity of active technical control measures over the required isolation periods. The choice of long-term storage does not reduce uncertainties. Postponement does neither, to the contrary, it violates the causality principle and cannot build on the know-how of future generations. These reflections influence the basis for an option analysis attempted in Chapter 12.

To handle uncertainty, science gradually develops more sophisticated technical models and methods [G182], institutionally it utilises peer-reviewing. In cases in which uncertainty is high, safety margins are built in; so-called conservatism compensates for ignorance, *i. e.*, insufficient knowledge. The public is in an even harder position with regard to the waste issue. They are confronted with a highly complex safety analysis methodology, a narrow expert community with clear interests and tasks (*e. g.*, legal requirement to prove safe disposal), stakeholders who have even more pronounced interests (find a quick and cheap radwaste solution to pursue the nuclear path), technical fixes, and the reproach to understand nothing but stay out of the way of the "solution". They are equally confronted with successful pressure groups who demand the "best" disposal sites at any costs and present critical experts and safety-compromising examples. In the media, people read and hear of incidents or accidents in nuclear installations, contaminated railway wagons, "systematically" falsified safety documents in nuclear installations, *etc.*

In such situations, it is difficult to compensate lack of knowledge (societal uncertainty) with confidence in the responsible bodies, a procedure usually chosen. Still, the public's basis for dealing with uncertainties is primarily process-based.

8.1 Empirical findings on decisions under uncertainty

According to NEA 1999 the "demonstration of safety" of disposal cannot be shown with a rigorous mathematical proof but "rather a convincing set of arguments that support a case for safety" [G182:11][G201:10*passim*]. Thus, the relevance of (the different types of) uncertainty is recognised. Relative safety, *i. e.*, "only" parametric uncertainty in the sense of *Figure 9-9*, is the case with technical barriers because components can be and are statistically well tested. If, as in the issue under scrutiny, long-term aspects dominate (over hundreds and thousands of years) it is risky to "increasingly sidestep to artificial barriers" as was proposed by Nagra and Federal officials in the beginning and mid-1980s [P160].

For the same reason, the environmental NGOs do no justice to the situation, as their 1998 "concept of controlled long-term storage" is exclusively or primarily based on technical barriers, for: "Technology (primary barriers) permits controlled storage in the long term (60–80 years)" [P27:32]. This idea is associated with the most uncertainty, *viz.*, scenario uncertainty, as the organisations, a year later, assume that "[o]ur conceptual idea does not ask for small modifications but for a new philosophy. Controllability over centuries/millennia has to be conceived of now, and corresponding structures making this feasible have to be established" [P257]. Such a development takes scenarios for granted with a continuous and stable societal and institutional order – much unsafer states than the already difficult predictions regarding the geosphere. Even if the intentions are laudable they are lacking in a comprehensive system and problem understanding, for "[o]ur conceptual idea wants to secure a permanent access of control to the facility surroundings for the generations to come so that a possible event of damage can be recognised early and prevented or limited, respectively. This may be most likely compared to patrols along a dam wall" [P257]. Controlling in a safety-oriented long-term repository goes far beyond the concrete physical presence of man as is shown in Section 13.2.

One may fall in with Breitschmid 2000, when he takes the view that "[t]he dilemmas cannot be resolved with a conventional scientific-technical procedure, but they [have to] be addressed by a prudent strategy in consideration of all possibly conceivable uncertainties and in a process-oriented manner" [P36]. But it is exactly the analysis of uncertainties which leads to the *opposite* conclusion from Breitschmid's: "… These insights oblige us to realise that future generations will have to deal with our radioactive waste in one way or another until radioactivity has decayed to a harmless level…. The optimum way will have to be elaborated in the future by each generation in a broad scientific-technical and societal discourse" [*ibid.*]. The basis for the converse conclusion is the analysis of the goal relations (Section 9.6). A proposal for a consistent procedure in the "dilemma" situation is put forth in Chapter 12.

9. CO-OPERATION PROBLEM

In an exacerbation of the classical prisoner's dilemma in radioactive waste governance, we deal with an "asymmetrical game" [D60:91]. The "players" neither are in the same situation nor do they have the same preferences. Representatives of the future generations do not sit at the table, let alone have a legal standing; their interests even cannot be forwarded (Section 6.4.2). *Inter*generational equity is not met. If we even export the waste from Swiss reactors, industry, labs, and hospitals we, in addition, violate *intra*generational equity issues. If we, at the most, try to co-operate with our successors, we are not allowed to benefit in excess from their (future) environmental contribution. We take advantage at any rate, for an anthropogenic waste problem is always an in-balance between something for nothing: The eventual environmental costs are borne by future generations without them having any benefit of the waste production.

Out of Franzen's 1998 proposals to solve the co-operation problem, [D23:26*passim*] one has already been adopted: to put pressure on a player after having announced his or her playing history, *viz.,* by requiring the track record – a proof of evidence – as a partial goal of the complex goal named "trust" (see Sections 6.4.4 and 9.6). The instruments of retaliation, external intervention, and moral appeal are time and again recurred to – sometimes, however, with effects contrary to intentions. Since there is no reciprocity with future generations, *i. e.,* non participating players (they cannot react to their ancestors' moves), the present generations are called upon for proceeding in utmost prudence. This prudent approach shall be developed by means of the concept of robustness (Chapter 13).

9.1 Empirical findings on co-operation problems

Co-operation and equity issues are touched upon with the supposed solution "abroad", with the recurrent theme of "NIMBY" or further inconsistencies of action (see Sections 6.4.1, 4.5 and 8.3). The principle that benefiting generations should bear the consequences of their action is shared by overwhelming public majority, in the European Union by 80 per cent of the respondents, according to the Eurobarometer 2001 [G133:10].

10. UTILISATION OF LATENCY PERIODS AS OPPORTUNITIES

In descriptive decision theory it was initially assumed that decision making is a sequence of choices. In recent years behaviour-oriented organisational researchers such as Cyert & March 1995 have adopted a modified view "challenging the first premise of many theories of sequence of choice, *viz.*, the premise that life means choice. They argue that living does not primarily mean choosing, but interpreting. In this sense, from a behavioral and an ethical standpoint, the results of a process generally are less relevant than the process itself" [D8:234-235, transl. suppl.].

Apart from this process-orientation and the importance of procedures (see below) we may also throw another, less negative, light on delays and retardations. At any rate, for (scientific) research, phases of latency – of seeming "unproductivity" – may be fruitful [D24:88*passim*][R96:178]. This may be the case in the phase of change waste management is experiencing, by considering claims for the development of disposition concepts. History shows that criticism has mostly, even if delayed, been taken up. In the setting of cost pressure due to the opening of energy markets, however, the contrary may take place – a deferral of decisions and a weakening of safety requirements for a disposal facility (see Section 12.7).

10.1 Empirical findings on latency periods

At Wellenberg, in 1994/1995, GNW and Nagra wanted to obtain the general licence for a LLW repository in just one step – even though they had been warned not to strive for the whole project but to proceed in stages. "A stepwise process with democratic attendance and a new philosophy, *viz.*, a controlled long-term storage, could gain acceptance, Leo Odermatt supposes. By assent, the people would let everything out of their hands" [BBr, 1995-6-14][115]. The applicants did not want to wait – and lost the referendum in June 1995. In a first reaction, they recurred to the usual suspicions: "The refusal must have been determined less by the cantonal opposition, but rather by the massive counter propaganda of the Swiss environmental organisations ... they succeeded in ... spreading ... fear and disbelief and, furthermore, in questioning the competence and the integrity of the Authorities ... apart from that ... it was less the project than rather the procedure which was rejected" [P216:1]. But on the grounds of the poll, which was immediately, in July 1995, commissioned by Nagra, it became clear that indeed the greatest motivation for denial had been the link between the exploratory gallery and the repository itself: "The people of Nidwalden would apparently like to be involved in the decision again as soon as further results are available after the gallery has been dug. The circumstance that the Cantonal Council would have been able to block the repository construction in the light of unfavourable findings made little difference.... Even though the wealth of favourable results from surface investigations and from test drilling hardly suggests any surprises to be expected, the people of Nidwalden evidently expect an additional basis of decision making from the gallery investigations" [P142]. Leo Odermatt "pointed out that, in the run-up of the referendum, his Committee had time and again proposed a phased approach. This strategy had, at that time, been vehemently rejected by Nagra" [BT, 1996-3-28]. The repeated defeat in 2002 will be addressed in Section 10.2.

The episode mentioned is just one among many – notions "as quick as possible", "without delay", *etc.* are associated with lamentations that one had a "backlog" of the programme. Such a behaviour is conjoined with unrealistic scheduling throughout – too optimistic in the case of LLW, dragging along regarding HLW (see Section 6.3.10).

---

[115] Odermatt was one of the leaders of the regional opposition in Nidwalden. Later on, from July 1998, he became a member of the Cantonal Council, the regional government of Nidwalden [TA, 1998-3-17].

Chapter 10

FINAL DISPOSAL SITING AS AN EXAMPLE OF SUB-OPTIMUM DECISION MAKING

1. GENERAL

Empirical content analysis shows that the discussion on disposal concepts in Switzerland was predominantly determined by external forces. After the principle of dilute and disperse (DD) had been abandoned ("to be washed down with plenty of water" [P266]) one envisaged "the construction of a big magazine" (for MIR waste) in the 1960s [P263, 2/63, Annex:1]. In 1968, the Federal Councillor Ritschard could announce: "In the mid-1960s, the Department of the Interior planned a building for storing waste. Shortly after, though, the dumping actions began" [P72:443]. From 1968, theoretically, the issue of high-level waste should have been on the agenda since the first NPP, Beznau I, was put into operation – this was not the case because the first reprocessing contracts did not call for repatriation of vitrified glass ingots. The Office of Energy in charge recorded "the following rule from 1969: The spent fuel elements ... were delivered abroad for reprocessing" [P185:51]. Two years later, the NPP Mühleberg made a statement to the effect that "the waste problem does not belong to the immediate concern of the nuclear power plants. The spent fuel elements are yielded to reprocessing plants" [P186]. Nagra was established in 1972 with the mission to build "below ground facilities" [P196]. From 1975 a small circle, in the KARA group, came up with the "national concept" of "final disposal" [P153][P154]. This committee was the first official group in charge of the issue and consisted, as mentioned, of government officers and representatives of the industry.

Thereby, "in concert ... it was refrained from appointing a Federal expert commission" and, with the Committee, "a flexible, less formal solution was found" [P153:3]. One collaborator within ASK, the predecessor of HSK, was in charge of radioactive waste alone.

Nagra had to attend to HLW only after the first concept of the energy industry of 1978. Initial optimism gave way to the insight that the issue had been underestimated. The attenuation of the requirements regarding the project "Guarantee 1985" followed up to the 1988 subdivision of the demonstration of disposal feasibility into a demonstration of technical (engineering) feasibility, of safety, and of siting feasibility [P76].

Claims to change the disposition concept were uttered in the 1970s already, by experts, affected communities and groups, as well as opponents, partly independently of each other. A debate was promised several times but never carried out (see Section 9.6.1). When the official actors chose final disposal to be the "national concept" in the mid-1970s, the associated project "Guarantee 1985" should have been realised in a timely manner, as the work schedule of the NPP owners documents [P283:6-61,6-64]. In the meantime, the Federal Office of Public Health backed off the issue by stating in 1975: "The definite solution is a long-term issue ... not urgent", according to [P92:10].

The issue was so politically charged that quests for conceptual modifications were blocked and guidelines (such as in "Guarantee 1985") were watered down; a debate along all dimensions (see *Figure 12-1*) and questions was even made impossible through the formation of factions. Nagra called disposal, not posing "particular requirements", "a subterranean facility, which would be around six times smaller than one tube of the Seelisberg tunnel", a road construction in Nidwalden [P209:30]. To the environmental NGOs, the monitoring task was "most likely comparable to patrols along a dam wall" [P257]. The federally appointed Technical Working Group (TAG) denied nuclear guardianship "which has to date in no country been seriously taken into account" and, in one breath, was in favour of "surface final disposal facilities" *(sic!)* which "continuously have to be monitored and if necessary maintained" [P271:7*passim*]. As equally mentioned, it also sanctioned Nagra's supposed compromise of 1998, the "adapted disposal concept", as corresponding to the "current international state of the art" [*ibid.*]. The political parties limited their contribution to financial aspects of disposal.

When the Federal Government drafted the Ordinance on Radioactive Waste Disposal Fund in 1999 [P269], it considered "the result of the technical-scientific discussion to date as politically not enough sustainable" [NZZ, 1999-6-8]. Instead of widely fostering this discourse, however, the Department in charge, DETEC, put reliance on political negotiation when it, in the same year, again set out to revise the outdated Atomic Energy Act of 1957:

"The preliminary draft to the Nuclear Energy Act shall start out from final disposal with long-time retrievability for the low- and intermediate-level wastes whereby one or another concept shall be the basis for the Statement to follow, depending on the results of the examination by the Expert Group [EKRA]. The disposition concept for the high-level active wastes shall be left open in the draft" [P276:1-2]. It was only the ensuing establishment of this Expert Group EKRA, which institutionally led to the substantive continuation of the discourse, according to the altogether positive echo in the media, evidently something the actors had waited for. The impetus for concrete modifications (in the Wellenberg project) was given by the Nidwalden Cantonal Government, which picked up the critical issues regarding the original disposal concepts and won over DETEC in 1995 to incorporate controllability and retrievability. It was also the government of Nidwalden which appointed the Cantonal Expert Group Wellenberg, KFW, in 2000, to get the conceptual modification accompanied (see Section 10.2). From the complexity of a disposal system explored in Section 6.2.2, it may be concluded as follows:

- Single criteria such as the half-life period of radionuclides, their origin, *etc.*, are not commensurate with the complexity of the issue, and should be substituted by a set of criteria of diverse indicators. There are no satisfactory individual positive "feasibility criteria" nor individual "exclusion criteria" as, *e. g.,* the hydraulic k value, which are representative to terminate a project.
- Similarly less successful is the strategy to await a "definite" safety analysis; the safety case has to be evaluated from several sides and perspectives and with a periodically adapted set of criteria according to the pertinent phase.
- The strict protection goals of the regulatory guideline R-21 [P133] evidently have to be complied with but, additionally, safety indicators (see *Figure 13-1*) have to be developed. These indicators should help to reduce uncertainties and to establish transparency.
- The selection of criteria and sites proposed, as well as the ensuing assessment and evaluation, have to be continuously traceable and plausible.
- The evidence of long-term safety of a deep geological disposal (repository) has to be supported by means of site-specific revised safety cases according to the state of the art. Since the natural barrier, the geosphere, is of decisive relevance, safety-related site characteristics are paramount in site selection. Accordingly, the Sections to follow, 10.2 and 10.3, explore the way how Nagra dealt with criteria such as "host rocks", "disposal design", exploration strategy", "exclusion cri-

teria" and finally came to a choice. The temporal conceptions, concerning schedule, are indicated as well.

The driving force behind the analysis is making the previous decision-making process transparent, in order to identify eventual deficiencies in methodology and possible insights for its development because the process of radioactive waste governance still goes on. For an adequate, "fair", appraisal, a contemporary-history perspective is needed that attempts to describe and reflect events "out of their time", out of their "contemporary reality" (see Section 4.3.2), but not *ex post*.

2. SITING OF A LOW-LEVEL RADIOACTIVE WASTE DISPOSAL FACILITY

On the establishment of Nagra, in 1972, their mission was to build "below ground facilities" for so-called "low active wastes" [P196], meaning that the then current investigations of NOK, the owner of the Beznau NPP, in the vicinity of their plant, was continued [P272:2*passim*][116]. They were enlarged to a gypsum and anhydrite programme with 22 potential sites in 1974 to 1976, whereby the selection criteria remained unclear and the programme was not implemented [P8:7][P151]. This was a Swiss variant of the international search for evaporites (above all, salt formations). Strong local and regional opposition was consequential to Nagra's harsh and rapid procedure, without local licences and without public consultation [P149][P150]. The test drilling in Bex, Canton of Vaud, was sunk without a cantonal licence [P151:9]. As in spring 1974 the first drilling was to be carried out in Airolo, Ticino, communal opposition stirred as soon as the project became public [P99][P134].

The authorities started out with a so-called model study in 1980 for a facility in the molasse of the Midlands, the lowlands of Switzerland [P23], between the Lakes of Geneva and Constance (see *Figure 10-1* overleaf). A year later already, and in a new survey area, Nagra narrowed down the circle of initially 100 to 20 potential sites: Among them there was no location in the Midlands, although the feasibility of the only quiet tectonic region in Switzerland was affirmed for a so-called "B disposal facility", *i. e.*, for low-level waste, including longer-lived intermediate-level waste [P226:vol. 1,41]. At that time, Wellenberg, running under the plot name "Altzellen", was not first choice: The geometric data, *i. e.*, the expansion of the rock host, were quali-

---

[116] The process and events are laid down in detail in Flüeler 2002e, main strings of arguments #1, 2, 3B, and 4B [G72:Vol. II].

Final disposal siting as an example of sub-optimum decision making 155

fied "medium" and their predictability "sufficient". One step below in ranking would have meant an elimination [*ibid.*:Annexe 5]. Other locations as well, such as Val Canaria in anhydrite in Ticino, having good characteristics throughout, were not pursued any longer [*ibid.*:Annexe 3].

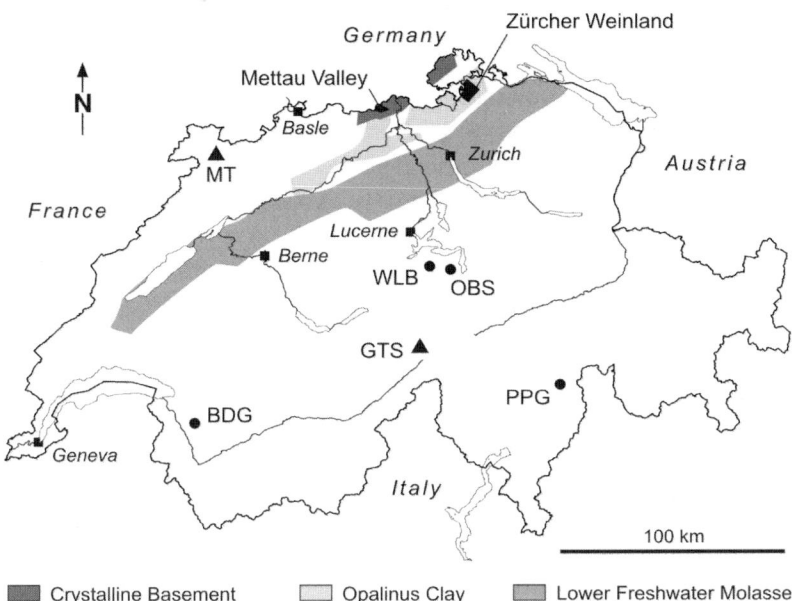

Figure 10-1. Locations of potential sites. The now-abandoned low-level siting programme contained four localities in the semi-final round: BDG Bois de la Glaive, PPG Piz Pian Grand, OBS Oberbauenstock, WLB Wellenberg (this one remaining in the final round and rejected twice by vote). The current high-level programme focuses on sediments: Opalinus Clay, whose top priority area according to Nagra is the "Zürcher Weinland" or Zurich Vineyard Region with the borehole of Benken, and Lower Freshwater Molasse as their 2nd priority region. The former focus was the Crystalline Basement in Northern Switzerland (lately with the Mettau Valley). Underground research laboratories are: GTS Grimsel (crystalline rocks), MT Mont Terri (sediments, Opalinus Clay). Source: Nagra.

Two years later, in 1983, there were only three locations left (*Figure 10-1*): Bois de la Glaive (anhydrite, Canton of Vaud, Western Switzerland), Oberbauenstock (marl, Central Switzerland), and Piz Pian Grand (crystalline formation, Grisons) [P200]. All of them are situated in the active and complex Alps and Pre-alps as regards to tectonics and geology. From the point of view of distribution, they belong to three regions of the country (West, Central, East) and they are peripheral with respect to the main electricity producers and consumers, *viz.*, the NPPs and the conurbation in the Swiss Midlands. Even if diversification is desirable, the problem of comparability arose in case the exploratory galleries were demised. The procedure went as follows: "For each of the three sites, Nagra will at first pursue their investigations in the same way [in parallel], they will set up working programmes for geological explorations and check on the location and the conditions in the host rocks by constructing exploration galleries provided that the detailed work does not come up with unexpected problems" [*ibid.*:V].

As a consequence, Nagra applied for test drilling in all three locations, in December 1993. In their September 1985 decision, the Federal Council rejected this proposal and demanded a fourth site, which should comply with the following five selection criteria [P75]:

"1. Good geologic predictability
2. Region with a low topographic relief [*i. e.*, a flat topography, tf]...
3. Simple and easily explorable hydrogeologic conditions
4. Subterranean water flow time to biosphere as long as is possible...
5. Tectonically and seismically quiet zone"

What Nagra offered in their application on Wellenberg in June 1987 did not fulfil these criteria. HSK decided upon Wellenberg in their expertise of 1986 and 1987: "[It] does not meet this goal" (full record in [P122]). But: In January of 1986 the former Cantonal Council of Nidwalden had invited Nagra for examinations [P22:6]. As a result, Nagra conceived that "an 'equal treatment' of all potential sites is not advisable and not possible in the exploratory phase" [P212:20].

Even the decision upon the project "Guarantee 1985" in June 1988, notably in favour of Nagra's Oberbauenstock, did not play a part anymore. In this policy decision, the Federal Government interpreted the demonstration of technical safety and of siting feasibility as proven [P76]. Criticism, also stemming from experts on behalf of HSK, did not change anything [P35] [P248].

Although, in retrospect, it is objectively not understandable any more, Nagra, in November 1988, again applied for exploratory gallery permits for all three original sites [P207]. They also decided to exclude the longer-lived intermediate-level waste from the LLW programme and to shift them to the HLW programme. This rearrangement led them to the argument that "the claim for an additional site [in the Midlands] is thus dispensable" [P271:19]. The envisaged radioactivity to be emplaced still was of considerably higher level than in other LLW facilities, *e. g.*, the Swedish SFR at Forsmark [G245:11]. In October 1990 the Federal Government determined not to decide "until for all three sites, as far as possible, comparable statements on geology can be made", after [P208:39]. This was the confirmation of a decision made in 1985 where already a "comparable interpretation of the work at the [three] sites" had been required [P75]. It was just Wellenberg that was added.

The Federal decision signified a *"de facto* link of all four sites" as Nagra rightly commented in their "Preliminary Report on Site Selection" of June 1993 [P211:11]. They did not abide by the guidelines of the Federal Council and, in this "Preliminary Report", made up their mind in favour of Wellenberg, which was "suitable with high probability" whereas the other sites were judged to be only "presumably suitable" [*ibid.*:52][117]. There is some evidence that Nagra had opted for one, particular, site, at an early date if one takes a look at the financial expenses: In 1991 and 1992 the co-operative spent 43 million Swiss francs for Wellenberg alone whereas the other three sites together amounted to only 5.2 million francs [P225, 1991:36, 1992:38]. It is not a definite analysis because corresponding data are not accessible. The reports issued in 1993, nevertheless, could not astonish observers: Wellenberg was selected [P210][P215].

The geologic committee KNE accepted the choice in 1994 though it pointed out – in case of failure – the "favourable" sites in the Midlands [P165:4]. As to the site of Bois de la Glaive, as part of the entire LLW programme, they made the following remark: "The aforementioned uncertainties in the evaluation of this site are predominantly based on the nature of the host rock and have been known since the programme started. They should have been partly eliminated by way of the originally planned oblique borings and underground galleries. This statement also holds for the other sites. Thus, it is deplorable that an important decision has to be taken today, without the results of such explorations" [*ibid.*:last part,8].

The quest for "geologic equal treatment" and "exploratory galleries at all sites" – as well as for monitoring/control and retrievability – had been raised before by environmental circles [P85][P87:10].

Although nuclear issues are regulated at the federal level, the canton seized the opportunity provided by cantonal mining law for a special underground concession. The mining licence was granted by the Cantonal Government in January 1995. This decision, however, needed to be endorsed by a cantonal ballot. In June 1995 the electorate of Nidwalden rejected the applicant's request after a general licence by 51.9 per cent. A survey at GNW's immediate behest revealed a month later that 65.5 per cent would have voted in favour of a submission for an underground exploratory gallery alone and

---

[117] Regarding the exploratory galleries, Nagra wrote in 1992: "There has never been the intention to execute all three [other] galleries" [P208:13]. To be fair one has to admit that the Federal Government's stand on this issue was inconsistent as well, in 1992: "The Federal Council did never assume that all four sites at stake were to be investigated by means of an exploratory gallery. Even … in 1989 the Federal Council had stated: 'For economic reasons it would not be bearable … to simultaneously sink exploratory galleries to the emplacement area at several sites…'" [P148:2-3].

61 per cent if the general concept had included controllability and retrievability [P142].

In response to the claims, Nagra presented a so-called "adapted disposal concept" in 1998, whereby "adapted" primarily meant to keep the shaft system open. Unlike reported in KASAM 2001 [G145:845], this modification was harshly criticised by the expert bodies (KNE 1998 [P167], KSA 1998 [P178]). Flüeler, in 1999, staked out the opinion that this "'would be the worst of all solutions'. For this variant would compromise the safety requirements because the operator could always revert to the fact that the facility be accessible and susceptible to repair work anytime" [P34].

The driving force for a productive continuation of the stalemated programme was the Cantonal Government of Nidwalden, where Nagra's favoured location, Wellenberg, was situated. It was not the Confederation, but the federally appointed expert group TAG that sanctioned Nagra's concept change and rejected exclusion criteria. The regional government established an expert group of their own in 2000, the Cantonal Expert Group Wellenberg (KFW) which had to examine the exploratory gallery project to be set up and the necessary disposal concept change according to the lines indicated by EKRA (summary in [P52:22-24]). The Canton established several conditions to be fulfilled before a mining concession could be granted, *e. g.,:*

- Restriction of the concession to the exploratory gallery;
- Definition of clear negative (exclusion) criteria for gallery results, leading to either continuation or abandonment of the project;
- Adaptation of the repository concept to monitored long-term geological disposal as proposed by EKRA;
- Clear definition of the waste categories to be emplaced (with an emphasis on the "short-lived" character of the repository).

On the basis of technical discussions, the safety authority HSK proposed exclusion criteria (primarily on water flow), which should be the yard stick for the new exploration programme at Wellenberg [P126][P127]. KFW considered that, given the existing studies and the work programme on the exploratory gallery, it was possible to step to the next phase since the geological evidence so far was positive [P157]. The group, however, prepared a report on the site selection procedure [P158] and put forward a series of proposals to improve the process, ranging from a clear separation of roles for involved stakeholders, funding issues, transparency, traceability of argumentation, transparent formulation of criteria, to controllability and retrievability, and stepwise approach. It also insisted on the need for a back-up alternative option in case of a failure in the single-option process.

After KFW had given their consent to the adaptations and progress work, GNW submitted a renewed and limited application just on the underground exploratory gallery at the beginning of 2001 [P110]. In view of the new context, the Cantonal Government granted the gallery concession in September 2001. The regional opposition was pleased with the participation of KFW experts who heard them, and reported their independent and critical views. It, however, opposed the governmental decision on the basis that the site was not providing the necessary guarantees to continue the qualification process. The funding received by the local community also raised concerns that the population was granted compensation for taking a risk. Finally, the part of Nagra having a leading role in the process was a source of distrust due to historical evidence. An additional matter for argument between proponents and opponents was that the result of the cantonal vote would be influenced by, or would influence, the oncoming federal referendum on the phasing out of nuclear energy in May 2003 (propositions, incidentally, rejected).

The Cantonal Government decision was to be submitted to popular vote in September 2002: It was negative again, this time by almost 58 percent [P231].

With no thorough analysis at hand (which sheds some light on the – missing – reflexivity of radioactive waste governance in Switzerland), and acknowledging the substantial improvements since 1995 as well as the anti-campaign of fear, one may postulate that four factors played a paramount role in the rejection [P52:24]:

1. The process for the selection of the Wellenberg site was not transparent and presented a hindrance to a well-founded decision because people could not understand "why just here?". Being in a type of catch-22 situation, the perceivedly affected voters felt the pressure of heavy investments on the applicant's side, which limited the power of exclusion criteria in a situation with no other alternative sites and options. This is the individual interpretation of the basic proposition that "[a] choice of one is not a choice" as the Canadian independent review Panel concluded in 1998 after an 8-year extensive examination and set of hearings to evaluate the Canadian AECL's disposal programme [G25:56].

2. The institutional system, with all its traditional rivals was, apart from the expert group KFW, still the same.

3. There was a lack of leadership in political governance, especially at the national level.

4. No tangible benefit in saying "yes" was perceived, and in this situation of doubt it was easier to say "no".

At any rate, the Cantonal Government denied further collaboration in the issue and concluded that, upon this repeated negative vote, Wellenberg is deemed "definitely out" as a repository for radioactive waste in Switzerland. Consequently, GNW was dismissed by November 2003 and their drilling locations sealed up by December 2003. The further rejection led to intense discussions at local and federal levels on most of the issues presented above. The responsible parties in Switzerland will have to re-evaluate the situation (see below).

Conclusions as to low-level and (short-lived) intermediate-level waste

The analysis of the process leads to the conclusion that

- the site selection procedure could – to a large extent – not be traced by non-participants;
- it was largely influenced by external factors;
- it did not experience broadly discussed concept modifications (from a facility deep underground to a less deep one, see below);
- the procedure as such did not lead to the latest site, Wellenberg [P158].

Nagra had already decided in 1981/1982 to base their work "on existing geologic maps and information" [P223, 82/9:2], a strategy which was sanctioned in 1984 by the oversight authority and which, indeed, was in contrast to the 1983 drilling applications of Nagra. In 1985 the Federal Government rejected the submissions for underground galleries which led to the fact that comparatively only little additional insight could be gained (from the surface). In spite of portents put forth by a regulatory expert in 1984 [P49], explorability of the host rock was no criterion for site selection. It was only the repeated demand of HSK for a deep underground facility (to deposit the long-lived intermediate-level waste, LLILW) that forced Nagra to examine more closely the extent of the host rock. HSK had identified, back in 1983, a necessary isolation period for ILW of "several 100,000 years" [P121: 13*passim*]. Nagra took its time until 1988 when, by reducing the inventory to so-called "short-lived" nuclides [P205:D1-7], they rendered the construction of a deep facility obsolete and averted to a site expansion to the Midlands.

Regarding exclusion criteria (for the site selection process), the waste community, *i. e.*, applicants and HSK, kept on referring to the protection goals of the guideline R-21 (respective calculations were the regulatory basis in favour of a deep facility for LLILW); despite the demand of the Subgroup Geology in 1984, the regulators never expected detailed criteria nor did Nagra propose any. The debate on criteria was confused with the debate on the "fourth site" (in the Midlands) and the political premise to proceed at three

Final disposal siting as an example of sub-optimum decision making 161

or four sites in parallel (which, rigorously, was not possible). It was only after the mandate for KFW established by the Cantonal Council of Nidwalden that an explicit examination of exclusion criteria by third parties was required. By that, the government took up requests voiced in the debate on the general licensing procedure in 1994/1995. With regard to the further process, it was envisaged to reduce the Wellenberg inventory for a second time [P159]. The Cantonal Government wanted a reduction of the LLW inventory to nuclides with a half-life of 30 years, the IAEA's definition of "short-lived" nuclides [G104:30][118]. This time, the design change should follow concrete geologic conditions and in a transparent and traceable procedure. It is typical of the approach that the KFW's homepage started with the motto of "open, transparent factual information as a contribution to the public debate" [P90] (dismissed after the negative decision in September 2002).

Two decisive framing conditions have changed in recent years, in spite of the latest rejection, at Wellenberg: The general conceptual phase is superseded by an implementation phase and the subsequent procedure is better planned and process-oriented, with the involvement of the main stakeholders. Transparency and public relations were part of KFW's mandate [P157:5], and the main critical group in Nidwalden, MNA, could follow up on the process and the arguments. The strategy of linear decision making (by authorities and operators) (*Figure 8-1*), eventually expanded by a more or less subtle risk communication (*Figure 8-2*) is gradually evolving towards a dynamic decision model (*Figure 13-5*, page 250).

3. SITING OF A HIGH-LEVEL RADIOACTIVE WASTE DISPOSAL FACILITY

In contrast to the relatively smaller issue of LLW the search for HLW sites is well analysed until 1988 [P41][P47][P48][P43][P44]. It is only with the present study, though, that a systematic examination of the argumentative patterns, according to uniform criteria, is undertaken[119]. In August 1980,

---

[118] From a technical standpoint, a less rigid but safety-oriented definition would have been more pertinent. In the "Wellenberg" inventory Ni-63 from decommissioning waste would have accounted for 90 per cent of the total activity, hence would have made obsolete the restriction to so-defined "short-lived" waste [P215:28-29]. Yet, the Cantonal Council's argument is understandable since Nagra – wrongly – had always, from 1988 beyond 1995, emphasised the "short-lived" character of the LLW project; this was repeated by administrators [P162:2] and the Federal Government [P277:24].

[119] The process and events are laid down in detail in Flüeler 2002e [G72:Vol. II], main strings of arguments #1, 2, 3A, and 4A1 (crystalline host rock) and 4A2 (sediments, after the Federal Council's decision of 1988) [*ibid.*].

a group of experts in geology raised a number of deficiencies in Nagra's programme, in an expert gathering with AGNEB and the co-operative: the criteria for the selection of the host rock types, the regions where to sink boreholes (without prior seismic investigations), the criteria for disposal at large. In addition it was complained that the research work had not been sufficiently published and that the schedule ("Guarantee" until 1985) was too tight [P8:Annex 5][P47:122*passim*].

According to federal appraisal and consent, Nagra had presented their demonstration of disposal feasibility ("Entsorgungsnachweis") in terms of construction feasibility and of demonstration of long-term safety ("Machbarkeits-" and "Sicherheitsnachweis", respectively). The empirical basis and geographic region of investigation was the crystalline basement in Northern Switzerland (see *Figure 10-1*). A suitable site could not be proposed, *i. e.*, Nagra failed to provide conclusive evidence for the demonstration of siting feasibility ("Standortnachweis"). It was supposedly difficult to find a sufficiently large and suitable host rock body in the heterogeneous and tectonically deformed crystalline basement [P121]. This basement, furthermore, was appraised to be too deep in the underground, under the newly detected Permo-Carbonic Trough.

For this reason, HSK proposed in their expertise on "Guarantee 1985" to expand the investigations to the more homogeneous and better explorable and predictable sediment formations [P122][P123]. This was endorsed by the Federal Council's decision of 1988 [P76]. Due to that, Nagra produced the so-called sediment study [P206]. This was appraised by the independent geoscientific Commission on Radioactive Waste Disposal, KNE, in a report of February 1990: "The selection of 2 formations (Opalinus Clay and Lower Freshwater Molasse) out of a range of sediments rocks occurring in Switzerland is not comprehensible as presented .... Nagra restricts itself even in the conceptual phase of the sediment study to just a partial area of Switzerland (NE Switzerland), singles out 8 siting regions as a further restriction, resulting during the further process in merely two siting regions" [P164:7]. This view was shared in a 1986 report by the Subgroup AGNEB, which, in 1989, was succeeded by KNE ([P274], also as reported in [P164:8]).

With the application of November 1994 Nagra intended to sink additional boreholes in crystalline formations as well as a deep calibration borehole in the Weinland (Vineyard Region), the northern part of the Canton of Zurich. This proposal was welcomed by KNE in 1995, not so though the drilling application for the crystalline sites of Leuggern/Böttstein in the Canton of Aargau: "It is with stupefaction that KNE observes how little Nagra considers their own investigation results in planning prospective explorations in the crystalline formations .... The chances of success [to discover, with oblique drilling, a rock body suitable in size for disposal] cannot be measured given the conditions reported" [P166:7]. Thereupon Nagra were not prepared to change their course of action, the KNE report had been generated without having consulted Nagra's experts [BT, 1995-7-4].

Subsequently the supervisory authorities established the "Working Group on Crystalline Formations in Northern Switzerland", consisting of HSK, KNE, and Nagra. This body agreed in June 1996 to relocate the investigations, for the time being seismics, to the west [P21]. In contrast to the debate on the notion of "guarantee" previously, it was defined what was to be understood by a key term, *viz.*, the demonstration of siting feasibility ("Standortnachweis"): "For further steps, including studies in sediment rocks, it is essential that a clear definition be spelled out of the terms set in the Federal Council's decision" of 1988 [*ibid.*] – in order to "prevent miscomprehension when interpreting the term 'Standortnachweis'" as was added later on [P125: 2-3].

After the seismic measurements in the west, in the Valley of Mettau, Nagra announced in summer 1998 not to carry out further drilling in crystalline formations for the moment, but to place "the centre of the studies" to the Opalinus Clay in the Zurich Weinland [press release of 1998-6-19]. On the evidence of the struggling and laborious development in the HLW programme, KNE pinpointed the following in 1998: "Although the issue of a safe disposal of radioactive wastes has been intensively explored in Switzerland for the last two decades and in numerous partial fields essential progress has been made, the results of the siting investigation/siting evaluation for a HLW repository are insufficient in comparison with programmes abroad. One of the reasons is that for too long only the problematic crystalline option was pursued and its difficult exploration was recognised too late, respectively …. Crystalline rocks 2001: Based on the current state of knowledge and an extensive assessment of the results obtained so far, there has to be taken a decision on whether to call off or to carry on the investigations into crystalline formations. Statements on their suitability must be documented with evidence-based results …. The siting search for a HLW repository in Switzerland shall be continued in a targeted and timely manner … . An eventual 20 year-long interim phase of 'doing nothing' in the HLW programme from 2001 raises doubts on whether a final HLW repository is really deployed in Switzerland. In KNE's view the maintenance of technical competence cannot be ensured given such a long interruption" [P167: 3*passim*].

In 2002, Nagra submitted the project "Entsorgungsnachweis" (demonstration of feasibility and siting of disposal) to the Federal Council. The documentation is aimed to demonstrate how and where spent fuel (SF), high-level radioactive waste (HLW) and long-lived intermediate-level waste (TRU) can be safely disposed of in Switzerland. The 2003 the NEA International Peer

Review concluded that "Nagra has presented a sound and practical disposal concept based on a specific realisation of the multibarrier concept" [P229:9]. After the extensive examination by the authorities and advisory bodies it is planned to open the documents for public scrutiny in mid-2005. It has to be pointed out that for this demonstration of feasibility no formal licensing procedure is stipulated by law, the more noteworthy is the increased allowance of public and stakeholder involvement in the process. Around 2006, the Federal Government, on the basis of this inclusive review, is scheduled to make a decision on the further procedure.

Conclusions on high-level and (long-lived) intermediate-level waste

With a glance to Sweden[120] Nagra selected granite as the host rock of choice for HLW way back in 1980, to produce the demonstration for "Guarantee 1985". Even after HSK's proposal in 1986 to "pursue disposal concepts in sediment rocks as well" [P122:85] the co-operative insisted on their crystalline programme. In 1996 still, McCombie of Nagra said that "[o]ne had posed the question in principle whether further programmes in the crystalline formations made sense. The positive conclusion allowed for a reference to Sweden, Finland, Japan, Canada, and Spain, all of which investigate their crystalline rocks for disposal" [TA, 1996-6-27]. The Federal advisory commission KSA demanded in their 1998 position paper: "The option of Opalinus Clay for the disposal of HLW/TRU shall be vigorously pursued. The option of crystalline rock calls for a conclusive analysis and a clear-cut decision on breaking off or continuing investigations. An option shall only be kept open if it indeed is actively followed" [P178:7]. As mentioned, also KNE observed Nagra's persistence in the once chosen but deadlocked programme.

For "Guarantee 1985", the year 2020 was determined to be the year of bringing the HLW facility into service [P199:28]. This was in accord with the Nordic programmes which, however, are still valid [G212][G235][121]. At any rate, the principle held that disposal facilities "are ready for operation at the accrual of respective waste categories" [P201:20] even if this 1985 for-

---

[120] For Switzerland, as well as internationally, Sweden has always been a reference. See in Flüeler 2002e [G72:Vol. II] under the catchword "Basismodell Schweden" (Swedish model for decisions).

[121] The basic idea behind the deadline of 2020 was that within 40 years after waste generation (around 1980) the heat production of HLW and spent fuel would be dropped to a level of acceptable safety. A longer (interim) storage would not be advantageous – certainly with uranium fuel which was decisive at that time – because the temperature curve flattens out [P91:49-50]. This assumption is supported by official reflections on the framework for waste financing [P28:3]. In their cost study of 2001 Nagra assumes an NPP operation period of 60 years [P288:2].

mulation was weakened by the phrase "only about from the year 2020" [*ibid.*:2].

In the disposal concept of 1992 it was likewise stated that the HLW facility would "be required only about from the year 2020" [P208:2]. Currently the co-operative plans to open such a facility "from 2024", the decision on the site would be "due only in about 20 years" [TA, 1998-9-9] or even: "reasonable around mid next century" as was released to the media by Nagra on 1998-6-19. The reactor owners assume an operation of the Central Interim Storage of 80 years – contrary to the 30 to 40 years originally envisaged. Such time deferrals do not comply with the causality principle. Spent fuel has been piling up since the 1980s, the Central Interim Storage is built. It is not decided how long the five nuclear power units will be in operation, though a claim for phase-out was turned down in a national plebiscite in May 2003. AGNEB has, in recent years, called for detailed scheduling [P20: 5] after they had refrained from such requirements for two decades.

According to statements by Nagra, the demonstration of siting feasibility seems to have just a formal status in the waste owners' disposal policy. One recalls the notion of a "sandbox exercise 85" the co-operative termed "Guarantee 1985" [P223, 8/81:2]. Seemingly a solution abroad is aimed at; with the operation of interim store facilities, no storage impasse has to be feared as was the case in the 1970s, and which led to the choice of reprocessing (see page 5). President Issler of Nagra was very transparent when he said in 1999: "The view across the boundaries is no question of trust but the consequence of opening markets and globalisation .... For countries such as Switzerland, with a small nuclear programme, a multinational solution is attractive" [TA, 1999-3-1].

Judging from the Federal response, the operators' strategy seems to succeed or, at least, have a parallel effect. In the 1999 preliminary draft to the Nuclear Energy Act, the Energy Department DETEC wrote: "For high-level and long-lived intermediate-level wastes, at present, it is not determined when the duty of disposal is fulfilled", for these wastes "a facility has to be in hand in the year 2020 at the earliest. Therefore, the disposition concept may today be left open" [P277:19]. HSK had prepared the terrain in 1997: "The deployment for the final disposal facility for high-level wastes, not required before several decades from now, shall not be initiated yet; the yet open issues, however, regarding the feasibility in principle shall be answered and possibilities of an international solution shall be explored" [P124:1]. The Director then in charge, Serge Prêtre, even meant in 1998: "The renouncement of Benken [in the Zurich Weinland] would not be tragic .... It is feasible that the Federal Council, some day, is of the opinion that the demonstration of siting feasibility for radioactive wastes in Switzerland is not necessary any longer and has become anachronistic. For this concept stems from the [19]80s, when one still assumed the construction of further nuclear power plants" [SoZ, 1998-9-6]. Even the Energy Minister and the media adopt the deferral of the deadline, partly the priority of a solution abroad [BBr, 1997-2-12]. There is some resistance, from the side of the Director of the FOE, Walter Steinmann, who declared before Nagra in 2002: "It must not be that the realisation of a HLW/TRU repository is deferred because this, firstly, is not necessary at present or, secondly, one anticipates an international, say, solution abroad" [P260].

It is paradoxical that it was up to the external bodies to exert pressure on a timely deployment of the HLW programme, against the resistance of the Confederation, the safety authority, and the applicants (or at least: the waste producers). All the more, this is not conceivable since one has to conclude that, from a technical point of view, the studies in the Opalinus Clay have rendered more positive results with a much lower effort in time and resources than the laborious work in the crystalline formations[122]. In 1998 KSA wrote in their position paper when they were alarmed by the struggling and disturbing development: "At any rate, detailed sites for repositories in Switzerland have to be demonstrated. The final disposal of HLW/TRU abroad shall not *a priori* be excluded, but shall only be pursued as an additional option next to the domestic disposal. Investigations into a disposal option abroad must by no means impair the search for a national disposal site and must not delay the deadline for disposal" [P178]. KNE seconded by criticising the "eventual 20 year-long interim phase of 'doing nothing' in the HLW programme" [P167].

The protest has been successful in so far as the Energy Ministry DETEC adhered to the duty to demonstrate disposal feasibility according to "Guarantee 1985" and, as it statuted, the disposal concept along the lines recommended by EKRA. In 2003, the FOE and AGNEB initiated a series of closed-door meetings of all federal bodies in charge (from the Administration to the advisory committees) to broadly discuss possible further steps, concerning the entire disposal programme [P20:5]. In the course of the review of Nagra's submission for demonstrating disposal feasibility of HLW in the Zurich Weinland, in 2003 the FOE established three bodies to accompany the work: a so-called Political Committee with representatives of the Cantonal Governments, a Working Group on Information and Communication, and a Technical Forum. Particular attention was paid to invite local communities, the regional opposition, as well as the neighbouring cantons

---

[122] No more stringent statement is possible due to the lack of available data. The information revealed in the latest – 2001 – cost study by the NPP operators confirms the assumption, though. Zuidema & Issler 2001 assess savings of 400 MCHF (in a whole of 1,900 MCHF) until "beginning of operation", as compared to the 1994 study: "this is primarily due to changes in host rock" [P288:3.2-3-3.2.4].

and the equally neighbouring state of Germany. Incidentally, the calibration borehole of Benken in the Zurich Weinland is less than 2 kilometres away from the German border (see *Figure 10-1*). The German Federal Office for Radiation Protection (BfS) and the adjacent Federal State, the Land of Baden-Württemberg, including its directly concerned administrative district of Waldshut, take part in the committees [P31].

4. CONCLUSIONS WITH REGARD TO SITING PROCEDURE

<u>General aspects</u>

The site selection can be visualised on three levels, each represented by two extremes (see *Figure 10-2* overleaf)[123]:

- On **Level 1,** there is the contrast between selection on a purely "technical basis", where **safety** as defined by experts is given top priority, and the selection through **volunteerism** where financial compensation is of importance.
- On **Level 2,** there is the contrast between the laypeople's demand for "absolute" safety (while still having to rest on **trust** in the experts) and the risk-oriented **expert** concept of "sufficient" safety.
- On **Level 3,** there is the insistence that the **"best site"** be identified as opposed to a site resulting from competing interests and financial **compensation.**

Hence, "absolute" statements exist with laypersons as well as experts. "Absolute" in this context means that in each judgement one dimension dominates all others (see Chapter 12). This is an observation throwing back the objection of experts against the "naïve" view of the impossible "zero risk" environment (see Wynne 1995, [R97:385]).

In analogy to the issue of knowledge generation, the following aspects are relevant to the site selection process: transparency, accountability and traceability of arguments, early involvement of the concerned stakeholders, iterative procedure, (mutual) trust in the stakeholders. In addition, it is crucial to define clear criteria beforehand and to stick to them (with regard to safety, ethics, *etc.*) (see Chapter 12).

---

[123] This approach was discussed and endorsed in a so-called Recommendation Group of the EU project COWAM, Community Waste Management [P52:34].

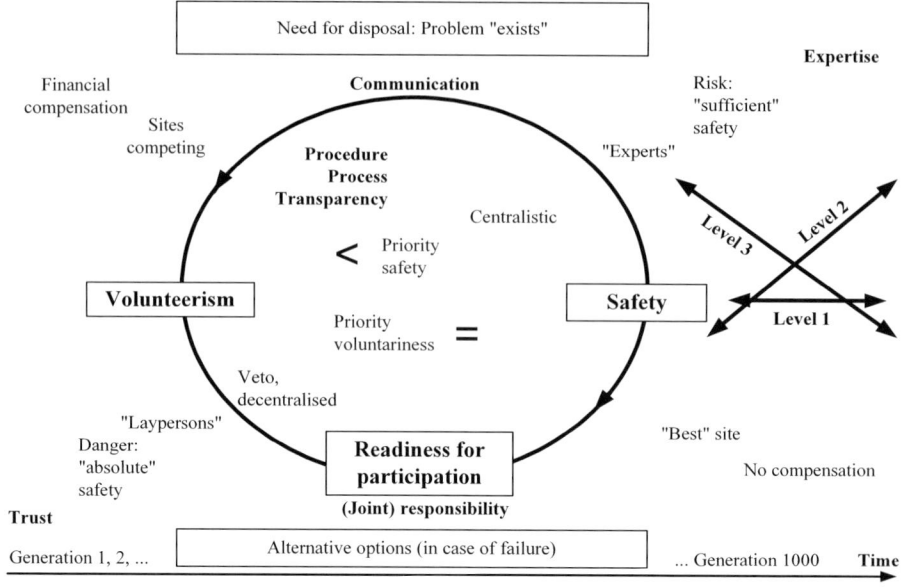

Figure 10-2. Approaches and mechanism of a siting selection procedure. Comments in the text.

When assessing site selections made in the past there should be distinctions made between methodology, the implementation, and result of the process. Methodologically it is of paramount importance to situate historical facts, decisions, and opinions – phenomena in general – in their temporal context without whitewashing them, *i. e.*, criticism must be formulated in a validated time-immanent manner rather than at the level of *ex post* arguments. In the specific case it has to be queried whether at a particular time additional knowledge might have been available, or whether all possible avenues for interaction and dialogue open to the principal stakeholders – mainly the applicant or the authorities – were adequately explored and implemented.

Changes in criteria and – more importantly – of the concept, have to be substantiated and carried out through consensus; bit by bit "closure" might be asymptotically achieved. All relevant partial steps have to be made visible and backed up by interim decisions. Otherwise – as has happened in nearly all national disposal programmes – delays are inevitable (though delays may still occur no matter what the process is). Since failure of the proposal is a possible outcome of the procedure and decisions need a choice of options, alternatives have to be considered as contingencies.

If an issue becomes too politicised, not even a minimum level of consensus – a "common ground" – between the main stakeholders can be reached

Final disposal siting as an example of sub-optimum decision making 169

(see Sections 12.4 and 13.3). As a consequence, there is a risk that demands to modify the concept are blocked and guidelines are watered down. Even attempts to instigate a rational (factual) discussion about all relevant dimensions and issues (see Chapter 12) may be seriously hampered through the formation of political blocs.

Compensation is a prominent topic in siting debates. The issue is not benefits or services directly related to a disposal project, such as the funding of communal liaison committees. Nonetheless, one may distinguish the following main positions on compensation:

- The case against financial compensation: Since every payment could be regarded as a sort of bribery, there is a strong argument against any compensation. For instance, in Sweden there is no discussion whatsoever about compensation for municipalities hosting a waste management facility. Although the waste management company pays taxes, these go to the State – not directly to the community. However, because of the contribution to the local employment situation and due to orders to local economy and services, even without "official" compensation schemes the region benefits from the facility.
- The case for financial compensation: According to the contrary argument, a host community/region provides a valuable service to the whole nation, and such service should be rewarded. Such a compensation is common in other areas, *e. g.*, hydropower plants pay local taxes and licence fees for the use of water and landscape to produce electricity; mining companies pay licence fees for the exploitation of underground resources; regional railway companies receive subsidies for providing necessary public transportation in areas where it is economically disadvantageous. In Switzerland, communities hosting (or in the vicinity of) interim storage facilities are reimbursed for so-called "gemeinwirtschaftliche Leistungen", *i. e.*, a service to the public; the Canton of Nidwalden declared the use of the sub-surface would be subject to a "mining licence" for which a substantial fee would have to be paid, in the event that the repository had been constructed. In addition to the NIMBY issue, there are real problems that a waste facility may cause to the hosting community. For instance, during the construction phase there will be additional (conventional) traffic and noise, or an otherwise picturesque landscape may become "industrialised" by the facility. This is another reason why compensation may be justified.

- Factual compensations in environmental context[124]: "Factual" means substantive, hardware, not "just" financial compensation. In principle, only those who have to bear a risk should be compensated, *i. e.*, the populations during the periods of greatest radiation risk from a disposal facility. If the risk is considerable, no financial discounting is allowed, and the fund for a "risk premium" should, consequently, be enormous; if, however, the risk is negligible, no compensation is necessary. Factual compensation in related domains (*e. g.,* the remediation of conventional contaminated sites, the inventory of wells and aquifers in the potentially affected area) would avoid the dilemma as well as provide benefits to present and future generations.

On the basis of the dimensional discourse in Chapter 12 I tend towards the third position. A payment must never be regarded as a compensation for any real risk beyond very strict tolerance levels. A repository must be safe and must not represent a real danger. However, even in the case when the facility is perceived as safe, the question arises "OK, it's safe. But why must it be here, and not somewhere else?" This, of course, is the usual NIMBY argument. Compensation may be considered as a means to overcome such NIMBY arguments but never to compromise on safety issues. This attitude should not impede activities towards "regional development" as proposed by AkEnd [G4:38-41].

Specific aspects

An analysis of the programme course and the decision-making processes in Swiss radioactive waste governance reveals that external factors were decisive – not just political factors, such as the link of the NPP licences to "Guarantee 1985", but also technical ones:

- The renunciation of sea dumping entailed interim storage and, later on, repositories for LLW.
- The shortage of storage capacity in the 1970s led to the (second, baseload) contracts for reprocessing (1978–1983).
- The threatening deadline for the repatriation of vitrified glass ingots from reprocessing made the Central Interim Storage necessary as

---

[124] In economics, factual services, as a compensation, are viewed to be inefficient because they cause transactional costs at the recipient, whereas – according to Frey & Schaltegger 2000 – monetary reward does not incur costs [M28:84]. The authors concede that risk perception (*sic!*) and, thus, the potential for conflict are very high in cases with asymmetric cost-benefit distributions [*ibid.*:51*passim*], such as prototypically with radwaste disposal. – Compensation may be counterproductive even to applicants [G26][G152][MA21][G243].

Final disposal siting as an example of sub-optimum decision making 171

well as – together with the blocked dumping – the definition of repository milestones[125].

This study substantiates the assertion that the radioactive waste proponent's, partly also the safety authorities' course of action, was neither transparent nor traceable for many years, even so for experts. The reasons for the initial over-optimism are manifold but three main aspects are paramount:

1. Well-known structural engineering was prioritised, whereas the decisive issue of long-term safety was underestimated.
2. Political considerations predominated the fact-based ones, whereby the waste issue was instrumentalised in energy politics by all parties.
3. Procedural issues of such a complex disposal programme were not duly followed, and if they were, they were dealt with as "political" aspects under point 2.

Setting up the programme radioactive waste management was considered primarily as a mining issue. Thus the tight deadlines are explainable even if the political premises of "Guarantee 1985" were not accounted for. The "deployment of respective repositories" was judged to be a "task to be accomplished in the medium- to long term" according to the waste owners' disposal programme of 1978 [P283:6-60]. The "steps for implementation" were assumed to be a dozen years until operation [*ibid.*:6-94]. From the current point of view this is a grotesque conception of a disposal programme. Maybe it was the fact that the legendary so-called "net plans" of the early 1980s [P10] never matched reality, which, in turn, led to not setting deadlines and milestones later on.

In assessing the selection procedure for HLW, HSK's view may be agreed on "[t]hat the procedure ... as a whole [complied] with the recommendations by [IAEA] as they are laid down in the Safety Guide: Siting of Geological Disposal Facilities" [G105][P128:5]. For this, the 1994 "Guide" is not useful with respect to a sufficient set of selection criteria, transparency, and traceability. In retrospect, it is recognised that Nagra's procedure corresponded to the then state of the art of site selection. Compared with major countries such as Germany, France, UK, or the USA it even was good practice; in leading industrialised nations eventually existent criteria were not considered or circumvented. Decision making was sub-optimum at an international level in face of the factual requirements of long-term safety and political acceptance, *viz.*, stakeholder involvement and democratic support.

---

[125] The issue of non-legality of storage after reprocessing in France is another topic (according to the French Law no. 91-1381 of 1991-12-30 [G227]).

This situation has lately been acknowledged by relevant parties in the international arena (see Section 11.3).

In the Swiss high-level waste programme, it was mainly the external review bodies that have promoted a consistent and continuous management. Particularly after the failure of the crystalline sub-programme, Nagra had the "option abroad" in their sights. With the low-level waste programme, the active role of the cantonal government in the potential siting canton, *i. e.,* Nidwalden, was noteworthy. Thus, the cantonal level took the lead in specifying and implementing the work until the further rejection of (GNW's second) application at Wellenberg in 2002 by the electorate. By taking up requests voiced in the debate on the general licensing procedure in 1994/1995, the small but sovereign canton of Nidwalden, with approximately 39,000 inhabitants, insisted on exclusion criteria being formulated (see Section 13.2) and the inventory being openly discussed. Indeed, the evaluation of a process-oriented set of criteria is more transparent (*Figure 13-3* and *Figure 13-4*), and thus more convincing, than the demonstration of compliance with a single, highly aggregated protection goal (see *Figure 13-1*). Conversely, the relatively passive role of the Federal Administration is conspicuous, in the alleged "national task" of radioactive waste disposal; this stands in contrast to their active role in the introduction of nuclear technology in the 1950s and 1960s.

Conceptually, it is recognised that the waste producers, the applicants, and the authorities have integrated many demands over the years, even though under pressure and only with reservations. Among the recent developments, it is envisaged to establish a pluralistic "Disposal Council", to revisit Nagra's position with respect to their owners, and to propose a novel site selection procedure [pers. comm., 2004-3]. Concerning institutional learning, the definition by Argyris 1982 is confirmed that organisational learning is a process of recognising and correcting errors [D4]. The forms of "adaptive learning" and "trial and error" dominate in the complex and politically charged radioactive waste governance arena[126]. Conclusions with respect to substantive, procedural, and institutional aspects are drawn in Chapter 13.

---

[126] Kissling-Näf *et al.* 1997 distinguish five learning types: "ordered learning", "instrumental learning", "trial and error", "lesson-drawing", and "learning in the lab" (pilot tests and social experiments) [D6:273][D47].

PART IV

CONCLUSIONS AND FURTHER DEVELOPMENT

Chapter 11

PATTERNS OF ARGUMENTS IN RADIOACTIVE WASTE GOVERNANCE

1. DOCUMENT ANALYSIS (OBJECTIVE 1A)

As outlined in Section 4.5, more than 2800 documents were sifted through and appraised and over 2400 entries were made, covering a time frame of over 40 years (1945/1956–2002), in order to reach Objective 1a according to *Figure 4-2*. As mentioned, concerning methodology, one has to note that many aspects could have been reasonably dealt with in another context. Protection goals and the definition of problems, *e. g.,* cover many or even most of the fields under scrutiny. Thus, continuous cross-references are required, inferred by the substantive, procedural and temporal interweaving of the complex system. It was emphasised that the method of content analysis was primarily employed to systematically investigate the data and not to classify documents into subcategories.

2. PATTERNS OF ARGUMENTS (OBJECTIVE 1B)

In order to structure the extensive material topical fields, so-called patterns of argumentation, were sifted (Objective 1b). The data were classified according to the criteria as laid out in Section 4.4, particularly from risk analysis and risk management, as well as risk perception and decision research. Substantive topics from Nagra's programmes (HLW and LLW), with related partial issues (isolation period, host rocks, conceptual notions, exploration strategies, *etc.*), were incorporated as well as procedural, project management, and organisational aspects. Recurrent themes were particularised (*e. g.,* option "abroad", "NIMBY", see Section 8.3). To highlight the contextualisation of radioactive waste management, brief information was listed separately, such as pertaining to the political context, technology policy, topicalities debated in each programme phase as well as the most relevant events on an international level.

2.1 Compilation of arguments

The itemised findings are reproduced in Flüeler 2002e [G72:Vol. II, tables 28-32], specific aspects in Vol. I, tables 16-25. The entries were arranged according to topics:

Risk analysis: protection goals, risk goals; principle of treatment: Confine and Concentrate (CC) or Dilute and Disperse (DD); robustness/ uncertainty;

(Risk) Management: disposition options; specification; time dimensions (deadlines); concept modifications (official); controllability; retrievability;

High-level waste (HLW): long term issue, need for isolation; host rocks; facility design; exploration strategy; exclusion criteria; site selection (traceability);

Low-level and intermediate-level waste (L/ILW): long term issue, need for isolation; host rocks; facility design; exploration strategy; exclusion criteria; site selection (traceability); "equal treatment";

"Solution abroad"

Problem view: problem "technical" *vs.* "political"; problem "existent" *vs.* "exported", problem "resolved/solvable" *vs.* "unresolved/unsolvable";

(Time) Management: premises, schedule, official; repatriation of waste from reprocessing; HLW repository, LLW repository (putting into service); interim storage; time requirement; "quick" solutions; "backlog", reference to – supposedly – good practices;

Nagra: premises, self-image and external image; pressure on Nagra, political pressure, substantive pressure, time pressure; information; scientific discourse; transparency;

Regulatory supervision: role of state; discharge of duties; personal resources; separation of duties; independent review; dismissal of Nagra *vs.* Federal solution;

Participation/procedures: information of the public; stakeholder involvement; legal means; speeding up proceedings;

Ethics: causality principle (violation): substantively, financially; goal relations; responsibility for waste, responsibility for energy policy; rationality; consistence of argumentation: Nagra, politics.

2.2 Findings

In Section 4.3.2 it was assumed after Menkes 1977 [M59] that a link between historic insights and current policy making can be constructed if

- "historical processes can be identified with or connected to policy decisions,
- the processes in contrast to events are transfe[r]able,
- the experience is generalizable, and
- parties at conflict can be identified and isomorphic situations can be modelled" [*ibid.*: 324].

Within the framework of the ongoing process in radioactive waste governance, partial processes including strings of argumentation, could be identified. The following *Table 11-1* (page 179) defines the pertinent decisions and stakeholders. Thus, certain statements are possible to make on the basis of historic insight. Above, the arguments were structured and commented according to particular aspects: risk aspects in detail in Sections 6.2.4, 6.3.10 and 6.4.5; decision aspects in Sections 9.2.1, 9.3.1, 9.4.1, 9.5.1, 9.6.1, 9.7.1, 9.8.1, 9.9.1 and 9.10.1.

2.2.1 Decision-making process

Following the assumptions and methodological guidelines in Chapter 4 it was possible to identify the different components and phases of decision-making processes with patterns or strings of argumentation. *Table 11-1* presents players (stakeholders) and decisional strategies in radioactive waste governance. For different periods I could identify different issues and interpretations of underlying goals (*e. g.,* from underground disposition via final disposal to monitored long-term geological disposal) and different types of problems (*e. g.,* from construction technology to long-term performance assessment including institutional aspects, see the two bottom lines of *Table 11-1*). Certain propositions can be made on the basis of the evidence obtained from historic insights.

The recording of the decision-related patterns (lines 4-6 of *Table 11-1*) emerged during the course of analysis. Partially using Böschen's[127] terminology we may discern a progressive widening and opening up of the debate: from the industry's expert organisation (*i. e.,* a "closed" expert) to an open expert community with so-called counter-experts. As a consequence thereof, the problem definition (line 7) extended from the internal technical problem of waste deposition in the 1960s and early 1970s to the conceptual debate (concerning issues ranging from final disposal to monitored long-term geological disposal) with the ensuing specification: Things and aspects under discussion became clearer and more outspoken. Whereas initially the resources were with the nuclear power utilities and, partly, with the regulatory body, mediation attempts were started after the electorate's acceptance of a moratorium of nuclear power construction in 1990, although these proved to be futile. Following the negative referendum on the proponent's (GNW's) application for a repository at Wellenberg in Central Switzerland in 1995, the trials were expanded to negotiations and, from 1999 onwards, to pluralistic expert discussions. Consistent with this, eventually the decisional conditions (line 6) widened from the insular existence of the construction technologists all the way to a strategic and inter-expert discourse; the power of defining the issue and framing increasingly moved away from the industry to include additional stakeholders (line 9). The problem horizon (second last line) initially was confined to construction technology, then developed to include long-term safety and, finally, has made radioactive waste governance both a technically and institutionally complex, long-term issue and programme. In summary, the progressive integration and generation of knowledge may be recognised as being characterised by an understanding of diverse perspectives, an openness in the ways of thinking and an exchange of ideas, which on the whole is appropriate for the complexity of the issue.

---

[127] Böschen's 2000 distinction between expert community ("expert public") and problem-related "public" cannot be endorsed because radioactive waste governance has always been a problem-oriented issue (of constraint) [M9:34-39].

Table 11-1. Patterns and strategies of decision in radioactive waste governance. KORA: Conflict-solving Group Radioactive Waste. For the other abbreviations and acronyms refer to the text (Cont. overleaf). Source: Flüeler 2002e [G72:166-167].

	– 1972	1975/76	1978/79	1980–1995	1995	1995–1999	1999–
(Main-) Stakeholders	Energy utilities (Local opposition)	Utilities Nagra Nuclear safety authorities (Regional opposition)	Utilities Nagra Safety authorities Science Regional opposition Env. organisations Parliament Public	Nagra, later on also GNW Safety authorities Advisory committees National opposition Env. organisations Parliament	Public (electorate of the Canton of Nidwalden) National/regional opposition Env. organisations Nagra/GNW Safety authorities	National/regional Opposition Env. organisations Advisory committees Nagra/GNW Safety authorities	Canton of Nidwalden (challenges the Federal administration) Federal Council EKRA/KFW Nagra Safety authorities Reg. opposition
Types of stakeholders	Engineers Construction technologists	Expert organisations Construction/ Geotechnology Authorities	Politics Closed expert community, earth sciences	Politics Partly open expert community, Counter-experts	Open expert community, Counter-experts	Dialogue groups (e. g., Energy Dialogue)	Regional government Committees Open expert community Pluralistic topical discussions
Resources	Industry	Industry Authorities	Industry Authorities Media	Attempt to mediate by Federal administration (KORA)		Negotiations (involvement of the env. organisations), Resource reduction at Nagra	
Decisions	Disposal as a duty Establishment of Nagra	Final disposal, Deadline of 1985 for "Project Guarantee"	Federal Decree to the Nuclear Act, "Guarantee 1985"	Commissioning of "Conflict solving groups" by federal administration (1990, after acceptance of the moratorium initiative)	Vote on the application for Wellenberg (rejected)	Intern. Waste Convention 1997 (national duty) – Search of industry for solutions abroad	Revision of the final disposal concept: Monitored long-term geological disposal

Decision types	Technical constraint	Conceptional issues Decide–Announce–Defend approach	Political premise Objective orientation	Political bargaining	Political premise, pressure on conceptual discussion	Economic argumentation (resource reduction, waste disposal fund)	Conceptual committee (EKRA) appointed by the federal administration Substantive negotiation Integrated process: widening of scope
Decisional condition	Insular	Corral	Fundamental positioning, Polarisation	Fundamental positioning, confrontation	Polarisation	Disintegration of traditional lines of position	Strategic and substantive discourse
Problem definition	1 (Disposition)	2 (Final disposal)	2a (Specification of 2)	(3) (Topics like controllability/retrievability)	3a (Topics of 3 officially set on the agenda)	3b (Discussion of 3)	4 (Monitored long-term geological disposal)
Types of problem definition and areas	Internal expert problem	Expert problem Proponent/authorities	Legal requirement Actual perception of public	Expert and political problem openly debated under pressure	Expert problem politically supported (topics of 3)	Concept issues openly debated	Expert issues openly debated
Power of definition	Industry	Expert organisation Authorities	Expert organisation Authorities Electorate	Authorities Expert community	Opposition Electorate	Industry (solution abroad, resource reduction with Nagra) Committees Intern. agencies (NEA, IAEO)	Committees International agencies (NEA, IAEO)
Problem horizon	Construction technology	Construction technology, Safety analysis	Long-term safety	Long-term safety	Long-term safety vs. retrievability	Long-term safety, retrievability, and controllability No "demonstration of long-term safety": "set of arguments"; chronic problem, flux	Institutional and objective long-term project Complexity
Main paradigm	Landfill	Final disposal	Final disposal	(Final disposal)		?-disposition	"Extended" final disposition

2.2.2 Improvement of system

It was also possible to identify topic-related patterns of reasoning. In summarising the complex interconnected strings of argumentation I may make the following statements:

Radioactive waste governance: a "technical" or a "political" issue?

- As a recurrent theme, the seeming juxtaposition of the "technical" (technically solved) and the "political" (*i. e.,* politically problematic) pervades all time periods. Implementers and authorities, on the one hand, have maintained that all technical issues are under control and the "reasons for delay" are merely "political"; on the other hand, the actors mainly in the NGO-oriented sphere have asserted that not even the "technical" basics have been solved to date. From 1957 to mid-2002 over 150 such examples were found. This phenomenon is not unique to nuclear issues though; in another context the Swiss constitutional lawyer Max Imboden in 1964 coined the following expression: "The essence of politics is based on trading off technical options" [P143:26)].
- The radioactive waste issue is misused as a political vehicle by both opponents and proponents of nuclear energy use: "insolvability" as an "argument" for phasing out *versus* "solution" as a "proof" of the legitimacy of a prolonged use of nuclear power.

Attitude towards technological issues (risk debate, long-term phenomena)

- As with other complex socio-technical fields, esp. with respect to technological constraints, the dimensions (see *Figure 12-1*) were often debated in reverse order: firstly, the technical and commercial aspects, followed by the political and economic, the social and, finally, the ethical aspects.
- Divergent attitudes towards technological issues become manifest in the differing attitudes towards safety and risk management. On the one hand, the nuclear waste community takes an inevitable "residual" risk for granted, which is to be reduced by way of risk assessment and concurrent mitigating measures. On the other hand, sections of

the opposition from the anti-nuclear movement demand "absolute safety"[128], "best solutions possible" and "best sites".
- Long-term aspects were not taken into account before the mid-1970s. To that, the waste community came up with the concept of final disposal, some other experts and the public repeatedly responded with the quest for controllability and retrievability. A conceptual attempt to integrate long-term aspects was made by an appointed external expert group, EKRA [P60][P61][P135].
- The insight that a rigorous mathematical proof of long-term safety is not feasible has led to the special emphasis of a stepwise procedure. The needed "set of arguments" [G182:11] referred to above comprises technological approaches (such as diverse and graded safety indicators, test facilities) as well as institutional components (quality assurance, reviewing procedures, see *Figure 13-1*).

Involvement of third parties and procedure
- Until 1980, "third parties", who did not belong to either implementers/proponents, regulators or hired experts, were not involved in the decision process. The linear model of "Decide–Announce–Defend (DAD)"[MA5:24][MA24] prevailed, in a seemingly contradictory mixture of Habermas's technocratic and decisionistic models [D33]. The "nuclear establishment" [MA23:21] of industry, the Federal Administration including the regulatory body, and politics was a closed circle. With the appointment of the "Subgroup Geology" of the Administration's strategic committee AGNEB in 1980 and their successor, the Commission for Nuclear Disposal (KNE), in 1989, this has gradually changed. In 1999 the conceptual committee EKRA followed, and in 2000 the Cantonal Expert Group Wellenberg (KFW), both predominantly resourced with external experts [G70].
- These aforementioned external bodies either acted as the initial impetus or were encouraged by others on decisive issues, *e. g.,* the "Subgroup Geology": criticism of programming, non-traceability of the siting process, extensive duty of publication, reviewing, participation of the public; KNE: criticism of the programme "crystalline formations", extension of investigations to sedimentary host formations, programme management; EKRA: acknowledgement of non-technical aspects to be justified, integration of controllability and retrievability as a part of the repository concept; KFW: exclusion criteria, discussion of inventory (for both see below), specification of

---

[128] Sundquist 2002 demonstrates that this notion of safety is evidently not confined to local and regional opposition as it determined the nuclear waste debate in Sweden for many years, until legislative changes in 1984 [MA43:75*passim*,111].

Patterns of arguments in radioactive waste governance 183

- the EKRA concept, transparency of reasoning, active involvement of additional stakeholders.
- The broad public forcefully obtained their "involvement" in popular votes and referenda. Thus, they could exert pressure on change and concept modifications.
- It was only after the nuclear "moratorium" vote of 1990 that representatives of "non-official" experts were admitted into the decision-making process (within the so-called "Action Programme 'Energy 2000' ").
- Political considerations predominated the fact-based ones. In Switzerland, many parties and stakeholder groups instrumentalised the waste issue in their energy politics. Procedural issues of such a complex disposal programme were not duly followed, and if they were, they were dealt with as "political" issues.

Concept and evolution of the disposal programme, learning processes

- The <u>concept, requirements and delays</u> have, time and time again, been changed – on some occasions considerably. It is only recently that corresponding revisions have been debated among a broader range of stakeholders.
- As mentioned, <u>political considerations</u> overruled the fact-based ones. The waste issue was in some respects misused by all major stakeholder groups, as a political battlefield or platform.
- With regard to <u>scheduling and factual issues</u> there was an over optimism particularly in the 1980s. Initially, familiar and well-known structural engineering issues were prioritised, whereas the critical issue of long-term safety was underestimated.
- Gaining a complete overview of the complex system of radioactive waste governance is almost impossible for all players. Thus, <u>proxies are defined.</u> Examples in the technical field are controllability, retrievability, exclusion criteria (see below), safety indicators (see *Figure 13-1*), traceability of reasoning. Separation of promotion and oversight, extensive publication of documents, stepwise and phased procedure, external reviewing serve as proxies for procedural issues.
- Regarding <u>learning processes,</u> the forms of "adaptive learning" and "trial and error" dominate in the complex and politically charged radioactive waste governance arena. They were induced by programmatic impasses, such as difficulties in the crystalline formations of Northern Switzerland with respect to high-level waste investigations, or political pressure like the rejection of applications as in the case of the potential low-level waste site in Wellenberg in 1995 and 2002. The range of instruments for learning is poorly developed (informa-

- The interaction of the stakeholder groups [D6: 272] suggests a partial convergence and gradual acquisition of values and notions such as controllability and retrievability or the expansion of the disposal concept.
- Resource exchange is hampered by the political instrumentalisation of the issue. Networks [D36:388] only existed among proponents and implementers as well as regulators (national supervisory bodies/NEA forum). Together with these networks, engineering experts, university institutes, and advisory committees constituted the nuclear community in a narrower sense (*e. g.,* forum of IAEA). Aside from the private National Cooperative for the Disposal of Radioactive Waste, Nagra, there is neither an independent research centre nor a real strategic body for radioactive waste issues in Switzerland. The Paul Scherrer Institute has a technical Waste Disposal Laboratory predominantly working for Nagra. On the one hand, this was obstructive to the process in the sense that relevant topics were only brought in and eventually adopted quite late in the investigated time period (*Figure 11-1*). On the other hand, in retrospect, it may have been useful as innovation and independence were possible in the research phase of radioactive waste governance. During the implementation phase, however, the lack of a connective network is fatal. Knowledge transfer and technically as well as societally reinforced programming can only be sustainable if sufficient resources are provided [*ibid.*:372].
- Leading role: In the high-level waste programme it is mainly the external review bodies that have promoted a consistent and continuous management. This is remarkably in line with findings in STS, such as by Jasanoff 1990 [M43:237]. Particularly after the failure of the crystalline sub-programme, Nagra had the "option abroad" in their sights. With the low-level waste programme, the active role of the cantonal government in the potential siting canton, *i. e.,* Nidwalden, was noteworthy. Thus, the cantonal level took the lead in specifying and implementing the work until the further rejection of (GNW's second) application at Wellenberg in 2002 by the electorate. By taking up requests voiced in the debate on the general licensing procedure in 1994/1995, the small but sovereign canton of Nidwalden, with approximately 39,000 inhabitants, insisted on exclusion criteria being formulated (see below) and the inventory being openly discussed. Indeed, the evaluation of a process-oriented set of criteria is more transparent (*Figure 13-4* and *Figure 13-5*), and thus more convincing, than the demonstration of compliance with a single, highly aggregated protection goal (see *Figure 13-1*). Conversely, the relatively

Patterns of arguments in radioactive waste governance 185

passive role of the Federal Administration is conspicuous, in the alleged "national task" of radioactive waste disposal; this stands in contrast to their active role in the introduction of nuclear technology in the 1950s and 1960s.

A schematic list of propositions and eventual integration of various substantive and institutional aspects is given in *Figure 11-1*:

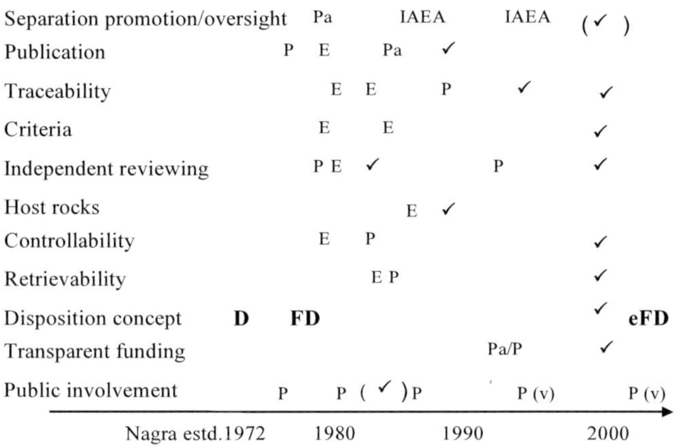

	Nagra estd.1972	1980	1990	2000
Separation promotion/oversight		Pa	IAEA	IAEA (✓)
Publication	P	E	Pa ✓	
Traceability		E E	P ✓	✓
Criteria		E E		✓
Independent reviewing		P E ✓	P	✓
Host rocks			E ✓	
Controllability		E P		✓
Retrievability		E P		✓
Disposition concept	D	FD		✓ eFD
Transparent funding			Pa/P	✓
Public involvement	P	P (✓) P	P (v)	P (v)

Figure 11-1. Integration of relevant aspects into the official disposition concept for radioactive waste in Switzerland. Note the time difference between first issue raising (P, E, Pa) and its consideration (✓). The empirical basis for the table is the content analysis of Flüeler 2002e [G72]. Pa Parliament, E Expert(s), P Public (v: vote at Wellenberg 1995/2002), IAEA International Atomic Energy Agency (recommendation; Waste Convention 1997), **D** Underground Disposition, **FD** Final Disposal, **eFD** extended Final Disposal/Monitored long-term geological disposal (acc. to EKRA 2000 [P60]).

3. COMPARISON WITH DECISION STRATEGIES IN OTHER COUNTRIES

3.1 Analysis by country

In contrast to Switzerland, other countries are documented with an imposing (Sweden, UK), even excessive, literature on radioactive waste governance (USA). The most relevant comparative studies are listed in the reference bodies "MA" (Section References, 2.5) and partly – especially from international organisations – in "G" (References, 2.4).

It is neither intended nor possible to conduct a comparative analysis on radioactive waste management. In the sense of validation (Section 4.6) it was checked whether, on the one hand, methodologically the chosen path was fruitful and whether, on the other hand, the patterns of argument and decision encountered in the present study appear or are discussed elsewhere. Furthermore, throughout the study, references were made to the international context and development.

It may be asserted that the constellation, as it appears in the radioactive waste "system Switzerland", is also shown in other industrialised nations analysed. In nuclear states obviously the links between military and civil technology is tighter than in Switzerland; Jacob 1990 even refers to a "nuclear establishment" in the USA case [MA23:21*passim*]. The mechanisms dealt with in PART I and PART II, however, are found everywhere, partly by virtue of facts (the characteristics of radioactivity and of the disposal system), partly by the circumstance that the management of the utilisation of radioactivity is harmonised worldwide via international organisations, *i. e.*, positive and negative connotations diffuse internationally.

Kemp 1992 [MA24] analysed the decision-making process in seven industrialised nations in a study which still is systematically pioneering yet as to the actual country-wise state outdated [*ibid.*:167]. According to that author, one extreme approach is DAD, "Decide–Announce–Defend" (see Section 8.1). Public involvement is minimised, the authorities in charge have the process under control, therefore it is labelled "centralised" and "closed". The other extreme is – in ideal terms – "devolved" and "open" planning with the steps: "Establish criteria–Consult–Filter–Decide". The public is involved at an early stage; "filtering", doubtlessly the act of trading off, takes place in a phased and transparent manner.

Evidently, there is no ideal solution. But "DAD is dead, and MUM is alive" as the Chairman of ICRP, Roger Clarke, coined it in 2003 [G74]; by "MUM" he meant "Meet–Understand–Modify". DAD is bound to be a failure [G163]. Many political and institutional boundary conditions play a part on the way to a "successful" strategy. The **Netherlands** officially favour long-term storage, controlled until the unforeseeable future, whereby the reasons ambivalently range from resource potential (plutonium, uranium) to risk uncertainties in the sense of minimum regret and progress in technology [G49][G9][G24][G273][G238][G272][G280]. Kemp locates the USA close to Switzerland, whereby there, one final disposal, WIPP, was commissioned in 1999 [G96], Yucca Mountain approved by the President in 2002 [G207], but not even a number of central interim storages were built [many G and MA]. Accordingly, Kemp does not judge the precise position of the respective state to be important on the decision-making process axis, but "the *direction* in which that approach is moving" [MA24:168].

In line with this, **Sweden** scrutinised decision strategies much more systematically than Switzerland and utilised participatory instruments on a larger scale [many G]. In a comparison of processes in Canada, UK, France, and Finland, KASAM 1998 reached the "conclusion … that the responsibility taken by the Government and the actual possibility for the local community to have insight and influence are two important factors for the credibility and results of the nuclear waste programme" [G143:14]. The Swedes have consented to a gradual phase-out of nuclear power in the popular referendum of 1980; even though it is not binding for governments it has been a yardstick in the discussion [G210]. The chances for success, to untie "the Gordian knot" (Kemp) are better in the north [MA24:127-128]. Internationally there is the mention of the "Oskarshamn model", named after a potential host municipality, where an extensive informal public involvement is linked to communal-based expertise and the formal procedure of an environmental impact assessment (EIA) [G5][G186:39-45][G145:39-40][G30][MA10:17-20] [M63]. Taxed as "unresolved issues" are "uncertainty of how to design the forthcoming planning and EIA-process, how to achieve confidence and legitimacy for actors and processes, and the long time span of the planning process" [G20:74].

Of particular interest are the comparisons with Sweden, which have been made in Switzerland, always as a model: from setting the deadline for "Guarantee 1985" via the choice of crystalline formations as the preferred host rocks (called in by Nagra and the authorities) up to the consistency and continuity of the Swedish disposal programme (consulted for the struggling HLW programme). Proponents and authorities do not put forward that Sweden has written down a ban on nuclear waste import – as most other EU Member States that cause considerable waste generation [G143:203*passim*]. Generally, there is an open discussion in Sweden about the nuclear waste issue; the R&D (and Demonstration) Programme by SKB is subject to a thorough review every three years. The advisory committee KASAM has contributed some fundamental ideas to the international debate, such as to integrate control functions into the final disposal concept back in 1988 [G139]. In their work "KASAM tries to cover a number of issues that are of special interest in the general debate and for which there may be a need for correct and readable information" [G142:14].

KASAM 1993 campaigned for the financial support of the host municipalities [G140]. As decided by the Swedish Government, municipalities involved in the feasibility phase of the site selection process are given 2 MSEK (around 200,000 EUR) per year from the Nuclear Waste Fund, paid by the NPP utility companies and state-administered. "This is used by the communities to build up competence, to inform the public and to employ experts to assist them in decisions they have to take", as Espejo & Gill 1998 wrote in a stimulating study on "The systemic roles of SKI and SSI in the

Swedish nuclear waste management system" [G56:5]. In 1999 the committee proposed to expand support for third parties, *viz.,* to environmental organisations at the national level "[i]n the light of the importance allocated" [G144:7].

It is evident that also the "Swedish way" is not free of flaws. In the 1970s the choice of disposal strategy took its course in much the same way as elsewhere. Sundquist 2002 [MA43:13/53*passim*]) plotted the course in remarkable similarity to Switzerland: forced upon the utilities by politics and society, linked with and subordinate to industry-state-defined technology and energy policy, supervised by "a small group of experts" and determined by a committee (whereby this, AKA, at least was established by the Parliament), in a speedy fashion (1976–1977), with overoptimistic timetables. The 1977 Nuclear Power Stipulation Act even required "absolute safety" albeit left to the Government to specify. From the late 1980s SKB changed their siting strategy from scientific-technical to voluntary-based claiming, in Sundquist's words, "that it should be possible to store [*sic!*] nuclear waste safely in most parts of Sweden" [*ibid.*:21]. SKB even actively approached the five Swedish municipalities with nuclear facilities, "local acceptance" being the primary criterion in the choice of feasibility study-locations [*ibid.*:124]. The site selection was criticised to be unsystematic, criteria (so-called "factors") to be non-existent. KASAM 1995 mourned that they "should be defined before the site investigations are started" [G141:5]. As of 2004, the municipalities of Oskarshamn and Östhammar have consented to accept site investigations in their territories [MA10:20]. With respect to the quality of the discourse, Sundquist 2002 viewed "both technocrats and democrats… hold dogmatic opinions, which support conflicting strategies", using the aspect of "uncertainty as a weapon" [MA43:223].

For a modification of strategy it is illustrative to analyse failures in national programmes including the learning effects gained. After the fiasco Nirex had experienced in the **United Kingdom** with their rock characterisation facility project in Sellafield (for LLW), the company launched a so-called Transparency Initiative in 1998. Subsequently it was completed revamped, as to personnel and organisational set-up. "We must learn the lessons", as David Wild of Nirex put it in 2001 [G276]:

- Related to *process* this amounts to determine a "clear up-front roadmap … with 'front end' consultation at key decision points", to include the authorities (up to then the disposal programme was left to the waste owners), to "include national consideration of approach to local issues", with a need for a "contract" with potential host communities, "an explicit recognition of veto and volunteerism", "planning gain" ("when is bribery not bribery but compensation?"), the previously "closed" site selection procedure should be opened up,

and "with an open application of criteria" at that, a published list of envisaged sites, and an "ongoing consultation".
- The *behaviour* of Nirex should be changed by "not driving the programme too quickly", work programmes should be open to influence from the stakeholders ("'add preview to review'"). An "absolute massive thing is involvement not information". "Respect for local view... and do not rely on Government policy".
- As for the *structure* a "separate organisation is required", separate from the waste producers, also a "separate long-term focus for thinking". Various and distinct skills are required (technical and scientific, social aspects, "public interest at heart").

It is judged relevant to formulate and discuss alternative options since in complex systems unexpected events (of technical or political nature) always have to be reckoned in. When setting the course, back up scenarios have to be elaborated so that impasses and failures do not bring the programme to a general halt. Further activities were carried out, such as BNFL's Stakeholder Dialogue [G36], followed by remarkable recommendations [G1], and the consensus conference in May 1999 [G239].

According to the Agreement between the Federal Government of **Germany** and the utility companies dated 14 June 2000 "exploration of the Gorleben salt dome shall be stopped for at least 3, at most however, 10 years" [G3, IV/4.]. At the same time, a Committee on a Selection Procedure for Repository Sites, AkEnd, was established [G186:105-108]. It was mandated to set up a proposal for a new siting procedure by the end of 2002: "For the intended new beginning it is of central importance that the selection procedure provides for the involvement of the general public and the dialogue with all those concerned right from the start" [G4:1]. They complied with the principles of decision theory in so far as they proposed a systematic yet process-oriented procedure with a set of comprehensive criteria, guiding principles on which to base the decision, a national debate on goals and uncertainties [G73:122-124]. The committee emphasised the public "willingness to participate" in contrast to the principle of volunteerism [G4:13], as a result of the alleged primacy of (passive) safety and of the factual constraint of existent waste. Addressing the issue of locally concentrated risk and, consequently, of compensation, they tabled the idea of a "regional development" instead of plain financial rewards to host communities (see also Minsch *et al.* 1998 [M61]). In the long-standing and fierce German debate on nuclear power the proposal was received with moderate consent [G132] but to completely restart the procedure is a contended issue [G151][G131].

France introduced the principle of voluntariness with the law of 1991 [G227], whereupon some 30 communities announced their interest for the establishment of an underground research laboratory (URL) to investigate into the disposal of HLW. Ten of them were selected on the grounds of technical and socio-political criteria, of which the waste implementer ANDRA picked eight. Due to heated regional and national opposition the French Government stopped the activities in June 2000 [G186:45-48,127-129][MA3] [MA10:21-22]. In 2003, one clay site – in Bure, Department Haute-Marne and Meuse in eastern France – was left for investigation while the search for a granite site failed in the absence of local support. In the case study undertaken in the COWAM exercise (see below) in 2003, "local people are concerned that the Meuse site in Bure remains the only [URL] in France and the implementation of the 1991 law is being delayed: because the Bure URL cannot be compared with another site, and the investigations are getting late, research activities are unlikely to provide definite results by 2006" [MA10:22], the date when the Law requires a milestone for Parliament to take a decision for further procedure.

Easterling & Kunreuther 1995 proposed "Facility Siting Guidelines" for a site selection procedure pertaining to environmentally relevant mega projects in the **USA**, together with representatives from politics, environmental organisations, citizens' groups and the industry [MA15]. The principles are the following [*ibid.*:168*passim*]:

"1. ESTABLISH THE NEED FOR THE FACILITY
   1a Achieve agreement that status quo cannot be maintained
   1b Choose solution that best addresses the problem
2. REDUCE THE PERCEPTION OF RISK
   2a Guarantee that strict standards will be met
   2b Use contingent agreements
   2c Work to develop trust
3. COMPENSATION
   3a Fully address all negative impacts of a facility
   3b Make the host community better off
4. EMPLOY A FAIR METHOD FOR SELECTING SITES
   4a Seek acceptable through a volunteer process
   4b Consider a competitive siting process
   4c Work for geographic fairness
5. INVOLVE AND RESPECT THE PUBLIC
   5a Institute a broad-based participatory process
   5b Seek consensus
6. KEEP THE PROCESS ON TRACK
   6a Set realistic timetables
   6b Keep multiple options on the table at all times"

Even if one does not agree with all captions ("Reduce the perception of risk"? or the "voluntary model", consensus on what?), some advice is helpful in setting a new standard in technologically-driven activities, such as the central involvement of concerned parties, independent reviewing and local control (local safety committee with influence). This power can be supported by way of "contingent agreements that spell out what will occur in case of accidents, interruption of service, changes in standards, or the emergence of new scientific information about risks or impacts" [*ibid.*:174]. Parallel to the "latency period" suggested here (see Section 9.10), Easterling & Kunreuther 1995 urge the promoters to "keep the process on track" while, at the same time, to allow "adequate time to consider the full range of options and to weigh technical evidence as it is gathered. In other words, it may be necessary 'to go slowly in order to go fast'" [*ibid.*:187]. Against the deadlock of the Canadian independent review Panel's remark of "[a] choice of one is not a choice" [G26:56] the authors emphasise the thorough examination of options, albeit not feasible "at all times" in the present context. As for the success, the proposal could be positively tested in 29 projects [G153], the real site of Yucca Mountain though was by far not chosen in compliance with the recommended criteria [many MA].

For that matter, **Canada** is a remarkable example of learning. In an extensive 8-year exercise this just-mentioned so-called Seaborn Panel Report identified the AECL concept of the 1980s as a major failure since it did not "demonstrate to have broad public support" (after [*ibid.*:53]), in the end, it did not address the issue sufficiently from the two complementary perspectives, technical and social. The Panel recommended to establish a new waste management organisation, "at an arm's length from the nuclear industry, entirely funded by the waste producers and owners … and subject to oversight by the government" [*ibid.*]. As a consequence, the Government of Canada enacted a new law in 2002, the Nuclear Fuel Waste Act, which was also the basis for the required new organisation, the Nuclear Waste Management Organization (NWMO). Its mission is "to develop collaboratively with Canadians a management approach for the long-term care of Canada's used nuclear fuel that is socially acceptable, technically sound, environmentally responsible and economically feasible" [G211:cover]. The novel attitude is already shown in the wording of NWMO's first "Discussion Document" called "Asking the right questions?": "It is not a journey that can be undertaken by a small, elite, isolated group. Rather, it is a journey that invites the perspectives of a broad cross-section of Canadians" [*ibid.*:10]. As opposed to the Swiss experience (see page 181), the approach deals with the dimensions in a logical and comprehensive way, starting with "overarching aspects", such as "institutions and governance", "engagement and participation in decision making", "aboriginal values", "ethical considerations", and "synthesis and continuous learning", followed by social, environmental, economic, and technical aspects [*ibid.*:8]. See Chapter 12.

Last but not least, the case of **Finland** is an exceptional one. This Nordic state is the only country so far to have commissioned a HLW repository, by way of the ratification in Parliament in May 2001 [G192:46]. The host municipality, Eurajoki, had consented as well as the safety authority STUK. After the construction of an underground research lab the repository shall be built from 2010 and go into operation in 2020 [G186:74-77]. Does this showpiece internationally serve as a model? As Mays 2002 observed in the 2001 workshop organised by NEA: "Often ... it was commented that the most intriguing features of co-operation among the Finnish actors simply had to be attributed to their 'culture' " [G192:113]. The pragmatic and straightforward way the Finns, esp. STUK, dealt with the issue, is impressive. One should be cautious though. Again, the host community is already the site of a NPP; one third of the municipal tax revenue comes from the respective company [G155:530][G192:51]. The Eurobarometer 56.2 as of autumn 2001 states that the Finns' public trust in government is – according to the survey – only 21.3 per cent, second lowest after the Italians' attitude towards (against) theirs [G133:16-17]. According to another survey (of 1998) half of the Finnish population is reported to believe that geology in Finland is not safe enough for disposal, 28 per cent think it is safe; 71 per cent hold that "radioactive waste is a permanent threat to the lives of future generations" [G204:522, transl. suppl.]. Thus, at least, there is a certain discrepancy between the perception by the overall population and the decision makers.

3.2 International organisations, research and development

It is amazing how the once technocratic, and pronuclear, international organisations **NEA** and **IAEA**[129] have evolved. Lately they have published a number of remarkable documents [see in G]. Internationally the situation is so muddled that non-technical issues are paid attention to even in the IAEA, such as at the Córdoba Conference in March 2000: "In almost all of the Conference's technical sessions, there was discussion of the need to involve all interested parties ['stakeholders'] in the decision-making processes related to radioactive waste management" [G118:vi]. Thereto, NEA established a so-called Forum on Stakeholder Confidence (FSC) in 2000 [G186].

The role of the international organisations (*Figure 12-3*, page 210) playing a part in the management of ionising radiation reflects the multiple function many state authorities have in the nuclear field:

---

[129] See, *e. g.*, footnotes 24 and 109.

- They were established as technical bodies, intrinsically having difficulties integrating non-technical aspects.
- On the one hand, they were founded as (technical) promoting instruments of nuclear technology (and by that, after Hypothesis 1, flawed with political implications).
- On the other hand, they act as advisory instruments regarding surveillance (radiation protection, reactor accidents, emergency) and the institutional oversight by the national bodies.
- Thereby, they do not virtually assert control (with the moderately successful exception of safeguard measures by IAEA), but give assistance and make recommendations albeit called "guidelines".

This explains not only their ambivalent role between promotion and control, but also why it took them a long run-up time to address the issues and criteria (such as information, transparency, openness, *etc.*) being looked into in this study. All Member States have to be considered, standards for harmonisation have to cover all states (worldwide in the case of the IAEA, all OECD countries in NEA). This inertia effect is not wholly negative for each position and change thereof has to be checked against a range of eventual impacts, increasingly now also non-technical ones. As an example the IAEA 1998 document on "Technical, institutional and economic factors important for developing a multinational radioactive waste repository" [G112] may serve, also valid for "national" projects: "A repository is, by definition, a long term project, extending over centuries for most LILW or even much longer periods for repositories in deep geological formations, receiving HLW with long lived radionuclides. A repository project involves a relatively long lead time (possibly more than 20 years for HLW or spent fuel) and is then anticipated to receive waste during several decades. After closing the repository, a surveillance and monitoring period will almost certainly be carried out even [*sic!*] for *shallow land burial* type repositories with LILW. This underlines once again the importance of the continuity factor not only from a contractual but also from a technical point of view (possibility/obligation to transfer/receive waste, waste acceptance criteria and quality of waste, control and monitoring, etc.). On the other hand, continuity is of equal importance for the proper functioning of the cost sharing arrangements and the respective payments" [*ibid.*:9].

During the 1990s, international nuclear agencies recognised that it is not up to them alone to decide on strategies with ethical, economical, and political dimensions. Thus the NEA stated in 1999: "Rather, an informed societal judgement is necessary" [G181:23]. Similar activities to the FSC were launched in the radiation protection field[130] by NEA's Committee on Radiation Protection and Public Health (CRPPH) through its Expert Group on the Processes of Stakeholder Involvement (EGPSI) and, *i. a.*, by way of a series of workshops [MA42][G17][M6]. For the radiation protection specialists the "fundamental question" was: "... is the issue one of integrating societal aspects into radiation protection decisions or, integrating radiation protection into societal decisions?" [G47:139]. The "broad conclusion was that radiation protection must adapt to meet the needs of society, and not the reverse" [G197:3].

For external parties this understanding may come in late, but the "prudent approach" is typical of technical institutions and, in a turbulent world of value and context changes, may be advisable. In their 1997 Conference on "Regulating the long-term safety of radioactive waste disposal" in Córdoba, NEA is afraid of, so to speak, too much public in the sense that "the regulators might be subjected to some pressure from the public in the conduct of their professional responsibilities, which might result in the risk of 'diluting good engineering practice to have better relations with the public'" [G199:245]. The dialogue, doubtlessly to be pursued, should not lead to a diffuse shift of roles with a "tendency among regulators to transfer to the public part of their responsibility" as Maurice Allègre, then President of ANDRA, put it [*ibid.*:249]. The agencies' approach, nevertheless, is still technology-oriented, not to say technocratic if one reviews the valid IAEA definition of the system "radioactive waste" of 1995: It is "the summation of all the individual components, for example body of laws, regulatory organizations, operators, facilities, etc. [*sic!*]" [G108:4-5].

Within the IAEA, tentative steps are undertaken with working groups [G121][G125], which are met with restrained positive response in the Board of Governors: "Action #7 – Develop a step-by-step programme of work aimed at addressing the broader societal dimensions of radioactive waste management, including an appropriate mechanism to advise on such a programme and assess its suitability and progress" [G122:7].

Also, at the international scale, we find the recurrent theme of the seeming juxtaposition of the "technical" (technically solved) and the "political" (*i. e.*, politically problematic) pervades all time periods. Referring to the "trans-scientific" nature of the issue, even in 1984 Parker *et al.* observed in a comparative study of countries which they did also for Switzerland: "... it is obvious that total consensus cannot be reached. Consequently, the only valid

---

[130] It is noteworthy that the expert group *par excellence,* the International Commission on Radiological Protection (ICRP), founded in 1928, started to put their drafts onto the Net for consultation (see www.icrp.org), and they lately recognised the need to include non-human species in their protection framework [G130].

course is to adopt a process that is as open, complete and fair as possible and to try to develop as broad a consensus as possible" [G217:101]. This recognition, and the role esp. regulators have to play in addressing uncertainty, is difficult to achieve. This idea is pursued in Section 13.5.

The locational solution by Parker is clear and, in an interesting way, cross: "... a 'solution' is to be found only in a national, not a scientific, context" [*ibid.*:108]. Sixteen years later, in 2000, Parker observed at the 44th IAEA General Conference: "For high level waste, the delays currently being experienced in many countries will negatively affect those countries moving towards establishing repositories unless the high level waste can be placed at or near presently existing waste disposal sites and nuclear energy facilities. Consensus is being reached that retrievability is essential for public acceptance though it violates the principle of this generation taking responsibility for its waste. Alternative methods (*e. g.* separation and transmutation) will be sought with limited success" [G218:2]. Without commenting on Parker's view on retrievability one may consent to the critical tendency of undermining "national" solutions, albeit the International Waste Convention of 1997 states that it is up to the signatory states to find a solution for this national task [G110:preamble xi].

At a **generic level** we are striving for a harmonisation of technical risks, based on the "New Approach"/"Global Approach" of the European Communities 1999/2000 [M10][M11] to deal with risks in an integral way. Some proposals were issued in this direction, *i. a.*, by the "International Risk and Governance Council" 2000 under the auspices of OECD [M40][M53] (established) or a "UN Risk Assessment Panel" suggested by the German Advisory Council on Global Change 1998 [M87]. There are first steps for implementation as recommended in the "Compass for Risk Analysis" [M47]. This is in line with a corresponding project in Switzerland from 1996 to 1999 which systematically analysed various technical risks [M26].

On a **programme level,** ideas were developed in 2002 by the **US National Research Council (NRC)** on a "staged development of geologic repositories for high-level radioactive waste" [G271]. HLW repositories are "first-of-a-kind, complex and long-term projects that must actively manage hazardous material for many decades .... As is the case for other complex projects, repository programs should proceed in stages" [G271:1]. In such an "Adaptive Staging" a "pilot scale facility" is envisaged, much the same as in the framework of EKRA (see Section 13.2), for monitoring and control [*ibid.*:174] in order "to apply this knowledge to affirm or modify the design and operations [*ibid.*:51]. For reviewing, *i. a.*, a local and regional "technical oversight board" is foreseen; as a "forum" of dialogue, NRC proposes a "stakeholder advisory board" [*ibid.*:66].

At the **process level,** and with an approach "from below", one has to mention the pioneering EU project **TRUSTNET** within the 5th Framework Programme, which is followed by TRUSTNET "in Action" (see Section 8.2). The "top down" paradigm is to be replaced by a "bottom up" paradigm (of "mutual trust"), whereby experts and other stakeholders shall enter into an intense dialogue. In eleven diverse case studies (from nuclear and chemical risks to genetically modified organisms and riverine flooding) it was attempted to discuss technical risks at local level and with broad stakeholder involvement [M18:3-9][G88]. So to speak, as a radioactive waste corollary the project **COWAM,** Community Waste Management Concerted Action, was carried out from 2000 to 2003 [MA33][MA10][G89]. In four highly interactive seminars the radioactive waste governance in Sweden, UK, France, Germany, Switzerland, Spain, and Belgium were discussed by means of concrete case studies, and mixed so-called recommendation groups laid down their views on the following topics: implementation of local democracy, access of non-experts to expertise in the local decision-making process, influence of the local actors on the national nuclear waste management framework, sustainable development in potential host regions, and quality of the decision-making process. A follow-up named COWAM 2 is to systematically scrutinise these specific issues. The objective of COWAM 2, until 2006, is to contribute to the actual improvement of the governance of radioactive waste by, *i. a.*, "better identifying and understanding societal expectations, needs and concerns as regards radioactive waste decision-making processes (DMPs), notably at the local and regional levels" [MA34].

The EU project **RISCOM II** (2000–2003) [G41] was to "support the participating organizations and the European Union in developing transparency in their nuclear waste programmes and means for a greater degree of public participation. Although the focus has been on nuclear waste, findings are expected to be relevant for decision making in complex issues in a much wider context" [MA1]. The undertaking provided a "'map' of values encountered in performance assessment, a review of dialogue processes and hearing formats, diagnosis of organizational structures and understanding of the organizational impact on transparency, consensus statements from a group of key actors, production and evaluation of a School's Web site" [*ibid.*].

In its final report RISCOM concluded in a comparison of their own project, COWAM, and the FSC exercise by NEA that "all three studies emphasise that radioactive waste management, due to its long-term nature, uncertainties, and emotive nature, is not the exclusive domain of technical expertise. Wider stakeholders' concerns should be addressed at the same level as technical issues. The decision-making process must be open, transparent, fair and participatory" [MA1:64,98-101].

Evidence that the **European Union** takes its commitment seriously to involve people is given in various fields, *e. g.,* with the White Paper on governance in 2001 [M12]. Three years before, the Committee of the Regions passed a resolution on "Nuclear Safety and Local/Regional Democracy" and held it indispensable to enhance transparency and involvement, to foster financial support and to execute an economic assessment of potential host sites [G40]. The Sixth Framework Programme of the EU launched in 2002 holds in its Priority Thematic Area 2.2 "Management of Radioactive Waste" that "[t]he absence of a broadly agreed approach to waste management and disposal is one of the main impediments to the continued and future use of nuclear energy". And: "Research alone cannot ensure societal acceptance; however, it is needed in order to … promote basic scientific understanding relating to safety and safety assessment methods, and to develop decision processes that are perceived as fair and equitable by the stakeholders involved" [G38:187*passim*][G59]. Apart from radiation protection [G60], this aspect is the connection to the new Priority Thematic Area 1.7 "Citizens and Governance in a Knowledge-based Society". Here, "an integrated understanding" shall be created "of how a knowledge-based society can promote the societal objectives of the EU … of sustainable development" [G58:44]. As a whole the present study also fits into the Sixth Environment Action Programme "Environment 2010: Our Future, Our Choice" [G37]. Having said this, one has to recognise that there is an urgent need for the EU to seek interaction with civil society for merely 11 per cent of the population trust the Union, according to the Eurobarometer 56.2 of October/November 2001 [G133:4]. In addition, whether the recent motion of 2002 by the European Commission to require Member States to decide on site locations for HLW with an "authorisation for operation of the disposal facility to be granted no later than 2018" [G39:41], for LLW by 2013, is backfiring or not, remains to be seen. At any rate, the German Atomic Forum's reaction was that the proposals "will infringe national competencies" and that "there is the risk that safety standards might be downgraded out of consideration for the candidate states (those applying to join the EU)" [G209].

To attempt an interim statement, one may draw some type of "learning curve" in the way the main institutions involved tried to integrate third party views (see *Figure 11-2* overleaf).

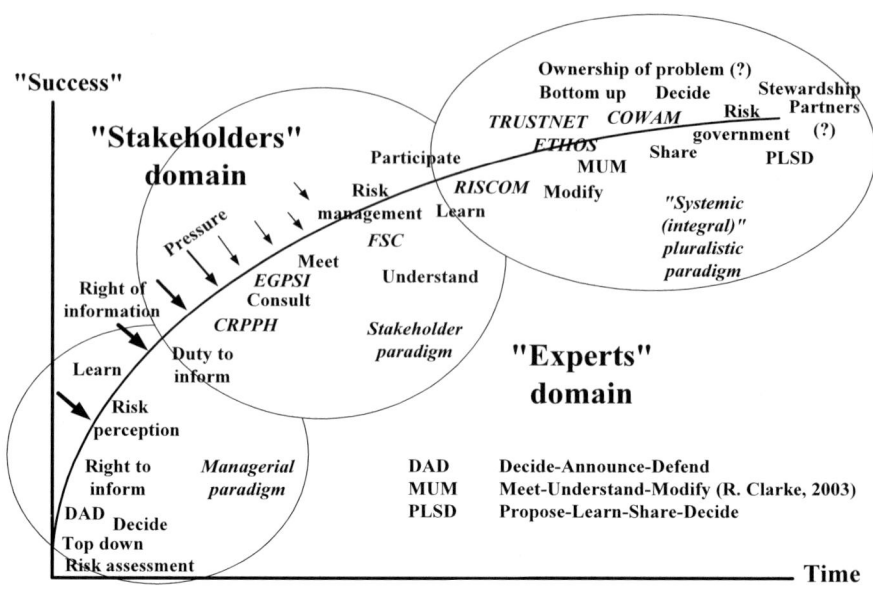

Figure 11-2. "Learning curve" in participation with respect to radioactive waste governance and research considering stakeholder involvement. Consult text in Sections 11.3.2 and 13.5, for risk management see Chapter 8. For abbreviations also refer to the text.

There is a definite tendency away from the top-down "Decide–Announce–Defend, DAD" approach, which is based on what could be called the managerial paradigm[131]. Both decision models as sketched in Sections 8.1 and 8.2 are based on the joint action by the expert community, administrative agencies and politics – presumably – *for* passive society, be it out of an arrogant or paternalistic background. A stirring experience to the radiation protection community – at least to many members – was the catastrophe of Chernobyl in 1986. In the European project ETHOS, from 1996, experts were confronted with the real fears and needs of people in contaminated regions of Belarus [G90]. Jacques Lochard of CRPPH, having participated in ETHOS, admitted at the end of the 2nd Villigen Workshop in 2001: "Everywhere in the world, the profession is confronted with the legacy of the past …. The time when it was possible to manage from inside the office is over. We need to go outside and interact with the population. We have to answer unexpected and difficult questions. We are facing values, concerns and emotions that were not part of our decision-aiding models. In fact, we are challenged at all levels of our expertise" and: "We have to listen, and

---

[131] This subsumes Habermas's 1971 "decisionistic" and "technocratic models" [D33] and resembles the "managerial model" as put forth by Beierle & Cayford 2002 [M5:2]

adopt a more modest and learning attitude as far as societal issues are concerned if we want to effectively take part as stakeholders in the decision-making processes in the future" [G158:263]. Keeping in mind the traditional expert-layperson dichotomy of acting experts *vs.* fearful populus, it is revealing how he continued: "This implies both a mourning process as far as our past position is concerned and some courage to overcome the fear of change" [*ibid.*]. This is where humble "MUM" emerges, "Meet–Understand–Modify", as Clarke put it at VALDOR 2003 [G35]. And this is where science has to understand the public, in reverse of Wynne's 1995 "Public Understanding of Science" [R97].

In view of the "co-production" of knowledge (by techno-science and society), this changing relationship also has a bearing on the once passive "stakeholders". They not only exert pressure on the "arrogant" expert establishment but – in participating – by and by take over some of the responsibility. Coming up the learning curve in *Figure 11-2* some conclusions on how to share decision making in long-term radioactive waste governance are drawn in Section 13.5.

Chapter 12

FUNDAMENTALS OF A COMPARISON OF DISPOSITION OPTIONS

1. VARIANTS AS BASES OF DECISIONS

The disposition of radioactive waste is decided on ethical grounds – and, in the end, has to be specifically designed according to technical criteria. Thereto, a study of variants is required. For a while the call for a comparison of options has been increasingly voiced [G140][G182]. NEA 1999 demanded: "Overall confidence must be developed in a much wider audience if a decision to implement is to be acceptable" [G181:8]. IAEA remarked a year before: "In view of the current questioning attitude of many people to the established view of experts … the possible alternatives: long term surface storage and disposal with the provision for retrieval, should be critically examined by independent international groups convened by the IAEA" [G113].

The central issue was put in a nutshell in the course of the examination of the Canadian disposal programme, when the government-appointed Canadian Review Panel concluded in 1998 after an 8-year extensive examination: "[a] choice of one is not a choice" [G25:56]. Decision theory, in principle, takes diverse options as a starting point being at the basis of a decision. Considering the analyses in Chapter 11 there indeed is a need for action to have a broad and open conceptual discussion. Back in 1996 IAEA wanted a "[c]omparison of different options for particular waste streams (e. g. shallow land burial or geological disposal for low level wastes)" [G109:17]. The NEA, in 1999, seconded in its "Review of developments in the last decade", and wanted to " … openly discuss the pros and cons of longer-term monitoring, reversibility and retrievability ... and of extended surface storage and of partitioning and transmutation", followed by corresponding "strategies and associated procedures" [G183:9*passim*]. The Agency has not acted accordingly, no resources have been released, no working group set up.

Below I attempt to lay out a framework to evaluate the management options for a specified concept of "sustainability"[132], and, by that, to attain Objective 3 of this study as set out in Chapter 2. Since, according to *Figure 12-1,* a variety of dimensions play a role, this framework has to be widely discussed. The aim is to obtain what Majone 1989 (and others) termed a "multiple evaluation" [M57:169], acknowledging the legitimacy of different criteria and perspectives as being more than the sum of partial appraisals.

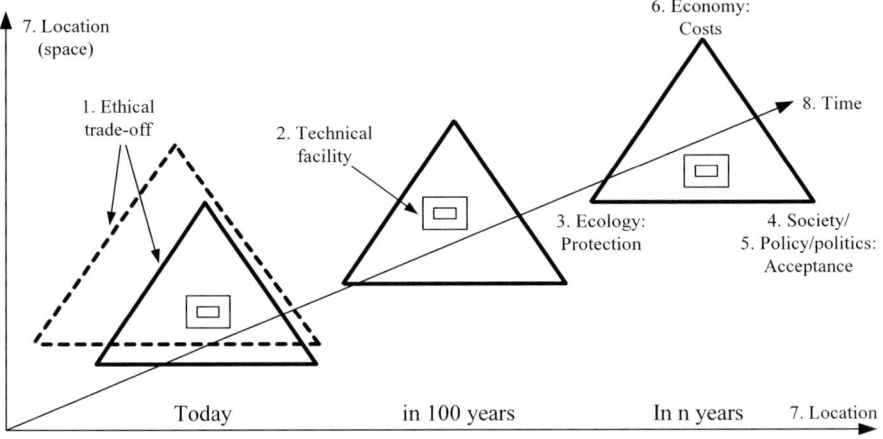

Figure 12-1. Sustainability of disposition systems. Eight dimensions – not just the three classical ones of the "magical" triangle Ecology–Society–Economy – have to be considered: An **ethical** trade-off takes place on the design of the facility (**technical** dimension), along the **ecological** dimension (protection of humans and the environment), the **social** and **political** dimensions (society and balance of power determine acceptance) as well as the **economical** dimension (costs of disposition including institutional control). This decision bears an eminent **spatial** (locality/location) and **temporal** dimension (period of isolation and concern). Source: Flüeler 2001a [G66:790].

The dimensions are briefly outlined below. Particularly in a complex case like radioactive waste governance it is advisable to proceed in a comprehensive, systematic and logical way. Regarding concept and practical approach, it is crucial not to lead the dimensional discourse in reverse order.

---

[132] A first outline was published in Flüeler 2001a [G66]. This contribution was based on Flüeler 1998a [G63], 1999 [G65] in the setting of a seminar organised by the US National Research Council 1999 [G21], and on a transdisciplinary seminar with a heterogeneous number of international stakeholders in 2000 as Flüeler 2000a [P89].

This was the case in Switzerland, in Canada, and apparently also in Sweden [MA43:4]. The trap questions might be: Shall one start off with geoscientific investigations or with a search for volunteering communities? What is more stable, rock or society? In Chapter 13 a proposal on the integration of the dimensions follows.

2. ECOLOGICAL, TEMPORAL AND SPATIAL DIMENSIONS

The dimensions of environment, time, and space/location have already been dealt with in Section 1.3. In addition to that we have to distinguish between the physically-induced – "objective" – temporal dimension and the project-related – institutional – temporal dimension. As mentioned: Had all four L/ILW waste types been deposited at the Wellenberg site as planned in 1994, after 600 years half of the radiotoxicity would still remain in the "repository for short-lived low- and intermediate-level radioactive waste" [P215:29]. The initial potential in high-level and long-lived intermediate-level waste is 500 times greater [P213:60]. Contrary to the DD strategy, of Dilute and Disperse, the CC strategy of concentration and isolation applied in disposal is inherently "undemocratic" [P170]: The hazard potential is raised locally, at a specific location, in order to not distribute it at the expense of entire humanity and the environment. This risk situation reveals right at the start that and how the dimensions are intertwined. The ecological long-term character of radioactive waste finds expression in temporal and geographic issues of distribution, which have to be debated as inter- and intragenerational equity issues in society and politics. Thus, the ecological and temporal dimensions set the tone for the remaining dimensions. The fact that the emergence of radioactive substances (the act of introducing them into environment) and the onset of their potential impact differ in space and – even more so – time, is of fundamental relevance for ethics. We are confronted, in Jonas's 1987 words, with the diachronicity of action and its effects, a novel situation to cope with [M46:84-85].

With regard to project management, a facility for radioactive waste is a long-term phenomenon as well (see *Figure 12-2* overleaf), as the IAEA correctly observed in 1998 [G112:9] (see above).

Another, so to speak, supradimensional, prerequisite in the dimension discourse, is the factual constraint that the existence of radioactive waste poses. All involved parties are confronted with this precondition. Unlike other controversial technical issues, "nuclear waste policy was", in Jacob's 1990 words, "not the engine that drove politics, but the product of political, economic, and social engines which drove the politics of nuclear waste" [MA23:22].

3. ETHICAL DIMENSION

The long-term characteristic entails the circumstance that the ones who benefit from nuclear energy (of which waste is one result) do – probably – not bear the risk associated with the waste; costs and benefits gape: a formidable ethical problem (*Figure 12-2*), *i. e.*, an issue of reflection upon values and norms. According to Beck 1988, "large-scale hazards of late industrialism" are characterised as follows [R6:76-77]:

- They cannot be delimited with regard to location, time and population concerned.
- Causality and liability cannot, in the long run, be attributed to anyone.
- The irreversibility of potential consequences cannot be compensated.

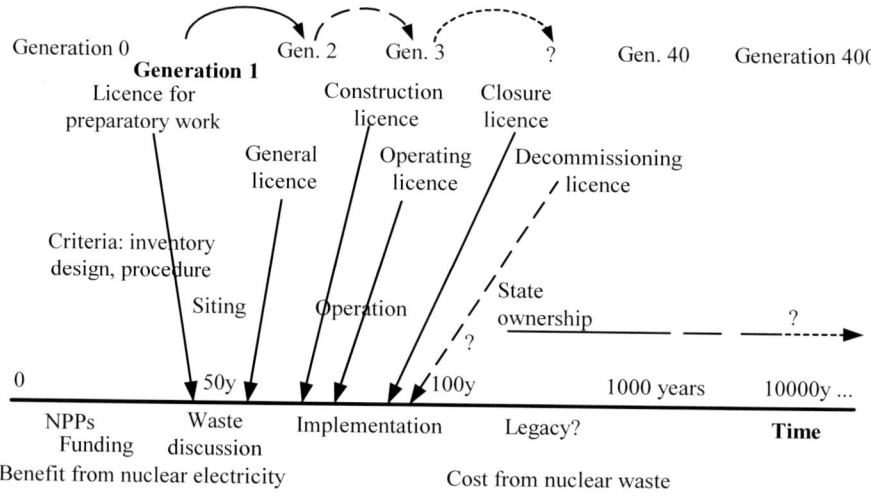

Figure 12-2. Radioactive waste management has a long-term safety *and* a long-term project character. It has to be backed up by the technical community, the political decision makers and the general public over decades. While still benefiting from nuclear electricity we, at present, are "**Generation 1**" having to start implementing radioactive waste management. Some duties – of monitoring, *etc.* – will have to be handed over to "Generation 2" being at the edge of merely bearing risks from waste. At any rate, duties of regulatory bodies stand out in long-lasting licensing.

Luhmann 1990 talks of "impositions of rationality" in the case when a transfer takes place from the (self-born, calculated) risk of the decision maker to the (imposed) danger for the people concerned and affected by the decision [R50:150]: "The risk-taking behaviour of one person turns into danger to another one, and the difference of danger and risk becomes a political problem" [*ibid.*:161].

The multidimensionality of the issue requires an appropriate reference system. Normatively, *i. e.*, with respect to values and ethics, the principle of sustainability seems to suggest itself. It facilitates a stepwise analysis according to various dimensions. Having been specified in the sense outlined – with the two pillars of passive safety and active control, see *Figure 9-8* – it builds on Jonas's 1977 imperative of responsibility and on precaution:

- Responsibility principle: "Act so that the effects of your action are compatible with the permanence of genuine human life ... are not destructive of the future possibility of such life"; "Do not compromise the conditions for an indefinite continuation of humanity on earth" [M45:11];
- Precautionary principle[133]: "[I]n case of uncertainty, act so as to avoid the worst possible outcome" [M15].

In the radioactive waste issue, as in all issues of technology, we have to do with consequences. It is not sufficient to maximise (today's?) overall public welfare (according to a utilitarian standpoint) or to have good intentions (as from a deontological standpoint). In teleological – or consequentialist – ethics it is essential what we know about technological processing and about impacts the use of technology entails. Responsibility ethics signifies that the ones, also institutions, who act (or forbear actions) have, firstly, to do this according to determined criteria, secondly, to justify their actions, and, thirdly, to bear responsibility for eventual consequences.

In contrast to my proposal, the opportunity has so far been missed by official actors to define the concept of sustainability more stringently. According to the Agenda 21 of 1992 the countries "should cooperate with international organizations to [p]romote ways of minimizing and limiting the creation of radioactive wastes ... [p]rovide for the safe storage ... and disposal of such wastes", and to "[n]ot export radioactive wastes to countries that prohibit the import of such waste" [G148:41, Section Two, Chapter 22].

Individual countries did not, until the present, specify the notion either. In the Netherlands, where the radioactive waste management is based on

---

[133] The European Union declared the precautionary principle the precept in tackling uncertainty [M11]. The EU environmental policy also requires preventive action, to rectify damage at the source and to apply the polluter-pays principle.

long-term storage, a debate has not taken place, though planned in 1989 [G49]. "Continuous monitoring" is not really based on an egalitarian concept of ethics, even if argued that way [G86] because we do not "equalise" future generations (with us being the deciding generation(s)) by forcing them to set up controls.

If we look at the **Swiss context** in particular, the productive concept of sustainability needs to be specified and implemented. The theoretically laudable idea of the Federal Department for Energy to base its ministerial strategy on sustainability and "to thereby disclose goal conflicts and substantiate the value judgements made" gets blurred if various "sustainabilities" are mentioned [P278:18]. The explanations are kept general: "Sustainability in energy matters means in detail: Ecological sustainability ... The safe disposition of nuclear waste ... economical sustainability ... The internalisation of external costs ..." [*ibid.*:3,18*passim*]. The debate initiated by the Federal Office of Environment on sustainability and sustainability research, respectively is no more helpful [P144][P252].

This situation is surprising considering that sustainability was codified by the Swiss electorate in the revised 1999 Federal Constitution as a state maxim of Switzerland: "The Confederation and the Cantons strive for a permanently sustainable relationship between, on the one hand, Nature and its potential for regeneration and, on the other hand, its utilisation by man" (Art. 73) [P40].

The nuclear community basically adopted one requirement of sustainability (protection) in the principles laid down in IAEA 1995 [G107] and in NEA 1995 [G175:13]:

"*Principle 1: Protection of human health*
Radioactive waste shall be managed in such a way as to secure an acceptable level of protection for human health."

"*Principle 2: Protection of the environment*
Radioactive waste shall be managed in such a way as to provide an acceptable level of protection of the environment."

"*Principle 3: Protection beyond national borders*
Radioactive waste shall be managed in such a way as to assure that possible effects on human health and the environment beyond national borders will be taken into account."

"*Principle 4: Protection of future generations*
Radioactive waste shall be managed in such a way that predicted impacts on the health of future generations will not be greater than relevant levels of impact that are acceptable today."

"Principle 5: Burdens on future generations
Radioactive waste shall be managed in such a way that will not impose undue burdens on future generations."
"Principle 6: National legal framework
Radioactive waste shall be managed within an appropriate national legal framework including clear allocation of responsibilities and provision for independent regulatory functions."
"Principle 7: Control of radioactive waste generation
Generation of radioactive waste shall be kept to the minimum practicable."

" ... the disposal concept requires that the presence of waste may safely be forgotten, after a period of institutional control to prevent early inadvertent intrusion. For the extreme case of retrieval from a sealed repository, engineering procedures might be difficult and costly, but not impossible, and somewhat analogous to the extraction of toxic mineral ores." NEA 1995

By that, a few other pertinent ethical principles were considered, such as the trustee principle, the principle of intragenerational equity, the principle of the permissible burden on future generations. Freedom of intervention by future generations is not fully excluded. But it was only the Swedish Consultative Committee KASAM 1988 that explicitly emphasised this second requirement of sustainability (controllability and, as an extreme variant, retrievability): "The choice of what to do must devolve upon the concerned [future] generation [*sic!*] in question and be based upon its own assessment of the advantages and disadvantages to be encountered .... These lines of reasoning lead to a double conclusion: A repository should be constructed so that it makes controls and corrective measures unnecessary, while at the same time not making controls and corrective measures impossible" [G139:15]. This approach has been adopted in the Waste Convention of 1997 [G110] by introducing sustainability explicitly (in Art. 1, Clause ii). In Art. 4, Clause vi, the principle of "equal opportunities of future generations" is followed by the phrase that "each Contracting Party shall take the appropriate steps ... to strive to avoid actions that impose reasonably predictable impacts on future generations greater than those permitted for the current generation" [*ibid.*]. Further documents issued by the agencies are no great help in developing reflection [G123].

If one considers basic principles of ethics, intragenerational equity might be met by thoroughly involving the concerned population around the waste site on the principle of informed consent [M81][G240]. Even though the risks are concentrated, people concerned can get a say in the decision-making process, which also leads to procedural equity (see Section 13.4).

Due to the fact that intergenerational equity – between generations – is inherently violated in the case of long-term radioactive waste management (see Sections 6.4.2 and 9.9), some proxies were sought, such as Kasperson's *et al.* 1983 "public defender of the future" [G147:366] or Posner's 1990 "Council of the Future" [G223][134]. Later on the idea was brought up by others but not conclusively specified. Following Dobson 1996 [M17], Fell and Fell (no year) favoured so-called "special representatives" to be selected "from the lobby which deals with sustainable development" [G61:156]. The issue of representativeness is not resolved [M1][135]. Since a just distribution of benefits and burdens according to Rawls's 1971 theory of justice [M73] is not possible, one might resort to the principle of retributive justice, which obliges benefiting generations to compensate potential risk bearers.

Being confronted with this ambiguous normative situation a possible track to follow is to simultaneously consult a *variety* of ethical and "pseudo-ethical" [M15] principles on condition that they are brought into line, *i. e.*, that they are at least compatible or complementary and, thus, add to robustness. This "ethical" approach is an attempt to reach what might lie in "the interests of the general public" [R31:179] whether present or yet to be born, without having to revert to short-lived surveys.

To mitigate the inherent "gap" between benefiting and risk-laden generations (see *Figure 12-2*) we might explore ways to institutionalise continuity along the principle of "equal opportunities of future generations" (in accordance with Rawls [M73:284]), the "chain of obligation" principle, and the principle of the "rolling present" [M67][G143:24*passim*][G203]. They have in common that continuous institutional provisions string the present and the succeeding few generations together, a claim which might be practical if sufficient political will and concomitant action exist. This idea is to be followed up in Section 13.5.

The issue of responsibility is a very complex one as has been indicated several times above. In the first place and according to IAEA 1995, "Member States shall ensure continuity of responsibilities" [G108:7], and based on the causality principle the implementers have to carry out a project. In the second place, today's society has to assume overall responsibility in benefiting from nuclear energy. In addition and surprising to some, if thorough "stakeholder involvement" is achieved, these other stakeholders start to own (a share of) the problem and, thus, a share of responsibility.

---

[134] More recently a variety of instruments are discussed in [M51].
[135] The intragenerational representativeness is not clear either. "Stakeholder involvement" *per se* does not necessarily result in involvement of "the public". Care for future generations is not secured either just by reverting to exhaustive present "stakeholder involvement" since surveys show that today's individuals are predominantly present-oriented [G46].

4. SOCIETAL AND POLITICAL DIMENSIONS

The concept of sustainability has a second advantage: It forces upon stakeholders, including decision makers, an examination of the dimensions and, consequently, it incorporates all parties' perspectives, needs, targets/goals, and knowledge systems.

Whether the stakeholders like it or not, ultimately, there *has* to be a partnership between the implementers, the public, politicians, political officials, and regulators, because all are interested in acquiring a sufficient safety level of the disposal system, even if for different reasons (see also Section 6.4.1):

- *Implementers* have the mandate to execute the project(s) based on given safety standards, *i. e.*, demonstrate safety, to fulfil the waste producers' responsibility to safely dispose of the waste, they want no delays, but want clear-cut requirements, at "reasonable" costs[136];
- *(Today's) public* wants safety, as well as control, transparency, and full participation;
- *Future generations* presumably put safety first, want no obligation to take safety measures, and no restriction in land use; but do want control in the case of grave system failure;
- *Politicians* want all requirements, depending on the political constellation (some favour delays according to a NIMTOO attitude, *i. e.*, "Not in my Term of Office");
- *Political officials (e. g.*, Department of Energy) have to accept liability in the long term and also have to integrate public requirements;
- *Regulators (safety authority)* have to supervise the demonstration of long-term safety (for "tomorrow's" risk bearers, in a trustee's role) and to define today's (and tomorrow's?) public requirements from a safety standpoint (that is, control may not entail a simple delay of closure but must also include instruments to additionally demonstrate safety).

Another type of sub-category of the social and political dimension is the judicial one. The legal requisite in Switzerland still is the "permanent and safe final disposition and disposal" [P39]. In addition, the country ratified the Waste Convention in 2000 and, thus, is obliged to find a "solution", the issue being an national task due to the formulation in preamble (xi) that the Contracting Parties are "[c]onvinced that radioactive waste should, as far as is compatible with the safety of the management of such material, be disposed of in the State in which it was generated" [G110]. As mentioned, this "national" character was repeatedly underlined by the Federal Government.

---

[136] The implementer's perspective may differ from the waste producers' (see 6.4.1 and *Figure 9-3* and *Figure 9-4*).

The causality principle must not conflict with this, from where the Federal Council concluded that "disposal is therefore not a task of the State" [P78:2747].

Since all objectives of all stakeholders can never be attained, prioritisation (according to goal relations mentioned in Section 9.6) and negotiation have to take place so as to adopt the stakeholders' respective responsibilities [MA30]. It would be daring – and utterly naïve – to maintain that their belief systems could be changed – certainly not in their core principles, but perhaps modifications could be made in their secondary aspects [D73:367-668,675] – in so far as the actors would identify some common interest or, in Carter's 1987 words, some "common ground" [MA6:427]. The yardstick proposed is, as mentioned, an enlarged notion of sustainability (see below). The element "society" in the sustainability triangle (*Figure 12-1*) addresses the participation of, and acceptance by, the public. The stakeholders involved in the Swiss radioactive waste system are plotted as a so-called policy field [D89:21-22] in *Figure 12-3*.

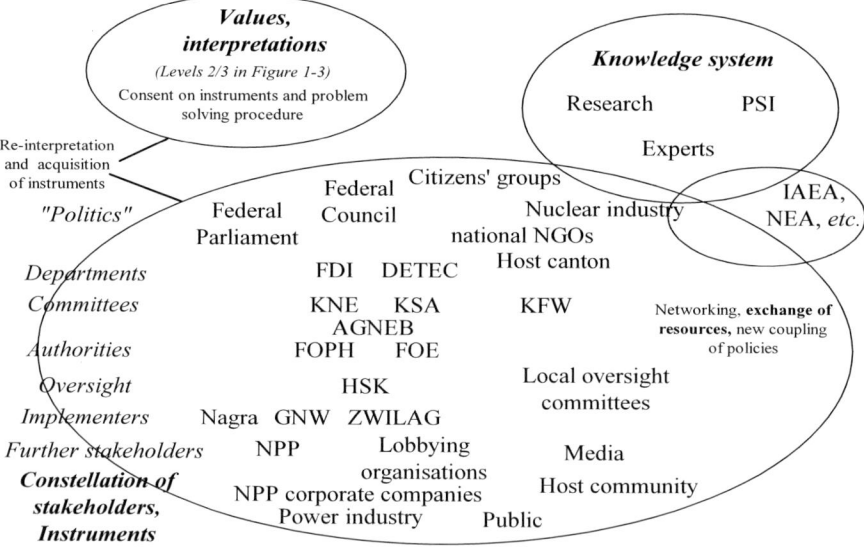

Figure 12-3. Major stakeholders in the Swiss radioactive waste system (simplified). The system also is a "policy field" [D89:21-22] with values/interpretations, a set of instruments, a constellation of stakeholders with types of co-operation and a knowledge system. Depending on the stakeholders, the perspective can be different (see *Figure 9-3* to *Figure 9-6*). According to the Swiss political system, the "Public" might be subsumed under "Politics" because goals and instruments are prone to change by way of initiatives or referenda. In contrast to the political science model, the knowledge system is not autonomous but part of the constellation of stakeholders (implementers, safety and other authorities contribute to knowledge generation, "experts", on the reverse, are also stakeholders).

The following features fall under a policy field: specific patterns of interpretation and values, a set of instruments, a constellation of stakeholders (actors), types of co-operation, and a knowledge system:

ad patterns of interpretation and values:

The concept of sustainability is to be checked for its various and diverse modes of interpretation, and the respective key notions such as controllability, retrievability, *etc.* It is not permissible to play barriers off against each other, as the saying goes, in Switzerland [P218] and elsewhere [MA43:14][G139][G240]: "What is more stable, society or rocks?" In my view, this is an utterly unproductive approach. According to STS [M43], technology culture as well as technical and scientific findings are part of society; they are – today and in the future – interrelated with society. Analytically, it would deny the fact of the institutional long-term dimension of the undertaking; even a "technocratic" approach assumes functioning institutions or, as Inglestam terms it: "institutional constancy" (after [G143:26-27][M39]). Conceptually, it would be a violation of the multi-barrier- or defence-in-depth approach upheld by the nuclear community [G117:85-89][G114:244]. Responsibilities, at that, must not be confused by juxtaposing "national" to "state" tasks.

ad set of instruments:

Together with the nuclear and radiation protection legislation, the basis is provided by the Guideline R-21 issued by the safety authorities [P133]:

"The overall objective of disposal is to eliminate radioactive waste in such a way that:
- human health and the environment are protected in the long term against the ionising radiation from the waste,
- no undue burdens are imposed on future generations." (Clause 4)

"Any measures which would facilitate surveillance and repair of a repository or retrieval of the waste shall not impair the functioning of the passive safety barriers." (Clause 5, Principle 5)

"The provisions for radioactive waste disposal are the responsibility of the present society which benefits from the waste-producing activities and shall not be passed on to future generations." (Clause 5, Principle 6)

"After a repository has been sealed, no further measures shall be necessary to ensure safety. The repository must be designed in such a way that it can be sealed within a few years." (Clause 5, Protection objective 3)

In addition to the requirements for protection, clauses pertaining to control and validation are included:

"Post-closure interventions into a repository system should not be hindered, provided this is compatible with the safety requirements. The retrieval of waste disposed of in accordance with current repository concepts is feasible in principle, even if the associated costs are high." (Clause 5, Principle 5)

"The applicant has to submit a safety analysis at each stage of the licensing procedure (general, construction, operating and closure license). Safety relevant informatio[n] on the repository system obtained from preliminary investigations should be supplemented by ongoing investigations during the construction and operation of the repository. The safety analysis for the post-closure phase should be refined in accordance with the improved knowledge of the repository system." (Clause 7.1)

"It ... has to be shown that the models used are applicable for the specific repository system (validation), taken both individually and as an overall model chain." (Clause 7.6)

These and other requisites have to be considered in the entire set of instruments by the stakeholders (*Figure 12-3*):

- IAEA, NEA and other international bodies: *e. g.,* Waste Convention;
- Federal Council, Parliament: concessions and licences; Atomic Energy Act/Nuclear Energy Act, Radiological Protection Act, ordinances, *etc.;*
- Siting canton(s): *e. g.,* mining law in Nidwalden;
- Oversight bodies: licences, clearances, special requirements, *etc.;*
- Committees: experts' reports, statements, recommendations, position papers, *etc.;*
- Environmental NGOs, citizens' groups; public: submissions, objections, complaints; popular rights (initiatives, referenda, petitions), *etc.*

Discursive methods of participation are discussed below.

ad constellation of stakeholders:

As may be clear from above, "the" actor constellation is not a constant but may change depending on the boundary conditions. The networks are dealt with below.

ad types of co-operation:

The mediation exercises, from the "Conflict-solving Group Radioactive Waste (KORA)" in 1991 to the "Energy Dialogue on Disposal" in 1998, carried out in Switzerland did not meet the standards of social science [M75]. Accordingly, Enderlin Cavigelli 1993 listed requisites to comply with in her evaluation of the Working Group on Reprocessing of "KORA", such as: "Every party disposes of sufficient resources for exchange ... access to information". She concluded from the actual process encountered: "There are justified doubts whether this condition... was fulfilled.... As for personnel and financial resources, the implementers clearly dispose of more resources for exchange" [P62:20]. The preconditions for mediation techniques are, as recommended by IDHEAP, listed in Section 9.7. Ideas for development are laid down in Section 13.3.

Back in the 1980s Parker *et al.* concluded that, in the end, one can only "… try to develop as broad a consensus as possible in support of the solution that is finally reached. It has to be recognized that there will always be an irreducible amount of uncertainty in the outcome of any solution" [G217:101]. By focussing on "common ground", rather than "consensus", it has to be emphasised that it is not intended to call for as many *voices* but for as many *perspectives* as possible so as to incorporate all relevant facets in the dimensional discourse: ethical, technical, ecological, economical, political, societal, spatial and temporal. Consensus, at that, would probably amount to majoritarian deliberation anyhow [G91:10]. This is not to avoid the issue of representativeness or, by no means, to devitalise claims for wider participation, but to focus on a dimensional discourse as inclusively as possible. In view of this multidimensionality, it is also an avenue to find society's way to some sort of sustainable "closure" of the issue [M4:12].

On this background, it is useful to specify what might be understood by "common ground". Trying to decompose ever-used buzzwords like "consensus" or "compromise" (see also Index) one may outline where and how "common ground" is likely to be achieved (*Table 12-1*). Keeping in mind the mental models of *Figure 1-3* it cannot be assumed to reach consensus "at heart", in the stakeholders' core beliefs (according to Sabatier 1987 [D73]). They must agree, though, on three levels:

1. Problem recognition (waste exists, problem to be tackled, "solved");
2. Main goal consensus (degree of protection and intervention);
3. Procedural strategy ("rules of the game").

ad knowledge system:

Experience shows that complex human systems such as society are not stable (as the fall of the Berlin wall strikingly indicates). But accordingly, the geological world view may also change as the increase in knowledge demonstrates when the Permo-Carbonic Trough was discovered (see Section 6.3.3).

Contrary to assumptions by political scientists [D89:21-22][D6:272], the knowledge system is not located outside the constellation of stakeholders but the technical research community is part of it. The Paul Scherrer Institute receives major contracts from Nagra [P220:23]; in a small country like Switzerland, researchers partly are decision makers. In addition to that, the radioactive waste issue is typical of a transdisciplinary constellation[137]. Scholz 2000 qualifies the transdisciplinary approach to expand knowledge

---

[137] See Chapter 3 for general, Section 13.4 for specific aspects, esp. *Figure 13-7*.

Table 12-1. Relations (and hierarchy) of consensus and dissent at diverse levels. The middle region is amenable to good chances for some "common ground", being above the non-negotiable core beliefs and below practical project management where a compromise is best to achieve. See text on previous page.

Level	State of agreement	Perspective/goal/fields (examples)
Secondary beliefs		
Implementation (dependent on policies, funding, authority)	Compromise	"Real" project/site
Procedure/methodology	Consensus	Siting, monitoring
Roles, decisions (instrumental and institutional goals)		Performance assessment, quality assurance, inclusive reviewing
Protection goals (passive protection, active control, involvement, power of decision) (= "success criteria")	Consensus	Safety and control goals
Factual constraints	Consensus	Waste existent
Concept of sustainability	Compromise ("weak" sustainability [M82][138])	Practical trade-off of dimensions (technical and social goals)
Core beliefs		
Attitudes of stakeholders	Dissent	Pro- *vs.* anti-nuclear
Models of rationality	Dissent	Technocentric/anthropocentric *vs.* biocentric or even ecocentric worldview

generation: "Transdisciplinarity aspires to make the change *from research for society to research with society* ... mutual learning sessions ... should be regarded as a tool to establish an efficient transfer of knowledge both from science to society and from problem owners (*i. e.* from science, industry, politics *etc.*) to science" [M80:13]. Nowotny *et al.* 2001 even talk of "trading zones of knowledge" [M65:143-147, after [M30]]. It is evident that a technical community has difficulty accepting that. The situation is particularly troublesome in the nuclear field, where the (Swiss) Confederation yielded complete responsibility to the waste producers, somewhat abusing the polluter-pays principle. Wälti 1993 described the Swiss situation as follows: "The mediation process is stopped by delegating implementation to a para-federal institution, at least in the planning phase. Nagra is largely autonomous in how to set up its planning. The Confederation's role is the control function. Enforcement by para-federal institutions generally is efficient and goal-oriented.... On the other hand, the

---

[138] "Weak" sustainability allows for substantial substitutability of resources, see page 217.

political monopoly by para-federal institutions weakens their ability to learn and adapt in a society, which puts high demands on participation... it is only by increased external pressure that [Nagra is] obliged [to integrate conflicting interests into its planning]" [P284:216-217]. Whether enforcement, *i. e.,* implementation, indeed is "efficient and goal-oriented" may be questioned looking at the history of radioactive waste governance in Switzerland.

Risk knowledge, monopoly on understanding, and resources were all with the waste producers until the 1980s, which impaired the position of other stakeholders, including the authorities, and restricted reviewing to the nuclear waste community. In the 1970s the way of thinking in the nuclear waste field was determined by the expert community of civil engineering. Only gradually the problem horizon widened to long-term aspects and issues like retrievability and controllability (in the current sense) were tackled in-depth. This was induced by a more critical public perception, pressure from independent experts, and legal pre-conditions (consult *Figure 11-1* and *Table 11-1*).[139] Progressively it is possible to broaden the basis for the generation of knowledge (see *Figure 13-7* on page 262).

Consistent with this, Minsch *et al.* 1998 [M61] propose four groups of strategies in their "Institutional reforms for a policy of sustainability":

- Strategies of reflexivity with "systems of reporting", expert bodies, improved structuring of information in decision-making processes as well as research, science, and education aimed at sustainability;
- Participatory strategies of self-organisation with rights to participate and discursive participatory models;
- Strategies to compensate and resolve conflicts with, *e. g.,* "advocatory" institutions such as a "Council of Sustainability" established by the Government, "services of sustainability" in organisations, free access to information, compensation for NGOs and controls of monopoly, "better integration of NGOs... into negotiation processes";
- Innovation strategies with, *i. a.,* a legal liability system, and a "co-operative development of the regional level".

5. ECONOMIC DIMENSION

The pre-eminent task is to balance the needs of ecology, society, and economy over the years (of potential impact). The element "economy" is not dealt with in a sustainable manner unless costs are internalised; *i. e.,* the

---

[139] This phenomenon is not unique to the nuclear waste field but is also observed with chemical risks [M9:281].

waste producers (and today's benefiting generations) have to accept responsibility for financing the "ongoing" safety management. The aim is not to impose a financial burden on future societies, in case of facility/repository failure.

Economics gains importance with the progressive penetration of the so-called liberalisation on the energy market, to which the quasi-monopolistic power industry, especially the capital-intensive nuclear energy production, has been exposed [G180]. This has an eminent bearing on radioactive waste management and governance. Pressure on the waste proponents and implementers is not only exerted by large parts of the public but also by some waste producers who own implementing institutions. The present constellation is characterised by globalisation, i. e., the opening up of the electricity market with concurrent increasing competition, and an enforced shareholder value perspective (lowest costs mean no further investment in the "back end") with parallel cut backs in Research and Development funds[140]. This becomes manifest in the search for "lean" solutions (e. g., on Pacific atolls and Russian dumps). In calling for cost reductions from both partisan sides, as is the case in Switzerland, one potential consequence is the building up of an "unholy" alliance as indicated: between cost-saving pronuclear shareholders[141] and antinuclear groups in favour of guardianship instead of final disposal repositories – this might lead to indefinite intermediate storage.

Previously, NPP operators were allowed to use the funds set aside for back end activities for other investments. A case of bankruptcy or liquidation might result in a lack of resources for disposal, a situation already criticised by the Federal Government in 1985 [P67]. A study commissioned by the Federal Government recommended to secure the funds under government control, which had already been suggested by a team of IAEA in 1999 [P141]. A similar fund already exists for decommissioning. But in the political negotiation, the joint and solidarity liability as a duty for additional cover was dropped in the Radioactive Waste Disposal Fund Ordinance, in force since 2000; no guarantee for regulatory safety research has been given by statute but, at least, an external administrative supervisory body was installed [P269].

The deployment of appropriate resources (goal-means relation) is vital for goal achievement. Long-time waste facilities require a sufficient support by proponents and operators for site characterisation, model verification, and safety performance validation. Monitored long-term facilities are bound to

---

[140] Such a tendency could be an "early sign of declining performance" in the evaluation of organisational safety culture [G185:15]. Cost reduction with resulting deteriorating safety would violate Article 11 of the IAEA Nuclear Safety Convention [G103].

[141] The argument of cost-saving has also been raised by economists, e. g., Nathwani 1993 viewed approaching "social risks" due to a supposedly "asymmetric and disproportionate allocation of scarce resources available to society today" [G171:1].

adequate resource allotment during the period of control[142]. Seiler 1986 asked to keep in mind that passing duties of surveillance to future generations was permissible from a legal standpoint only if the surveillance was ensured during the period required [P253:134]. It was inconsistent to base the Radioactive Waste Disposal Fund Ordinance on the conception of final disposal but, at the same time, to "keep the concept open" for HLW as DETEC decided in 1999. The Federal Council "considers the result of the technical-scientific discussion to date as politically not enough sustainable" [NZZ, 1999-6-8]. In the meantime, the EKRA concept of "monitored long-term geological disposal" has been chosen as the basis of the new Nuclear Energy Act, in force from 2005. In their plea for controlled and continuous storage, Damveld & van den Berg 1998 pertinently comment: "An important boundary condition, however, is that sufficient money is set aside to pay for the future storage costs" [P54:2]. This is in agreement with Rawls 1971, demanding that each generation must also put aside in each period of time a "suitable amount of real capital accumulation" [M73:284].

Let us consider the following analogy: The current utilisation of non-renewable resources is permissible on terms of sustainability if, according to Erdmann 1995, "the associated losses to future generations are compensated – primarily in terms of increased know-how to substitute resources" [M22:42]. Based on this, it would be justified that present generations would be "allowed" to impose an increased risk and increased control, respectively, on future generations on condition that sufficient resources were ensured, *i. e.*, also a guaranteed transfer of know-how and technology. It is true for radioactive waste management what Erdmann holds for the energy issue: "The critical bottleneck is that the current civilisation perhaps is incapable of sustaining and/or unwilling to sustain the learning and innovation process at the necessary extent and speed. Sustainability is only achievable if and as long as the problem solving capacity proceeds faster than the continuously emerging and newly created problems which are to be solved by the human mind" [*ibid.*]. The above-mentioned, negative socio-economic stimuli may paradoxically serve as a prerequisite for social learning [D71:359].

Specifically with regard to the cost of disposal, Zuidema & Issler 2001 identify the following main factors [P288:3.2-6]:

---

[142] Swahn 1992 calculated the simple case of monitoring against proliferation in Sweden. For a 24-hour surveillance by one person, a staff of four people per shift would amount to 400,000 USD per year [G255:158]. A surveillance period of 250,000 years entails costs of 100 billion USD. The Swedish Nuclear Waste Fund is to be provisioned by the waste producers today, partly via real interest rate of the capital of 2.5 per cent. But a net interest assumes an effective corresponding economical growth in the same order of magnitude, *i. e.*, over 250,000 years – this has to be reckoned as impossible.

- Legal and regulatory boundary conditions, regulation of responsibilities, strategic premises, such as a parallel procedure of various options, licensing procedures, decision paths and courts including possibilities to appeal, financial state contributions top R&D);
- Technical framing conditions (size of the nuclear programme, relevance of waste producers other than NPPs, geological conditions, spatial planning, surprises, and difficulties);
- Procedural strategy (site selection procedure with number of milestones, *etc.*, host rocks explored, choice of disposal concept);
- Societal and political influence ("societal will to solve disposal, ... acceptance, willingness to make a decision in politics and with authorities, ... exertion of influence by the anti-nuclear movement ...").

In an analysis of the Finnish case (see Section 11.3.1), the authors locate "the following points which facilitated the great and lower-cost progress" if perhaps different conclusions might be drawn: "The way to proceed where it was avoided to sort out technical issues (*e. g.,* siting investigations) with a high effort, before the societal and political decisions in principle... had been taken, and the discipline of all parties involved to stick to the time schedules once agreed on" [*ibid.*:3.2-7]. A systematic procedure is proposed in Section 13.4.

6. TECHNICAL DIMENSION: IMPLEMENTATION

Various technical options have been suggested to date to cope with the sustainability requirements of protection and intervention, although each has had a different emphasis, especially with respect to overall goals. This is a striking illustration of "interpretative flexibility" [M13], the way different actors interpret facts, artefacts and concepts. For an overview see *Figure 12-4* overleaf:

Fundamentals of a comparison of disposition options 219

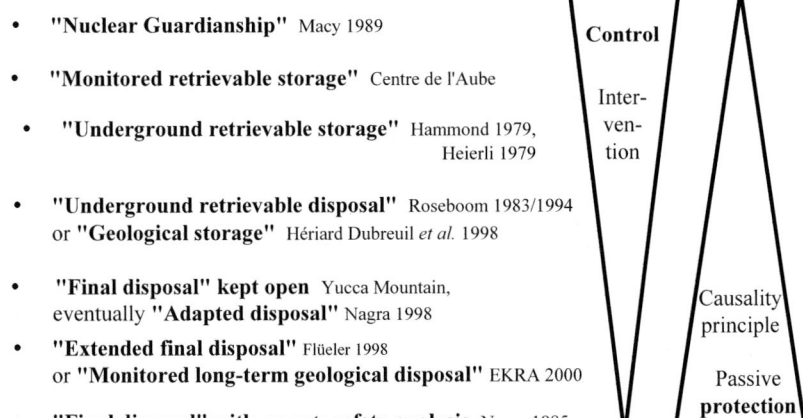

- **"Nuclear Guardianship"** Macy 1989
- **"Monitored retrievable storage"** Centre de l'Aube
- **"Underground retrievable storage"** Hammond 1979, Heierli 1979
- **"Underground retrievable disposal"** Roseboom 1983/1994 or **"Geological storage"** Hériard Dubreuil *et al.* 1998
- **"Final disposal"** kept open Yucca Mountain, eventually **"Adapted disposal"** Nagra 1998
- **"Extended final disposal"** Flüeler 1998 or **"Monitored long-term geological disposal"** EKRA 2000
- **"Final disposal"** with *ex ante* safety analysis Nagra 1985

Figure 12-4. Various recommended options to comply with the main objectives of sustainability: Protection and control/intervention.
Sources: Macy [G159], ANDRA's Centre de l'Aube [G7], Hammond 1979 [G87], Heierli 1979 [P118], Roseboom 1994 [G232], Hériard Dubreuil *et al.* 1998 [G90], US DOE's Yucca Mountain [G34][G14], Nagra's "adapted disposal" [P219], "extended final disposal" [P88], EKRA's "monitored long-term geological disposal" [P60], Nagra's traditional concept [P202]. Source: Flüeler 2001a [G66:791].

To a certain extent, the concept options may be combined. The internationally accepted standards do indeed allow space for interpretation. It is, *e. g.,* crucial how the following NEA/IAEA/CEC 1991 [G201] requirements for long-term disposal are met:

"… it is recognised that the future behaviour of the disposal system must be understood well enough to assure that no harmful releases of radioactive substances to the environment are likely to occur…. Safety assessments must proceed iteratively with disposal system siting and development, to determine if further information is needed and, if so, what type of information is needed…. Confidence is also built through the process of assuring (or validating) that the predictive models used in safety assessments adequately represent the behaviour of the real system…. To obtain such assurance indirectly requires a systematic evaluation of modelling results against data from experiments in laboratories and in the field…." [*ibid.*:10,13]

The International Waste Convention of 1997 points into the same direction in Art. 16iii: "For a disposal facility the results … obtained [by operation, maintenance, monitoring, inspection and testing] shall be used to verify and to review the validity of assumptions made and to update the assessments [of safety] for the period after closure" [G110]. The Convention additionally statutes in preamble xi) that the Contracting Parties are "[c]onvinced that radioactive waste should, as far as is compatible with the safety of the management of such material, be disposed of in the State in which it was generated" [*ibid.*]. To do that they have to ensure the necessary qualified staff, including funding, in order to guarantee the continuity of the institutional control and surveillance measures, "... to be continued for the period deemed necessary following the closure of a disposal facility" [*ibid.*].

The primary goal in radioactive waste governance is the stability of the disposition system as it performs: The protection against harmful emission of radioactivity is to be continuously ensured. The secondary, or complementary, goal is flexibility, defined as the potential to intervene (for what reason has to be determined, reasonably in advance). Thereto, a representative monitoring and surveillance programme has to be specified, comprising full publication of the work, an intensive reviewing, and corresponding quality assurance.

Based on the arguments explored so far, *Table 12-2* gives an overview of some of the consequences that can occur if contradictory sustainability objectives (with the corner points "Control" and "Protection") are prioritised in deciding the preferred option. It portrays a continuum in which the decision makers have to speak up and lay open their arguments for one option or another.

Table 12-2. Disposition conceptions compared. Implications of the dominance of the two juxtaposed sustainability objectives (reversibility/retrievability is looked at as an extreme variant of "control"). Depending on the specification of the disposal concepts in question, convergences are possible (Cont. to page 222). Source: Flüeler 2001a [G66:792].

Prevailing objectives Criterion	**Control** [degree of importance]	**Protection** [degree of importance]
Protection	Low (in view of the possibility to repair, safety standards might be reduced in the construction phase)	Higher (primary objective)
Short-term safety	High (depending on expenses)	High (depending on expenses)
Long-term safety	Lower (secondary objective)	Higher (primary objective)
(Possible) system flow	Fast movements of societal/technical properties, not stable	Slow degradation of geological systems, natural phenomena partly predictable
System change	Possibly abrupt	Gradual
Safeguards	Bad (relatively easy recuperation)	Good

Fundamentals of a comparison of disposition options

Prevailing objectives Criterion	**Control** [degree of importance]	**Protection** [degree of importance]
Potential to intervene	High (primary objective)	Low (secondary objective)
Flexibility	High (intervention: modification of technical barriers) Danger of proliferation (primary objective) Danger of postponement	Low (secondary objective)
Constraints	High (control)	Medium (depending on the quality of the safety assessments)
Uncertainty 1: vagueness	Medium (technical barriers)	High (deviation, sets of data)
Uncertainty 2: insecurity (see *Figure 9-9*)	High (period of effectivity of control system; unwanted intervention)	Medium (depending on the robustness of the scenarios)
Experience	Short, bad	Almost none
Belief in technology **(progress dependability)**	High (future "solutions," investment into nuclear research provided)	Medium (existing "solutions")
Resources (valuable substances)	At hand, only feasible with continued and enforced utilisation of nuclear technology (partitioning/ transmutation, fast breeders, advanced reactors, *etc.*)	Difficult to recover
Causality principle	Violated	Adhered to in principle
Decision makers	Legitimated parts of present society representing interests of present and future risk bearers	
Collisions in interest (equity aspects)	Present generations: partly benefiting and partly risk bearing (intragenerational equity) Future generations: risk bearing (intergenerational equity)	

Prevailing objectives Criterion	**Control** [degree of importance]	**Protection** [degree of importance]
Externalities/expenses (human, technical and financial resources)	Very high, burden to future generations (higher degree of discounting), higher belief in technological and medical progress	High, mainly burden on benefiting generations (lower degree of discounting), adherence to present technology (i. e., at disposition up to definite sealing of system)
Strategic background: variety of possible "hidden agendas" (no hierarchy)	1. Long-term commitment of society to radiological burdens ("guardianship")	1. "Dispose and walk away" mentality
	2. Willingness to recovery actions in case of major system failure (one interpretation of the notion of "freedom of action")	2. Dominance of passive safety, no burden on future generations (another interpretation of the notion of "freedom of action")
	3. Resource storage for a second nuclear era (retrievability)	3. Willingness to execute the polluter pays principle (to force demonstration of long-term safety by the producers)
	4. Political opportunism (highest chance of receiving wide acceptance)	4. Fear of "walk away" mentality of waste producers due to cost-cutting and "lean management"
	5. Argument for "least-cost solution" for waste producers and present generations	5. Argument that controllability may compromise safety standards and concept quality (eventually leading to "modified open-end final disposal")
	6. Argument to phase out nuclear power (no "permanent solution")	6. Argument to continue nuclear path (waste "solution" presented)
	7. Fear of "technical fix" of final disposal (reversibility)	7. Fear of "political fix" of never-ending interim storage
Extreme option variant	**Surface storage shelter**	**final geological repository**

7. DISCUSSION: ON THE SEARCH FOR A WELL-SUPPORTED DECISION

Figure 12-4 spreads out the variety of options in the technical field, *Table 12-2* spells out the criteria to be checked in the assessment of options. The empirical content analysis and the literature study in Section 11.3. reveal that controllability, retrievability and procedural aspects like transparency, traceability of arguments, and stakeholder involvement are key elements in appraising options. In the end, the matter is to find a sustained "solution" to the problem. A corresponding decision is well supported if it integrates relevant parts of both the problem and solution ranges of the main stakeholders (see *Table 9-2*). For this very reason a broad discussion is called for on fundamental principles. As may be gathered from Section 9.6.1 goal discourses have been initiated in Switzerland very early in the process, time and again, from different directions and out of different interests. A discussion on the management concept was repeatedly promised but – until recently, with EKRA in 2000 – never carried out. On the contrary, demands or guidelines were blocked and attenuated.

It is decisive not to aim at "quick" technocratic or political "solutions" (technical or political "fixes"). An increased and stepwise "control" by third parties is essential – by strengthening the regulatory bodies, intensifying the review process, and involving stakeholders hitherto excluded or not treated equally.

As a counter, proactively and in self- and external reflection [R97: 381-382], all partners have to discuss eventual inconsistencies and contradictions (and there are some as *Table 12-2* shows). They also have to adequately consider the time dimensions (construction, impacts of the facility). This has to be done in respect of other, conflicting opinions. The aim is to reach a sustained minimum consensus as indicated above. With increasing knowledge, and after joint interim decisions, the options might be narrowed. By that, the various and diverse options should be "conceptionally comparable" as was claimed, by proponents and opponents, in the final report of the "Energy Dialogue Disposal" 1998 and by the environmental NGOs [P27:10].

The vehement and persistent controversy over controllability and retrievability demonstrates that the complementary goal flexibility (potential of intervention) of sustainability must be integrated into the disposal concept. But the ultimate overall goal is the permanent protection from ionising radiation. On that, the major stakeholders, in Switzerland and internationally and also over time, do agree. Accordingly, the NGOs wrote in their contribution "Perspectives of action (position paper of the environmental organisations)" to the "Energy Dialogue Disposal": "The measure of general safety (demonstrated permanent protection of humans and their environment) must be increased compared to the current final disposal project", in characterising the time dimension they added "[a]ccording to the period of potential threat" [P27:33]. Damveld & van den Berg 1999 correctly observed: "According to the principles of sustainable development, the needs of the future generations may not be endangered by the present one. Therefore, the storage of nuclear waste has to be designed in a way that no harm will occur in future"

[P55:5]. By this yardstick, primarily, the options according to *Table 12-2* have to be judged:

"Quick fixes", presumed adoption of misgivings

1993 Nagra: "Regarding technical matters, erecting a disposal facility does not pose particular requirements. In the case of a repository for short-lived wastes it comes to, for example, a subterranean facility which would be around six times smaller than one tube of the Seelisberg tunnel", a road construction in Nidwalden [P209:30].

1993 Greenpeace: "the stores ... will have the effect of bunkers[143] from war times to future generations, as a memorial of the misdirected technology of lunacy" [P113].

1995 Greenpeace: "The solution is impressively simple, reasonable, and economically efficient. It has been propagated by Greenpeace for years.... No further atomic sites come into being by storing waste in or at the atomic power plants = less sources of danger in Switzerland" [P115].

1996 Issler, Nagra, after the voting defeat at Wellenberg: "We have already modified our concept. The wastes are anytime controllable and retrievable during the period of concession [a licence issued by the Canton, tf, comment added]. The future generation [*sic!*] shall decide whether to keep open the facility further on" [P70:20].

1999 Nagra: Their "adapted disposal concept" ("facility kept open") is interpreted as a "geological facility" with the benefits of a "facility at the surface" [P221:15].

Inadequate time dimensions

1999 Environmental NGOs: Their "concept of controlled long-term storage" is primarily based on technical barriers, for: "Technology (primary barriers) permits controlled storage in the long term (60–80 years)" [P27:32].

2000 Breitschmid: "... These insights oblige us to realise that future generations will have to deal with our radioactive waste in one way or another until radioactivity has decayed to a harmless level .... The optimum way will have to be elaborated in the future by each generation in a broad scientific-technical and societal discourse" [P36].

---

[143] This is a choice of words from previous days: According to Swiss media in 1976 the US Energy Research and Development Administration (ERDA) "puts the policy of its predecessor, *Atomic Energy Commission (AEC)*, on ice, to store the high-level wastes preliminarily in concrete bunkers on the surface, and, instead, directly heads for solidification and subsequent final disposal in geological formations" [NZZ, 1976-3-27].

Inadequate prioritisation of goals

1998 Damveld & van den Berg: "The present generation places great demands on the disposition of nuclear waste, which should also be true for future generations to avoid negative effects. Permanent retrievability can comply with that. Thus, with permanent retrievability [each generation to come] gets the possibility of controlling the waste and taking commensurate measures" [P54:7].

1998 French Government, regarding retrievability: "It is decisive that future generations are not tied down by decisions previously taken and that they may change the strategy, in view of the technical and societal developments having occurred in the meantime." "The duty of today's decision makers" rests with the "preparation of all possible paths of research" [G85, transl. suppl.].

2000 GNW: "GNW ... takes note of the report by ... EKRA ... with satisfaction.... The Working Group juxtaposed the new Disposal Concept '98, elaborated by GNW and combining the benefits of controlled storage with the benefits of permanent final disposal, to be confronted with the simple long-term storage without geological safety barriers. The comparison came indisputably out in support of geological safety and the claim for reversibility and, thus, in favour of the GNW concept" [P108].

2000 Breitschmid: "The central ethical premise ... is the demand that each generation remains capable of acting in all conceivable developments and disturbances .... If [actions] are forborne for whatever reason, the technical and geological barriers of utmost optimum [*sic!*] still offer a certain defence against radiation-induced damages .... All systems of control and surveillance can, however, curb the effect of geological barriers. On behalf of long-term safety with appropriate room for manoeuvre this disadvantage has to be put up with..." [P36:14*passim*].

Ill-conceived conclusions and "conceptions", inappropriate comparisons

1997 Kreuzer, in his journal "nux": claim for reversibility of decisions in the long run [P176].

1998 Forum vera: "When implementing the repositories ... no irrevocable decisions may be taken and no constraints may be created. The permanent sealing of the facility may be left to our descendants" [P95:2].

1998 Rechsteiner, representative in the National Council: "In order to fill in the financial holes even the junk reactors are not stopped .... If the market performed correctly, one would close down [the NPP of] Leibstadt [with the highest debts] and to declare oneself bankruptcy ... Without having a say the people have to incur additional safety losses. It would be more honest to admit the mistakes made, even if it amounts to costs for us .... The Federal

Council's announcement to orderly retreat [from nuclear energy, as of 1998-10-28] could be converted to beneficial political decisions: – resignation from the international atomic energy organisations (savings 10 to 20 million francs) .... Technical progress cannot be blocked ..." [BaZ, 1998-11-25].

1998 Nagra: presents the so-called "adapted disposal concept" seemingly adopting claims from the run-up to the – lost – referendum at Wellenberg: keeping open the access shaft (named "GNW 98" later on), "[t]he adapted disposal concept equally meets all the ... demands [passive long-term safety, causality principle, and freedom of action for future generations]" [P219:I].

1998 Environmental NGOs: "Typology of facilities: ... design ... geographical site ... *e. g.,* cavern in dry rock ... tectonically undisturbed zone ... possibly the vicinity of atomic power plant" [P275:26].

1999 Environmental NGOs: "The observed weaknesses of the final disposal concept [ignoring absence of knowledge, obligation for transfer of information, no possibility to intervene, too great risks] make it indispensable to search for an alternative. Controllable and retrievable long-term storage, in contrast, offers the requisite increase of safety" [P256:[3]].

1999 Damveld & van den Berg: " ... the storage [*sic!*] of nuclear waste has to be designed in a way that no harm will occur in future" [P55:5].

1999 Environmental NGOs: "Reversibility is the pivotal principle of our conceptual idea. Reversibility is not commensurate with the final disposal concept. Our conceptual idea wants to secure a permanent access of control to the facility surroundings for the generations to come so that a possible event of damage can be recognised early and prevented or limited, respectively. This may be most likely compared to patrols along a dam wall" [P257].

2000 Breitschmid: " ... These insights oblige us to realise that future generations will have to deal with our radioactive waste in one way or another until radioactivity has decayed to a harmless level .... The dilemmas cannot be resolved with a conventional scientific-technical procedure but they [have to] be addressed by a prudent strategy in consideration of all possibly conceivable uncertainties and in a process-oriented manner. The optimum way will have to be elaborated in the future by each generation in a broad scientific-technical and societal discourse" [P36].

Promotionally effective but inconsiderate statements such as the ones cited are attractive to politics, at first glance even plausible – at least for the targeted clientship –, and they reinforce groupthinking in their corporate reasoning [D41]. Behaving like this, on the one hand, the stakeholder groups isolate themselves, whether intentionally or not. On the other hand, it may lead to the situation that criticism (also in-house) is not permitted anymore, and that no common solution range may be discovered or created.

Under the sign of goal discussion, controllability, and thus surveillance, should be ensured *prior to* retrievability, namely to check whether the main goal, long-term safety, is met or not. Retrievability and retrieval, respectively, can eventually serve as a control of second order on condition that the controlling results call for it. Or: Retrievability serves to reclaim resources, this being an instrument of quite a different type (see *Figure 6-1*). If after Damveld & van den Berg 1999 "the concomitant effort will increase since continuous storage has to be sustained" [P54] the ethical principle of equal treatment (of present and future generations) is violated. As the proposal of a pilot facility by EKRA [P60] illustrates (Section 13.2), the issue of controllability under (geological) disposal conditions is the real crucial theme. It was KNE which pointed out back in 1998: "Lately the argument has come up that one may not impair the scope of action of future generations by irreversible steps. This claim is met in the current disposal concepts foreseeing stepwise construction, operation and closure in several phases. The waste principally remains retrievable over a long period of time" [P167].

Analysing the reasons for storing reveals different possible arguments behind them: on the one side, the misgiving that so-called permanent solutions according to "out of sight, out of mind" might harbour unforgivable and irremediable defects [P114], on the other side, the hope might emerge, from an energy political controversial viewpoint, that facilities could be used as a low-cost resource deposit once the nuclear path revisited a revival. In both cases, extremely cost-intensive surveillance systems would have to be set up and maintained (on safety and safeguard grounds). They might be so costly that their funding could not be provisioned by the waste producers, but would have to be provided by the future actors. In order to – in both cases! – not infringe the principles of the "rolling present" and of the "chain of obligation", nuclear technology must forcefully be maintained and promoted to secure the vital know-how and funds at least to a certain extent. In essence this would amount to a convergence of Macy's "nuclear guardianship" and Weinberg's nuclear "priesthood", a situation where both parties would surely not feel at ease. At least, in the case of the protection-oriented (and/or anti-nuclear) argumentation, this position seems paradoxical and inconsistent.

In the perspective of decision theory, prolonged storage is the option of non-decision. Often out of opposed positions[144] it is associated with the, intuitively attractive, claim for more research. This attitude was well taxed

---

[144] Transmutation (and separation) has been promoted over time by pro- and anti-nuclear followers, *e. g.*, [P26][P177][P277] *vs.* [P232], which could be similarly traced on an international scale. This has been, time and again, put into perspective by experts or at least made dependent on major prerequisites [P192][P167][P163][MA35][G270][G264].

by the Director of the Finnish waste implementer Posiva, Veijo Ryhänen in 2000: "Non-implementation of final disposal (zero alternative) would in practice mean that interim storage is continued indefinitely. As to decision-making[,] the alternative based[,] on nuclide separation and transmutation[,] would in practice revert to the zero alternative, because the required technology is not available today. In addition, even this method would not eliminate the need of final disposal ... it is unavoidable to choose between development of final disposal and postponing of decision (continued interim storage) ... final disposal ... will leave more freedom of choice for the future generations than the zero alternative; spent fuel can even be retrieved from the deep repository if required" [G235:[3]].

It is particularly questionable that the non-decision statement was reiterated in DETEC's preliminary draft of the revision of the Nuclear Energy Act in 1999, and again in 2000 in the final draft: "At present it is not determined when the duty of disposal is complied with for high-level and long-lived intermediate-level wastes .... An interim storage ... is conceivable in the case when a geological disposal is possible but not advisable (*e. g.,* careful use of resources, further technical possibilities of disposition such as transmutation)" [P277:19][P279:19]. Partitioning and transmutation as valuable strategies to partly decrease the hazard potential assume dramatically improved reprocessing, a well-functioning fuel cycle on an industrial scale, as well as a clear commitment to nuclear R&D and a nuclear-based energy policy. DETEC violated, however, the above mentioned ethical principles and the legally defined causality principle, because neither ways were laid out, nor programmes suggested nor research intensified. To the contrary, no deadlines are indicated, the Energy Research Commission (CORE) intends to cut back the minimum research funds in the disposal field and, according to the draft, reprocessing was to be banned. The revised law, Nuclear Energy Act, in force from 2005, imposes a 10-year moratorium on new contracts for reprocessing only, but does not specify the remaining requirements.

As to the issue of non-decision, Greber 1995 of the Canadian AECL observed from an ethical point of view: "Decisions on disposal have to be taken with due regard to an unavoidable degree of uncertainty. Absolute certainty in assessing the very long-term effects of disposal is not possible. Nor is there certainty with respect to actions or inactions on the part of future societies ... there is a responsibility to do the best that can be done on the basis of available information. The ethical approach if there are doubts about the future impacts of our decisions is to err on the side of prudency. However, it must be recognized that taking no action when dealing with nuclear fuel waste is a de facto decision to pass on to future generations the burden of dealing with the waste that this generation has produced" [G82:143].

After screening international literature, and on the principle of sustainability, the arguments in favour of deferring the conceptual decision, with concomitant continued storage, are not imperative. As the few respective surveys show, people have difficulty grasping long-term aspects, incidentally in conformity with the industry (see *Table 6-2* and [G169]). But the prolonged hazard potential and the time limits of institutions as the sanctuaries of control and technology development diverge so dramatically that the causality principle does not admit further delay. In addition, US studies reveal that the alternative of on-site storage (at the NPPs) is not favoured compared to final disposal [G15].

To conclude, IAEA pertinently observed in its notable paper called "Regulatory decision making in the presence of uncertainty" of 1997: "It is desirable to engage the applicant, the regulatory body, and other stakeholders in a continuing dialogue from which the level of assurance to be required in demonstrations of compliance will emerge. Such a dialogue may result in changes in the methods and approaches used by the applicant and in the expectations of the regulators. The principles on which the safety assessment is based need to be robust and readily communicable to a wide range of audiences …. As a corollary, delaying a decision to advance in the process to finally dispose of radioactive waste in a geological facility on the grounds of incomplete knowledge may be inappropriate, because alternatives can only be interim solutions and irreducible uncertainties will naturally always remain" [G111:30]. The following Chapter attempts to set out a workable path to tackle the complex issue.

Chapter 13

INTEGRATED RISK ANALYSIS: OUTLINE OF AN OVERALL SYSTEM ROBUSTNESS

1. NEED FOR AN INTEGRATION OF ASPECTS: GENERAL REMARKS ON DEALING WITH DISSENTING VIEWS

In a review of the developments in the last decade, the NEA Radioactive Waste Management Committee (RWMC) 1999 recognised a better integration of the main technical challenges of deep geological disposal projects, such as the design of engineered systems, the characterisation of potential disposal sites, and the evaluation of total-system performance [G183:7][145]. The international nuclear community, however, recognised that it is not up to them alone to decide on such complex issues, thus the Nuclear Energy Agency, in 1999, stated: "Rather, an informed societal judgement is necessary" [G181:23]. A year later, NEA conceded that "radioactive waste management institutions have become more and more aware that technical expertise and expert confidence in the safety of geological disposal ... are insufficient, on their own, to justify to a wider audience geologic disposal ... the decisions *whether, when* and *how* to implement geologic disposal will need a thorough public examination and involvement of all relevant stakeholders" [G186:3]. Accordingly, in line with efforts of the IAEA [G118][G122], NEA established a so-called "Forum on Stakeholder Confidence" (FSC) "... to

---

[145] Fundaments to this Chapter were presented at the ESREL 2001 Conference [G67], the 2003 HSK/IAEA/NEA Workshop on Regulatory Decision Making Processes [G71], the VALDOR 2003 Conference [G74] and the PSAM7–ESREL'04 Conference [G75].

facilitate the sharing of international experience in addressing the societal dimension of radioactive waste management" [G154:258]. The Sixth European Framework Programme for Research and Technological Development, launched in 2002, concedes that "research ... alone cannot ensure societal acceptance; however, it is needed in order to ... promote basic scientific understanding relating to safety and safety assessment methods, and to develop decision processes that are perceived as fair and equitable by the stakeholders involved" [G59].

Since trade-offs are inevitable in decisions on radioactive waste governance (e. g., in the triangle of sustainability, see *Figure 12-1*), a concept of an "integrated" – technical and societal – robustness is proposed. With such an approach, attempts are made to minimise negative side effects resulting from a long-term disposal system.

In general, a system is robust if it is not sensitive to significant parameter changes, such as from external impact, and if it rests within well-defined boundaries [M86:33]. Robust procedures, as defined in a narrow sense, can only be achieved when the problem at hand is strictly technical. The system characteristics of radioactive waste are such, though, that with respect to long-term safety it "is not intended to imply a rigorous proof of safety, in a mathematical sense, but rather a convincing set of arguments that support a case for safety" [G182:11][G201:10,13]. Therefore, even technical robustness cannot be treated as in "conventional" technical systems. Nevertheless, it is precisely the robust control systems that are designed to manage the above-mentioned manifold types of uncertainty.

The aim is to attain, on the one hand, a conservative, passive, and stable system with, on the other hand, control and intervention mechanisms built in. The underlying assumption is that dealing with a complex socio-technical system, such as the disposition of radioactive waste, needs an integrated perspective [G219:265][M7]. Much of the widespread blockage faced in this sensitive policy area for decades may be ascribed to the neglect of looking at the various dimensions involved. Only by such an approach is it possible to advance to level 3 in *Figure 1-3* and to, in a technical perspective, such a blurred concept like "social rationality" (Section 8.3).

Applied to the radioactive waste field, it means that the system calls for technical barriers against releases of radioactivity, as well as societal checks to achieve and sustain confidence in technical assessments and, hence, acceptance. It is, in fact, an integration of societal aspects into the defence-in-depth strategy familiar to radioactive waste-performance assessments [G117:85-89].

The conducted analysis of the decision-making process in radioactive waste governance virtually is its embedding into a "political dimension" of radioactivity and facilitates a more extensive perspective than that obtained so far; though it does not resolve the initial question on "technical *vs.* political issues" of Section 1.2, the aspect is raised to another level. The integration of insights from risk perception, decision science, STS, management, and institutional learning can be a stimulus for broadening the approaches, *e. g.,* about aim, goal, and scope of risk analysis. Thereby, I strive for an "integrated robustness" of the system under study and an "integrated risk analysis" to investigate this system. And there is a need for clearing up the issue. This goes along with IAEA's – unhonoured – claim to link together defence in depth with robustness in 1998: "The key principle … is the concept of defence in depth …. One of the important aspects is the evaluation of the robustness of repository systems …. One of the main issues is obtaining a better understanding of the meaning of principles such as defence in depth in the context of waste disposal" [G114:244]. It has to be emphasised that this is not an attempt out of a deterministic point of view by just building as many barriers as possible "around the waste" but the notion of robustness adopted encompasses conceptual issues and is process-oriented. Neither is the approach "objectivistic" by serially switching technical and societal "robustness" or putting them on the same level.

2. TECHNICAL ROBUSTNESS

There is wide international consensus that long-term radiation safety should not depend unduly on active measures. Hence, protection should be implemented primarily at the design stage [G114:243]. Due to the required longevity of the disposal system, "[t]he aim of the performance assessment is not to predict the behaviour of the system in the long term, but rather to test the robustness of the concept as regards safety criteria" [*ibid.*:245].

As regards technical robustness, NEA makes the following distinction [G182:30-37]:

- Engineered robustness: "[i]ntentional design provisions that improve performance" such as overbuilding of barriers, waste conditioning in a stable matrix, and physical separation of waste into packages of limited size;
- Intrinsic robustness: "[i]ntentional siting and design provisions that avoid detrimental phenomena and the sources of uncertainty" such as siting in sedimentary layers deep underground, with self-healing properties and an uneventful geological history, away from potential natural resources;

- (Technical) system robustness: combination of siting and design provisions supplemented by peer-review and quality assurance procedures.

Prediction of detriments to health over very long time periods is critical. Therefore, it is useful to consider safety indicators other than dose and risk criteria [G106][G126]. This approach leads to:

- Performance robustness: comparison of the anthropocentric criteria individual human dose and risk with waste and environmental safety indicators, which are ruled by less uncertainty than the aggregated radiological protection goals.

The technical components of robustness, as envisaged by the international nuclear community, are depicted in *Figure 13-1* overleaf.

Claims for revisiting final disposal have already been raised in the 1970s, shortly after the respective decision by the nuclear power industry and the authorities. The spectrum of proposals ranges from the "expansion" of the concept all the way to permanent surface storage in bunkers (*Figure 12-4*). The demand after an "extended final disposal" by Flüeler 1998 [P88] and the "Monitored long-term geological disposal" presented by EKRA in January 2000 [P60][P135] distinguish themselves as much from the so-called "Nuclear Guardianship" of indefinite control [P175][G159] as of the surface deposits of the sort at the French Centre de l'Aube, Spanish El Cabril, or British Drigg[146]. These official and implemented "engineered surface facilities", planned to be controlled for around 300 years, are sanctioned as (final) disposal by the IAEA [G101:11,13] as much as the foreseen underground facility at Yucca Mountain in the US state of Nevada. The access shafts "could ... be ... kept open for hundreds of years" [G263:17], the caverns are not to be backfilled according to normal operation [G34][G14]. The US Geological Survey 1999 held, though, that "little substance has been given to monitoring .... Design of a substantive monitoring program is needed both to assuage public fears regarding 'out of sight, out of mind' and to ensure that our descendants will have the proper data to decide whether and when to seal the repository" [G266:17].

---

[146] Refer to Buser 1998 for a critical review of the guardianship idea in the Swiss context [P45][P46].

Integrated risk analysis: outline of an overall system robustness

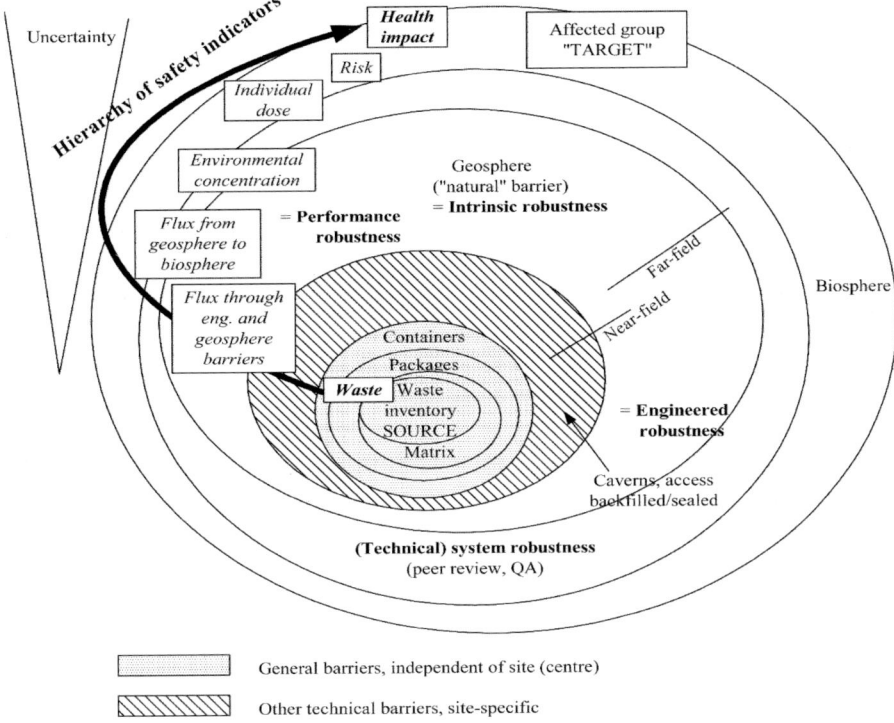

Figure 13-1. Technical robustness. Long-term safety of long-lived radioactive waste is primarily based on passive technical (general, site-specific) or natural barriers. Technical robustness should be obtained through a careful selection of the site (intrinsic), prudent design of the disposal system (engineered), technical system robustness (reviewing, quality assurance, QA), and performance robustness. This last type is characterised by safety indicators other than dose and risk criteria (these are associated with higher uncertainties than waste or environmental measures).

The idea and proposal by Flüeler and EKRA, respectively, also include the traditional international, also Nagra's, final disposal concept, according to which it was merely designated to optionally set up environmental monitoring on the surface, and that under the caption "post-closure phase". GNW suggested in 1994: "After closing down the repository for LLW and sealing the access- and linking shafts, repository surveillance may be continued by control measurements at the surface …. Because surveillance does not have to contribute to the long-term safety of the repository, the stipulated control and surveillance measures are not specified any further in the present study" [P106:83]. This was notably concerned with the technical safety report of the general licence application for Wellenberg.

Nagra assumed, and in principle still assumes, "a definitive removal without the intention of retrieval": This retrieval "albeit always possible in principle, becomes more and more costly, in technical as well as economical terms, with the progressive implementation of the graded measures of confinement, and concomitant with an increasing radiological hazard for the operating personnel" [P201:15]. Precisely to meet this aim, however, risk control and retrievability must be integrated into the project design, a claim already put forward by Wakerley & Edwards 1986 [G274] and part of the EKRA concept. After its defeat in 1995, the GNW offered a delayed closure in its "adapted disposal concept" for Wellenberg, an idea that violates the overriding goal of passive safety though, as was repeatedly criticised.

A demonstration of long-term safety has to be provided even if the requisites are inherently unfavourable[147] for design and safety analysis for repositories are founded on model conceptions and experiments in the field or in the laboratory. In order to re-examine the system's long-term behaviour, the concepts must be verified and, above all, validated during all disposal phases and *in situ*, particularly under final disposal conditions during the post-closure phase. By this we do not understand "drillings on the surface as well as drillings to be carried out below ground", as GNW tried to make readers believe under the heading of "final long-term safety analyses" following "repository operation" [P106:82]. This would indeed amount to the situation that surveillance is "technically possible, but at the expense of long-term safety" [P192:23].

After having disturbed the geoscientific equilibrium by destructive testing and construction, with phased closure, the repository will supposedly regain a balanced state in the long run, which shall correspond to the model assumptions in the safety case. According to the EKRA concept, it is conceived to erect various shafts and galleries for validation- and surveillance purposes in order to detect eventual changes in the near-field of the facility system and to monitor the vicinity of the host rock towards the biosphere (see *Figure 13-2* overleaf). The measuring period depends on the new stable state and on the results during the monitoring period. Based on that, the waste is retrieved in the case of a grave system failure, or the facility is sealed if the "ultimate" safety assessment is positive.

---

[147] Rometsch of Nagra remarked in 1985: "The models imparting the decisive predictions on the future behaviour of the repository are products of the human mind. There is no direct way for them to be validated with experiments for the time periods to be covered are about two hundred times longer than the experimenters' life expectancy" [P245]. The technical community attempts to cope with this particular with conservatism [G162].

Integrated risk analysis: outline of an overall system robustness 237

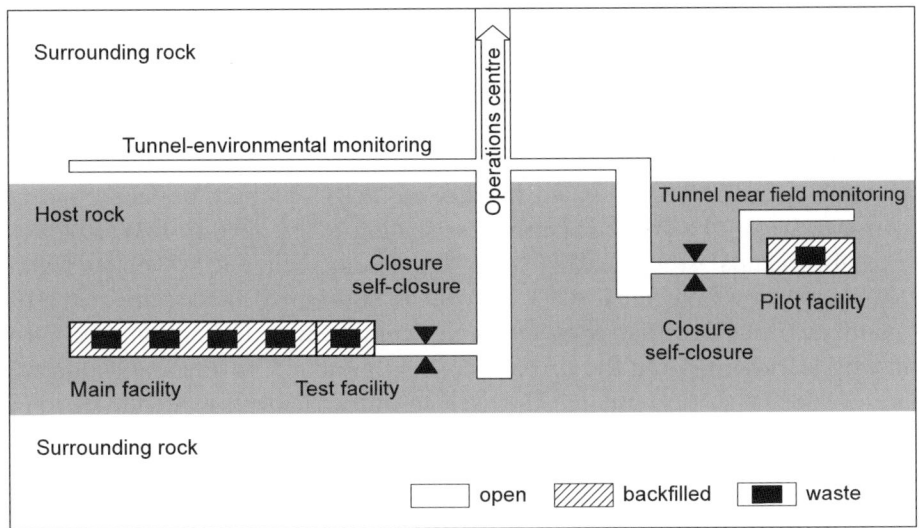

Figure 13-2. The key elements of the so-called "monitored long-term geological disposal" proposed by the Swiss Expert Group on Disposal Concepts for Radioactive Waste (EKRA) [P60] are three facilities: a Test facility, a Pilot facility, and the Main disposal facility. The Test facility shall be erected during, or shortly after, site characterisation in the host rock, and serves as a rock laboratory for the investigation of safety-relevant processes, to specify the safety analysis and to adequately plan the design of the Main facility. The Test facility could be situated at the entrance to the Main facility, whose caverns, containing about 95 per cent of the waste (black boxes), are to be backfilled as soon as the waste is deposited. The Pilot facility is hydraulically isolated; the intent is to load it with a, say, 5 per cent representative sample of the total waste activity. The Pilot facility shall function as a so-called "demonstration facility" to assess the long-term behaviour of the technical barriers and the near-field of the entire disposal system. It is conceivable that the Pilot facility could be kept open after the Main/Test facility has been closed by backfilling. A carefully selected cavern structure facilitates retrievability from all facilities without compromising safety. For validation and surveillance purposes, several tunnels and shafts are foreseen to survey the facility near-fields and to carry out environmental monitoring in the surroundings of the host rock toward the biosphere. After the observation phase (period to be determined), the waste is retrieved (in case of system failure) or the facilities are sealed (in case of a positive final safety analysis). A passive self-closure mechanism is provided for, in case rapid sealing becomes necessary. Schematic graph from EKRA 2000 [P60:55][P135].

Control, thereby, is also aimed at minimising irreversible consequences of a repository failure, according to the minimum regret principle. Evidently, under the eyes of the authorities, it has to be deployed by the repository operator, and not "until our grandchildren have been convinced by our meticulous work and decide to backfill and seal the access shaft", as GNW vaguely planned according to their brochure "The repository at Wellenberg: answers to your questions" [P107:[7]], until they entered discussion with KFW.

In the context of *Figure 13-1*, the term robust or robustness is used in relation to technical and environmental risk analysis. In general, a system is robust if, as mentioned, it is not sensitive to significant parameter changes.

Such a concept of control is explosive by nature, by the fact that it was picked up in 1998 by the Forum vera affiliated with Nagra: "When implementing the repositories ... no irrevocable decisions may be taken and no constraints may be created. Permanent sealing of the facility may be left to our descendants" [P95:2]. This is a striking coincidence with demands from NGOs as shown in Section 12.7. What is subsumed under "the room for manoeuvre must remain open" does not comply with the notion of control having led to criticising the first Wellenberg project: The constraints already are created, and the polluters themselves must establish evidence of long-term safety; this responsibility must not be passed on, not even in parts. The legal situation in Switzerland is unambiguous: "Once having sealed the repository, the owner ceases to be liable. For damages occurring after sealing the Confederation is exclusively liable and to an unlimited extent .... Because sealing is decisive for supervision and liability, it may only be commissioned if the repository owner has delivered proof of long-term safety" [P3][P2].

What is outlined here has been debated internationally for some time [G165][G10][G205], also in working groups convened by IAEA [G109]. KASAM 1993 suggested a retrievable "demonstration-scale repository" to SKB with a share of 5 to 10 per cent to be erected prior to the main repository. Its time horizon, though, was only several years and extensive long-term investigations, such as those according to EKRA, were not foreseen [G140:12-15]. A consensus has not been established yet due to the sensitive trade-offs – control measures must not impair the passive safety barriers. In 1997 NEA was afraid of the idea that "regulators might be subjected to some pressure from the public in the conduct of their professional responsibilities, which might result [in] the risk of 'diluting good engineering practice to have better relations with the public' ". The dialogue should not lead to undermining responsibilities, and unambiguous legal stipulations – as the "rules of the game" for licensing – were deemed to be necessary [G199:245-246].

The EU project "Thematic Network on the role of monitoring in a phased approach to geological disposal" (2001–2003) strived for a better understanding of what may be meant by monitoring and its possible contribution to decision making [G42][G51][G8]. The following passage of its final report reflects the cautious, and timid, view towards the issue: "There is a range of approaches to monitoring. It is important to understand the reasons for these differences and the role played by monitoring within any safety and repository implementation strategy. The extent of monitoring should be limited to that which could reveal useful results for the decision-making process or for the confirmation of safety ... it is important not to give the impression that such monitoring indicates a lack of confidence in the safety of the disposal system" [G205:ii]. Apart from the informative annex on country states with regard to long-term confirmational monitoring it laconically states that "[s]pecially instrumented test or pilot facilities may be used in a research mode if this is thought desirable" [*ibid.*:53].

The EKRA concept was the basis for discussions with KFW, from mid-2000 to mid-2002, within the LLW programme [P110], as well as Nagra's submission for the demonstration of HLW disposal feasibility in late-2002 [P222:III,1]. This is in compliance with the new Federal Nuclear Energy Act stipulating "deep geological disposal" in Art. 31 on the basis of EKRA [P78: 2718,2746], as well as a concomitant "observation phase", being a "lengthy period during which a geological deep disposal facility can be monitored prior to sealing and the radioactive wastes can be retrieved without great effort" (Art. 3 lit. a) [P264]. It is a matter of discussion, however, whether Nagra's pending proposal regarding the disposal feasibility of HLW is "fully compatible with the concept of monitored long-term geological disposal", as maintained in the safety report going with it [P222:1]. In a conceptual phase, one has to sufficiently define core topics. They are, *e. g.*, the "observation phase" mentioned, or controllability and retrievability, so that their importance – for the safety case – can be appraised. It is particularly the pilot phase "to test predictive models and to facilitate the early detection of any unexpected undesirable behaviour of the system" [*ibid.*:24], which should be adequately qualified, for example, on:

- the definition of the Pilot facility, including so-called test niches;
- specific criteria for the choice of location;
- criteria for the waste selection representative of the Main facility;
- criteria and requirements of applicability of the measuring data (with respect to waste, facility, and geosphere; interactions);
- statements on monitoring targets, measuring strategy and procedures, measuring parameters, set of instruments, monitoring period.

As mentioned, the robustness of a system can only be tested and audited if its parameters are clearly defined, and if it is ensured that the system rests within well-set boundaries [M86:33]. With respect to Wellenberg this specifically meant to the expert body KFW that "small modifications of boundary conditions must not result in dominant dose shares of single nuclides close to the radiological protection goal" [P156:11]; the prevailing protection goal is 0.1 mSv/a as an individual dose, according to the guideline R-21 [P133].

For Nagra as well, the notion of robustness is central to its HLW demonstration of feasibility [P222:40,44,46]. It is utilised close to a hundred times in the safety report, albeit with different meanings, *e. g.*, [*ibid.*:XXI, XXIII,12,22,25,27,34-35,39,43]. I consider it of paramount importance that its use – and usefulness – be examined and separated from other concepts such as reliability, resilience or, *vice versa,* vulnerability:

- Residual uncertainties may not impair the system (is a cumulation of conservatism admissible/necessary/inacceptable?); how and where are probabilistics applicable? Shall there be a time-cut, and where [G194]?
- Fundamental mechanisms must function continuously (such as, *e. g.,* diffusion).
- Multiple barriers have to mesh, barriers should not be unnecessarily used but spared so that they indeed can fulfil their functions.
- Basic assumptions have to be validated.
- Safety functions have to successfully complement each other (*e. g.,* careful site selection).
- Special attention has to be given to interfaces which – understandably – have no process owners, *i. e.,* for whom responsibility is not clearly assigned (particularly pertinent where technical meet non-technical spheres, see *Figure 13-6* or *Figure 13-8*).

In summary, many questions remain to be scrutinised, even at the level of technical robustness. The newly proposed concept of "expanded final disposal" must undergo extensive reviewing if it is to serve as a convincing reconciliation of passive "safety" and "control". Otherwise it would be left open to attack by people saying it would be another, more sophisticated, attempt to persuade the wider public of a non-credible technology. It has been picked up in the international discussion [G271:70-74][G33:167] yet not thoroughly examined. The NEA International Peer Review on Nagra's HLW Programme in Opalinus Clay, in 2003/2004, did not take the opportunity, nor was it given by Swiss mandate, of analysing the truly novel proposal by EKRA [P229:29]. Instead, it is falsely equated to "keeping the disposal facility open for an extended period of monitoring" as planned in Yucca Mountain [*ibid.*].

<u>Excursus: connection of long-term safety and transparency as exemplified with the inventory issue of radioactive waste</u>

The proposed process-based link between engineering design and consideration of stakeholder notions (arrows in *Figure 13-1*, *Figure 13-6*, and *Figure 13-8* below) is illustrated hereafter[148]. The question of inventory is closely associated with the (progression of) safety analysis for radioactive long-term disposal. In response to requests for comprehensibility, it is proposed that a stepwise plausible procedure on the way to a consensual "end product", the "demonstration" of long-term safety, be developed. This proce-

---

[148] For details see Flüeler 2002e [G72:204-208].

dure should be traceable for non-specialists. Since the mid-1980s, the topic of waste inventory, including definition and allocation to eventual facilities, has concerned interested parties in Swiss radioactive waste governance (see, *e. g.,* page 95). A consistent waste classification has also been called for internationally: At the IAEA's 3rd Scientific Forum in 2000 it was complained in the official document that "[t]here is an air of mystery attached to radioactive waste, which is at least partly the result of the complex and sometimes obscure terminology used by specialists" [G119:5]. Consistent with this is a 1999 recommendation by an International Regulatory Review Team (IRRT) to the Swiss authorities [P141:45].

In conformity with international standards, the nuclear community has, until recently, refrained from defining the inventory before the construction of a respective facility. While acknowledging the intrinsic complexity of giving conclusive evidence of long-term safety [G182[G193], it is suggested here to augment the presentation of the safety case as follows:

- The evidence of long-term safety of a deep geological disposal facility has to be supported by means of revised safety cases according to the state of the art.
- The strict protection goals of the regulatory guideline [P133] evidently have to be complied with but, additionally, safety indicators (see *Figure 13-1*) have to be developed. These indicators should help to build up multiple lines of arguments, to reduce uncertainties and to establish transparency.
- Single criteria, such as the half-life period of radionuclides, their origin, *etc.*, are not commensurate with the complexity of the issue and should be substituted by a set of criteria of diverse indicators.
- The selection of criteria and sites proposed, as well as the ensuing assessment and evaluation, have to be traceable and plausible.
- In the end, a "robust" disposal system, in the sense suggested above, has to be guaranteed.

In arguing for the extension of the safety case, the following reasons are put forth. Thus far, all stakeholders, with the exception of the few existing performance assessment analysts, have been referred to as "end products". On one of two levels, the technical level, they are given safety reports providing information about meeting the protection goals; on the other level, the legal one, they face the general licence application, at best the application for the concession for a facility. In technical matters, the strategy selected according to legal requirements takes the safety of humans and the environment as a basis. It is called Strategy **A** in *Figure 13-3* overleaf and is exclusively product-related to outsiders. Strategy **B**, though, in the legal procedure, facilitates participation even today and is therefore process-related. It is granted that the decision here is "open", but the fundamentals to the decision (from the technical level according to Strategy **A**) are not presented prior to the final process phase.

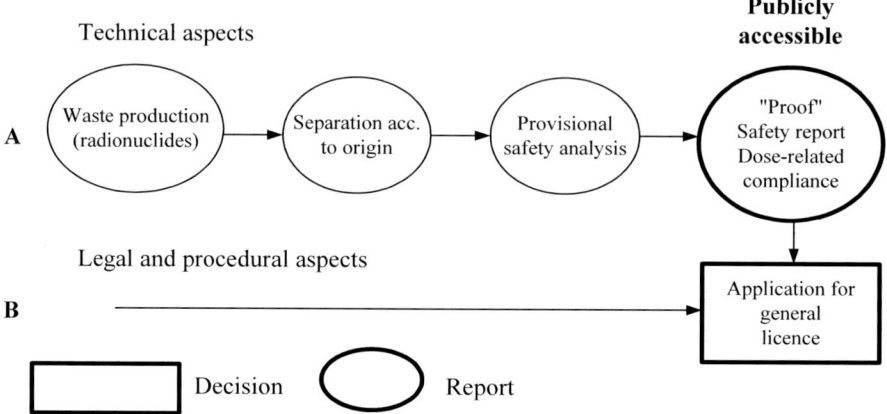

Figure 13-3. Decision process according to the current legal procedure. In product-related Strategy **A**, only experts are involved. In the legal procedure of Strategy **B**, however, the involvement of concerned parties is foreseen, whereby their decisional basis emanates from Strategy **A**, and this not before the final stage. The technical level is related to the legal level merely unidirectionally.

An expanded procedure that would be perceived as stepwise by the public, even in technical matters, would relate the waste and the siting discussions to each other via the criteria discussion, in a recursive manner using strategy **B**, as depicted in *Figure 13-4* overleaf.

In discussions with KFW, from mid-2000 to mid-2002, Nagra and GNW responded to requests and adopted the requisites by EKRA as specified by KFW [P157][P109]. GNW submitted a revised application in 2001, this time only for an exploratory gallery as agreed in the discussions, notably with the safety authority HSK and with self-restricted participation by the regional opposition in Nidwalden. As mentioned, the second application, of 2001, was rejected as well.

Demonstrating the stepwise procedure using Wellenberg as an example, the safety authority HSK, the proponent and the Cantonal advisory body KFW, agreed on so-called exclusion criteria [P157][P127]. Their purpose was to support the assessment of observations made during the preparatory phase by means of an exploratory shaft. They would have allowed a decision on whether to continue or abandon the project[149].

---

[149] For a short summary of the approach in English see HSK [P129:7,58][P130:65-66]. Also consult www.energie-schweiz.ch/internet/00118/index.html?lang=en.

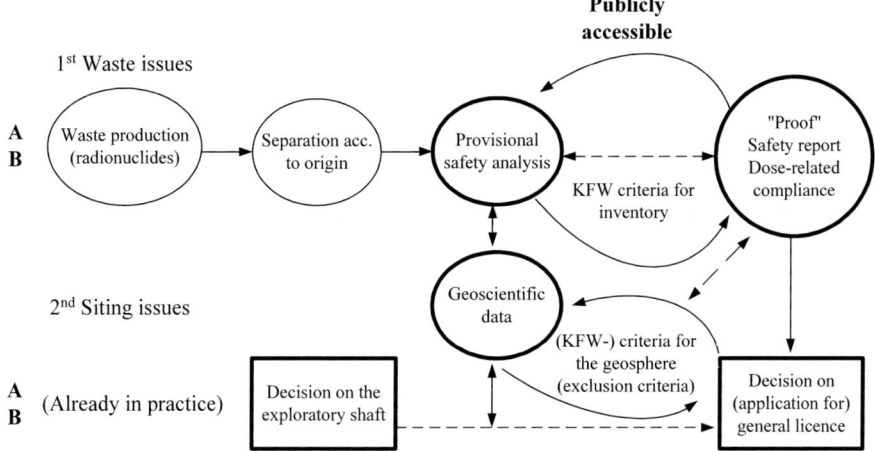

Figure 13-4. Proposal of an integral and recursive decision-making process. Technical and legal levels as well as waste and siting issues, respectively, are interconnected, via the introduction of criteria. The circle of involved parties in the development of the safety case is enlarged. The strategy is process- as well as product-related. The procedure shall contribute to the robustness of the overall system.

Paralleling these consensually achieved criteria in the geoscientific field, a set of criteria for inventory and waste classification ought to be developed. The independent expert body KFW presented the initial ideas for a respective conception [P159]. Subsequently, the Federal Interdepartmental Working Group on Radioactive Waste Management (AGNEB) appointed a Subgroup on Inventory Issues. Such efforts regarding the diversification of safety indicators and the improvement of transparency, as well as stakeholder involvement, are in accord with international discussions of late[150] and the recent research endeavours of the European Union [G59][G57].

3. SOCIETAL ROBUSTNESS OWING TO AN EXTENDED DECISION MODEL

The systems approach spells out the fact that the various project phases of a facility site require at least several decades, from characterisation, design via operation to monitoring and closure/sealing (see *Figure 12-2* and *Figure 13-9*). Consequently, management fundamentals and concepts are liable to an integration into a consistent and stepwise decision-making process. Their instruments have to be designed in a dynamic, adaptive, even experimental manner [G43][M2:375], but the ultimate goal remains, *viz.,* the

---

[150] [G106][G109][G114][G118][G120][G126][G182][G186][G193][G271][G4].

passive protection of present and future generations and environments. According to Rip 1987, a system is "socially robust" if most arguments, evidence, social alignments, interests, and cultural values lead to a consistent option [D71:359]. Therefore, the concerned and deciding stakeholders have to eventually achieve consensus on some common interests, along the lines suggested in Section 12.4.

"Overall robustness", in a way, is a fuzzy notion, but it recognises the complex socio-technical character of the issue and has the potential to stepwise and iteratively integrate structural and procedural/dynamic elements – as well as various and diverse types of uncertainty – into radioactive waste governance. By means of (individual) risk perception criteria and (institutional) decision-making criteria, the content analysis facilitates to empirically sift and sort out relevant demands and requirements by society for radioactive waste governance. Over time, various stakeholder groups have raised critical and crucial topics which, eventually, were adopted by the institutions in charge. They are, *i. a.*, the separation of promotion and oversight in the Federal administration, adequate funding according to the causality principle, extensive duty of publication, traceability of reasoning, transparent formulation of criteria (for siting, inventory, *etc.*), controllability, retrievability, extensive independent, *i. e.*, pluralistic, reviewing, stepwise and phased procedure, participation of the public (*e. g.*, also in oversight committees).

This result concurs with the insight gained by the international nuclear community in recent years, as elucidated in Section 11.3.. This goes so far that the regulators' organisation, NEA, started to integrate the public and opposing groups quasi as clients "with a legitimate interest" into their quality management programme schemes [G189:15]. Adopting an inclusive perspective, one may also note the EU Framework 6 Priority Thematic Area 1.7 "Citizens and Governance in a Knowledge-based Society", where "an integrated understanding" shall be created "of how a knowledge-based society can promote the societal objectives of the EU ... of sustainable development" [G38:44][G58]. Under the caption 1.7.2 "Citizenship, democracy and new forms of governance" the aim is "to support the development of forms of multi-level governance which are accountable, legitimate, and sufficiently robust and flexible to address societal change ... to assure the effectiveness and legitimacy of policy making" but also "to support the development of institutional and social capacity in the field of conflict resolution, identify factors leading to success or failure in preventing conflict, and develop improved options for conflict mediation". In the long, run the issue is "to promote citizens' involvement and participation in European policy making, to understand perceptions and impacts of citizenship" [G58].

In *Table 13-1* the milestones of participation in the Swiss process of radioactive waste governance are listed. Indicators with participatory

Integrated risk analysis: outline of an overall system robustness 245

relevance are highlighted in *italics,* and only when they first appear. Obviously the aspects have to be appraised in the context and not in isolation.

Table 13-1. Corner stones of third party involvement. All stakeholders, except for authorities, proponents, and their experts, are judged "third parties". **Bold years** depict **popular votes.** Indicators for *participation* are in *italics* (Cont. to page 248).

1957 *Popular vote:* 77 per cent favour a constitutional article on atomic energy.

1959 Atomic Energy Act: parliament rejects *instrument of concession,* solely regulatory approval needed for licences.

1961 First project of an national "store" for radioactive waste: "The *problem* being posed is much less of a technical than a *psychological* one" [P263,13/61:14].

1964 Imboden, constitutional lawyer, with a nation-shaking book on the "Helvetic malaise": "The essence of politics is based on trading off technical options" [P143:26].

1964+ Decisions to construct nuclear power plants in Switzerland.

1967 Federal Commission KSA with respect to the NPP project of Kaiseraugst: "The issue of the definitive deposition of solid waste ... in KSA's view ... is ... *technically solvable."*

1968 Project of a "definitive depot" at Lossy abandoned after local protest (as well as in Lucens).

1969 First *political resolution* against NPPs (Cantonal Council of Basel-City).

1970 Courvoisier, ASK (today HSK, safety authority): "The waste problem does not exist" [P247:44].

1972 First *parliamentarian inquiry* at Federal level after the 1957 debate on the constitutional article.

1974 Federal Councillor Ritschard: "Our great problems are the radioactive wastes. Therefore, a commission is working on it. Resistance against disposal, against these wastes, is substantial. But we still hope to find a solution" [P250:1667].
– First test drilling in Airolo, Ticino: *Communal opposition* stirs as soon as the project becomes public.

1975 Hunzinger, head Radiation Protection in FOPH: "The ultimate solution is a long-term issue ... not urgent. The technical problems are solvable. The representatives of the Confederation appealed to the opponents against the atom not to instigate ... against the investigations by Nagra. In order to examine the geologically interesting information certain psychological preconditions have to be there" (hearing on Kaiseraugst, Sept. 18–21) [P92:10,51].

1977 Rausch, administrative lawyer: first legal expertise, *judicial gaps* existent (participation of the parties of the proceedings not guaranteed) [P239].

1978 DETEC: "... worldwide it was missed ... to demonstrate a practical method of elimination in time (*i. e.,* when there scarcely was opposition); one ought not merely trust the authorities but "follow the principally quite trivial considerations"; "opposition has to be broken ... in case of need" [P66].

	– Determination of project "Guarantee 1985" by the Federal Administration and the power industry: *no publication, no public scrutiny of the concept.* – Establishment of the Interdepartmental Working Group on Radioactive Waste Management (AGNEB).
1978	Federal Decree on the Atomic Energy Act, duty and mandate for "the permanent and safe final disposition and disposal of the … radioactive wastes" [P39].
1979	The official Federal brochure on the Federal Decree promised: "Furthermore, a safe disposal of the radioactive wastes has to be ensured before an eventual licence [for a new NPP] may be issued" [P74:11]. – Having this promise in mind: The **first people's initiative** on phasing out nuclear is rejected (by around 55 per cent). – Positive national vote on the **Federal Decree** (69 per cent); a following on survey reveals that 86 per cent of the voters hold that "safe disposal … must be guaranteed in the longer term" [P103:16-17]. – Criticism raised by the Swiss geoscientific community, formal discussion with university geologists: Criteria and guidelines are missing, need for a National Geological Institute, *international review, publication.*
1980	Appointment of the Subgroup of AGNEB, the first geoscientific expert body: demands "a categorically necessary public involvement and publication" [P8: Annex V,1]. – Rausch, administrative lawyer: "… on the other hand it has to be noted that even today's practice according to the Atomic Energy Act does not comply with the standards as they are stipulated by the Federal Law on the Protection of Water Bodies not only for large facilities but also for each single family house: no construction licence without previous guarantee of sewage purification" [Art. 19 and 20]" [P240:188-189].
1981	AGNEB: The definition of "Guarantee" is a "political reaction to the question whether an unsolvable waste problem might arise by the generation of nuclear energy" (*sic!*) [P9:Annex III].
1982	Rometsch, President of Nagra: "… one is looking for technical solutions for a psychological issue, which of course is impossible" [P223, 8/82:3].
1983/ 1984	*Massive opposition* in Ollon (Canton of Vaud), Mesocco (Grisons), partly in Oberbauen (Uri), demands referendum and retrievability of waste.
1984	Parker *et al.*: "…. a 'solution' is to be found only in a national, not a scientific, context" [G217:108].
1984	**Second people's initiative** on phasing out nuclear is rejected by 55 per cent.
1985	Cantons potentially hosting a site: demand for a "geological equal treatment" and for exclusion criteria. Establishment of the regional opposition group in Nidwalden, MNA: demands for a mandatory vote on licences (cantonal popular initiative).
1986	Parliamentarian motion for *full access* "to all documents … in the interest of …

Integrated risk analysis: outline of an overall system robustness 247

widely supported democratic decision making" (House of Representatives) [P189].
– Seiler, administrative lawyer: "In ... the field of disposal there has never at all been a procedural involvement of third parties in so far as this entire topic was ignored in the issuance of operation licences ... nor when fixing the deadline of [Guarantee] 1985 were third parties given a stake" [P253:31].

1987 Nagra: "In each country and in the end, the core of the waste disposal issue lies in the relationship of trust between the population and the technical persons responsible in the implementing organisations" [P204:Annex 3].
– Nagra: "In view of the political origin of the demand for ['Gu]arantee[']is seems to be crucial that a consensus among all concerned parties be elaborated on how the work shall and can be continued with respect to the realisation of final disposal" [P204:accomp. letter] (definition of "all concerned parties" was left open).

1989 Establishment of KNE, the Commission on Radioactive Waste Disposal, as the successor of the Subgroup of AGNEB: expert body with representatives of the Swiss universities and consulting companies.
– Parliamentarian motion in favour of "permanent interim storage" at the sites of the NPPs [P236].
– From 1989 more attempts were made to *speed up proceedings* in nuclear legislation, revoked in 1994/1995 due to harsh opposition by the cantons and the Senate.

1990 The **third people's initiative** on phasing out nuclear is rejected, whereas the **fourth initiative** on a 10-year ban of construction is adopted (by 65 per cent). A majority of the (winning) opponents of nuclear energy holds the "view that the wastes shall be deposited in Switzerland if suitable sites are found" (62 per cent, 87 per cent of the nuclear proponents underscore this statement) [P101:31].
– Largely unsuccessful *mediation exercises ("Conflict solving group")*, with implementers, authorities, and the national environmental NGOs, take place [P168].

1995 The application by GNW for a **general licence at Wellenberg** is rejected by 52 per cent of the voting Nidwalden electorate.
– First comment by Nagra: "The close result suggests that ... it was less the project than rather the procedure which was rejected. If this interpretation ... holds to be true, a revised application for a partial concession might be appropriate" [P216:1].
– Most relevant reason for rejection: connection of the exploratory gallery with the repository as such. "The people of Nidwalden would apparently like to be involved in the decision again as soon as further results are available after the gallery has been dug. The circumstance that the Cantonal Council would have been able to block the repository construction in the light of unfavourable findings made little difference" [P142].

1998 The *"Energy Dialogue", another mediation,* again with implementers, authorities, and the national environmental NGOs, takes place [P275]. "Both" parties, pro- and anti-nuclear, insist on their maximum demands (keeping the nuclear option open and phasing out, respectively).

2000–   Discussions in the expert bodies EKRA (national) and KFW (cantonal) are opened
2002   to the environmental NGOs and the regional opposition groups, respectively.
2002   The Nidwalden electorate refuses to grant a licence for an **exploratory gallery** alone at Wellenberg by almost 58 per cent [P231].
– The Parliament passes the revised Atomic Energy Act, now called *Nuclear Energy Act,* prohibiting a final vote by an eventual host canton but stipulating that "the Department [Ministry] gives a share to the host canton as well as to the neighbouring cantons and countries in direct vicinity with regard to the preparation of the general licensing decision" [P264:Art. 44].
2003   The **fifth people's initiative** on phasing out nuclear is rejected as well as the **sixth** one on the continuation of the winning moratorium in 1990 (by around 66 and 58 per cent, respectively).
Nagra submits the project "Entsorgungsnachweis" (demonstration of feasibility and siting of disposal) on HLW to the Federal Council.
– The FOE establishes three bodies to accompany this work: a so-called Political Committee with representatives of the Cantonal Governments, a Working Group on Information and Communication, and a Technical Forum. Particular attention is paid to invite *local communities, the regional opposition,* as well as the neighbouring *cantons* and the equally neighbouring state of *Germany.*

The enumeration makes clear that the dialogue with the citizens was not cultivated for years. The broad public forcefully obtained their "involvement" in popular votes and referenda (see Section 11.2.2). Until 1980, even other "third parties", who did not belong to either implementers/proponents, regulators or hired experts, were not involved in the decision process. Even in 1980, the Subgroup Geology of AGNEB, the first geoscientific expert body, felt obliged to demand "a categorically necessary public involvement and publication" [P8:Annex V,1].

Stimuli for a revival of the concept discussion were provided by external expert bodies in the HLW programme (first the Subgroup Geology, then KNE), in the case of the LLW programme it needed the voting defeat of Nagra/GNW at Wellenberg in 1995. The Cantonal Government, up to that point notedly in favour of the project, changed their minds and reconsidered their role of a trustee for their population and their environment in terms of safety. They adopted criticism expressed by the 1995 opposition and appointed an expert committee of their own (KFW). By that, they initiated a transparent, open and stepwise process. This was, apart from the purely conceptional EKRA at the national level, a novel situation in the history of Swiss radio active waste. The mandate for KFW, as issued in June 2000, was: "Review of the exclusion criteria, ... of the waste inventory, ... the plan for the exploratory gallery and the revised disposal concept on the grounds of the EKRA report, factual information as a contribution to the public debate, controlling at the location, further work as commissioned by the Canton" [press release of 2000-6-21]. Apparently now, the conceptual revision was not enough for the electorate of Nidwalden as the one and only site already was determined.

Seemingly striking is the fact that the issue of stakeholder involvement has gained terrain at the international scale as well. Some type of paradigm shift seems to happen, from information and education to dialogue and involvement of the public (see Section 6.4.5, Chapter 8, and Section 11.3). Continuing the reflections in Chapter 8, a model of an inclusive participation of stakeholders may be laid out within the framework of a (from now on) integrated risk analysis.

It is incontestable that non-experts show perception "anomalies", *i. e.,* identical or similar factual circumstances are perceived differently on the ground of non-rational considerations, such as radiation risks with radon *vs.* industrially caused low-level radiation (see Section 6.3.1). It must be pinpointed, though, that the technical community is subject to anomalies as well, due to their narrow (quantitative) definition of risk. They are traditionally not receptive to an "extended" risk understanding (Sections 6.2.1 and 8.1). To attenuate this phenomenon and to approach broadly based decision making, it is proposed that an expanded risk analysis (*Figure 13-5* on page 250), and, consequently, a comprehensive decision model, are characterised by the following attributes

Process:

On the initiative of a conceptual body (Section 13.4), with a broad composition of societal concerns, a discourse takes place on principles and alternative options based thereon. Proponents, authorities, and this body reach a consensus on a conceptual framework. This agreement undergoes a wide public debate. Following an inclusive consultation, the proponent elaborates a project (previously step **1** in *Figure 8-1*). The authorities review and commission expertise. Eventually the project is revised. Finally, the concerned region (the concerned public) decides on the project.[151]

The process is characterised by dynamic mutual understanding and learning, as well as drawing up of the programme and project(s), respectively. A transparent decision base, with all relevant advantages and disadvantages of the options, must be provided for and openly discussed. Modifications and changes in time and context have to be considered, *i. e.,* decision criteria may change in the course of a project, such as the appraisal of monitoring of

---

[151] According to Swiss legislation, an optional national referendum is foreseen, whereas the hosting region is granted some participatory right "insofar as [considering the concerns] does not disproportionately impair the project" [P264:Art. 44].

long-term disposition facilities. Contradictory goals with all their pros and cons have to be fully put on the table. Such a procedure corresponds to so-called "double-loop learning"[152] proposed by organisational learning theory, whose requisite, however, is, *i. a.*, institutionalised programme evaluation [D52:3*passim*].

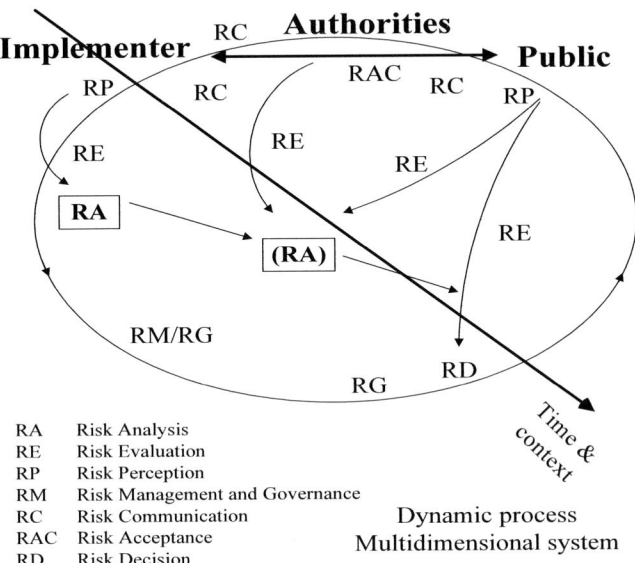

RA	Risk Analysis
RE	Risk Evaluation
RP	Risk Perception
RM	Risk Management and Governance
RC	Risk Communication
RAC	Risk Acceptance
RD	Risk Decision

Dynamic process
Multidimensional system

Figure 13-5. Dynamic and pluralistic decision making. Multi step-3-actor-model: All stakeholders are equal and, in theory, learn from each other. "Public" stands for third parties, *i. e.*, stakeholders not pertaining to implementers or authorities. There are no "better" or "worse" types of perception of reality or of rationality. The process is dynamic, iterative and stepwise, and multidimensional.

Mental models

- "Absolute" rationality for risk analyses in a narrower sense;
- "Bounded" rationality as the heuristical approach by laypeople;
- "Social rationality" (see Section 8.3) for an "extended" risk evaluation to be integrated into the implementer's safety case, based on, *i. a.*, risk criteria under scrutiny in this study and validated within the Swiss context.

---

[152] Argyris & Schön 1978 [D5][D4:449*passim*] distinguish between "double-loop learning" and "single-loop learning" where "the primary strategies are to control the relevant environment and tasks unilaterally and to protect themselves and their group unilaterally" [D3:368] and "minor, reversible changes of a behavioural and instrumental type are made" [D19:59].

Techniques

Risk management and risk governance is an inclusive, stepwise, and recursive approach, including the partial steps and instruments such as risk perception, -analysis, -communication, -acceptance, -evaluation, and -decision. It is vital to divide a project into phases and concomitant stages with (interim) decision points [G271:32].

Actors/stakeholders

The self-evident but not always present fact is acknowledged that all actors are subject to value-loaded risk perception. There are no "better" perceptions of reality or notions of rationality [R97:382][R30]. In plain words, this amounts to accepting the public, in their diversity, as an equal partner, equal partners, *i. e.*, risk communication does not take place just in the final stage of a project. Apart from some common "basic knowledge", the base of knowledge exchange (see below) is some minimum comprehension of other people's perspectives. All stakeholders act according to their responsibilities. An inclusive trade-off must take place, in which each stakeholder group can bring in their competences. In simple terms, this means that an "informed society" would not favour the "elimination" of radioactive waste in volcanoes, even though in this case the issue would be "solved" "quickly", "feasible", cheaper, highly "acceptable" and far away from strenuous domestic programmes. What this means in practice, focussed on radioactive waste governance in Switzerland, is outlined in *Table 13-2, Figure 13-6* and *Figure 13-9*.

Aim

Beyond the indispensable assessment of risk targets and the instrumental consideration of non-quantitative risk perception aspects, we strive for an all-embracing governance of risk. It includes "rules, processes and behaviour that affect the way in which powers are exercised ... particularly as regards openness, participation, accountability, effectiveness and coherence" as is stipulated by the respective European White paper [M12:8, see footnote 1].

I do not propose a new method, but the expansion of the current mental models and approaches: In addition to the "absolute" and the "bounded" rationality, the "social rationality" is recognised (see Section 8.3 and [D67]). It assumes, on the one hand, an extended notion of risk and facilitates, on the other hand, a more comprehensive risk analysis, within the framework of a defined overall system. An inclusive participation and integration of concerns is vital, especially in the conceptual phase, because this is the only feasible way to, in the long run, legitimise well-supported decisions on issues with far-reaching consequences. Besides, Wynne 1989 pointed out that information (of any type) only has informational character to those who share the underlying assumptions, otherwise it is (de-)valuated as an artefact (after [R59:40]). This dynamic approach with mutual learning is indeed time-consuming but likely to be, in the long run, more effective as well as more efficient than previous approaches.

Since not all goals of all stakeholders can be achieved, as mentioned on page 213, they have to be negotiated according to their responsibilities [MA30:134]. No change of stakeholders' belief systems is envisaged, at least not in their core principles, but modifications in their secondary aspects [D73:367-668,675]. They are to identify some common interest or, in Carter's 1987 words, some "common ground" [MA6:427] (see Section 12.4).

Rip's 1987 cited definition of "social robustness" (page 244) sheds light on the circumstance that "process" is not just an issue of participation (the more "stakeholder involvement" the better), but that many aspects from diverse perspectives (and fields) are expounded. The inherent emphasis on the process is a productive approach to eventually integrate different aspects and perspectives (see research potential within the European 6th Framework Programme [G57]):

Long-term safety

As a matter of course, robust procedures, as defined in a narrow sense, can only be achieved if the parameters are clearly defined and if it is assured that the system rests within well-set boundaries [M86:33]. Yet, the system characteristics of radioactive, and, for that matter, chemically toxic, waste are unique and technically complex. Once defined, the waste has to be stored in a safe way, since it emits hazardous ionising radiation. Depending on the hazard potential of the waste in question, its isolation period from the biosphere ranges from 100's to 100,000's of years. The main mechanism in an underground (geological) site is a low-level but long-term, chronic release into the environment; it is a slow degradation of an open system with concurrent large uncertainties (see Section 1.3). These system characteristics lead to the admission that the required long-term safety "is not intended to imply a rigorous proof of safety, in a mathematical sense, but rather a convincing set of arguments that support a case for safety" [G182:11]. Nevertheless, it is precisely the robust control systems that are designed to manage the mentioned uncertainties (an important aspect not dealt with here). "Confidence" in performance assessment is only attained step by step [*ibid.*].

Long-term project

Radioactive waste governance is not only confronted with the "objective" long-term dimension, but also with institutional longevity. From site selection via characterisation, design and operation to surveillance and closure, the various phases will last for decades. And so will the corresponding project management, which necessarily entails entwined and tangled processes. A (disposal) project, laid out for long time periods, has to be backed up over decades by the technical community, the political decision makers and the general public [G182:11]. It is to be implemented in a phased approach, with feedback for recourse, and interim decisions [G271].

Sustainability

The multi-dimensionality of the issue asks for applying the principle of sustainability. Normatively, it seems to suggest itself as a reference system: On the one hand, it facilitates a stepwise analysis according to various dimensions (see *Figure 12-1*), on the other hand, it forces us to integrally examine these dimensions and consequently perspectives, needs and knowledge systems. In its dimensional approach, the concept inherently is system- and process-related.

In view of the objective longevity of the hazard potential, the primary goal of the entire undertaking has to be the long-term safety of humans and the environment, whereas the secondary goal is flexibility, defined as intervention potential (controllability, retrievability). (For a prolonged discourse of the reasoning see above.) However that may be: The decision has to be taken by society.

Decisional issue

A decision is more than the preference of an option – it involves the decision-making process with problem definition, judgement, choice and implementation [D48:388*passim*]. So it is not "only process" that counts .... "Good" decisions are always goal-related decisions: "good" with respect to what? (see Section 7.6). "Good" decisions imply good processes (which do not necessarily result in good decisions, though). A multitude of stakeholders and perspectives are involved. Therefore, particularly due to the complexity of the issue, the process and procedure, and not only the result, are vital for the decisions to be taken. It is a stunning interaction to face.

The following are attributes of a "good" decision-making process:

– *Stepwise:* planning phases with milestones;
– *Periodic orientation, reviewing and interim decisions:* for technical and political back-up;
– *Open and comprehensive option analysis;*
– *Iterative, with opportunities for recourse* (and mutual learning);
– *Reliable, accountable:* unambiguous rules to be complied with (only modifiable by prior consent);
– *Consistent, minimising conflicts:* technical and non-technical sets of criteria;

– *Coherent, continuous:* for sufficient trust in "the system" (see below);
– *Traceable:* arguments and reasoning to be fully comprehended by interested parties;
– *Transparent:* in broad discussion forums aspects put up for discussion at early stages;
– *"Fair" procedure and treatment of the intra- and intergenerational equity issues*, taking into account the twofold – spatial and temporal – asymmetry: benefit of nuclear electricity is broadly distributed whereas cost/risk of waste disposal locally concentrated and transferred to future generations.

To reach well supported and stable decisions an "informed consent" is needed which, in turn, requires a demonstration of (all, most) possible tracks and consequences of actions [D21].

Risk debate

The discourse on risk issues is factual, and political bargaining about what are conceived as "risks" is done by the actors engaged: "Risks ... are to be understood as consequences of decisions on conditions of uncertainty, social action and situations of interest", as the sociologists Nowotny & Eisikovic 1990 describe the negotiation of interests [R59:13]. To repeat it, it is not aimed at changing belief systems, but at finding a common factual basis. If some common basis can be found, we may speak not only of negotiation but of deliberation.

Technology and progress debate

Scope and orientation of technology policy are publicly and politically contentious issues that are continuously and always debated [D15:247 *passim*]. Assumptions on advances or re-directions of technology directly impact on the choices of disposition (disposal, storage and transmutation) (see [G175]).

Integration of technical and political aspects, learning process

Credible and sustained compromise can only, if at all, be achieved if collective learning takes place, if authorities turn their backs to technocratic planning, stop separating technical from political issues and involve the concerned stakeholders in all relevant planning phases [P280]. All stakeholders have to realise that, in the end, effectively sustainable radioactive waste management only results from transdisciplinary "mutual learning", learning from each other. "Transdisciplinarity aspires to make the change *from research for society to research with society*" as Scholz 2000 puts it; " ... mutual learning sessions ... should be regarded as a tool to establish an efficient transfer of knowledge both from science to society and from problem owners (*i. e.* from science, industry, politics *etc.*) to science" [M79:13].

According to Gibbons *et al.* 1994, problem-driven, dynamic transdisciplinarity is one attribute of the so-called "mode 2" of knowledge production. On top, or beside, "mode 1" of conventional disciplinary scientific knowledge acquisition, this newer strategy, to gain insight into the world, is produced in the process-oriented context of application; it is marked by new, often transient forms of organisation with members of heterogeneous experience: "The experience gathered in this process creates a competence which becomes highly valued and which is transferred to new contexts" [M31:6]. By integrating a number of interests, also so-called concerned groups, mode 2 knowledge gains more social accountability and makes all participants more reflexive. Gibbons *et al.* even argue that "the individuals themselves cannot function effectively without reflecting – trying to operate from the standpoint of – all the actors involved" [*ibid.*:6].

Responsibility, project implementation and mediation

Often authorities put off crucial decisions and break them down into a number of partial decisions; political opposition is forced to obstruct each step, which in turn leads authorities to consider themselves being in the role of defenders [P284:216]. Mediation builds on the assumption that consensus is possible and, in the long run, of some use to all involved [M83:232].

Trust in the system via procedure

Procedures symbolise the continuity of similar experience and may add to actors retaining and gaining trust in the political [R48:199]. In the context of laypeople and probabilistic analysis, Sowden 1984 referred to "process utilities" [D78:297]. This is endorsed by experience in environmental impact assessment procedures [M63][D80].

Performance via networks

Organisational theory assumes that performance in companies is attributed to their system characteristics rather than to interests or intentions of individual actors [D67]. Cyert & March 1995 even hold that achievements are generated through intricate networks within and between companies. Networks are characterised largely by interactions and processes [D8][D58].

Modern management principles

The current ISO 9000:2000*passim* quality management standards are guided by dynamics, stakeholder involvement (from worker to client), system thinking and continual improvement [M41][M63]. The insight into the need of involving stakeholders hitherto excluded goes as far as to recognise "Concerned Action groups", *i. e.*, opponents, as "clients" in the NEA 2001 proposal of a "Quality Management Model" [G189:15]. As to a comprehensive quality control, quality is determined by a wider set of criteria than, *e. g.*, peer-reviewing in Gibbons *et al.*'s mode 1 [M31:6*passim*], since additional issues of cost-effectiveness, competitiveness, social acceptance, *etc.* are raised. Such an approach enhances the openness of the process and the reliability and robustness of the decisions.

Interdisciplinary convergences as mentioned, regarding processes in complex issues, such as radioactive waste governance, allow the conclusion that for the overall decision not only the results are relevant but also the procedure how to reach them. Thereby, radioactive waste governance may, and does have to, develop a satisfactory degree of maturity.

Despite the emphasis on process-orientation, this is not a call for arbitrariness; it does not amount to "anything goes". In view of the sustainability goal relation "protection *vs.* control" and process- *vs.* outcome-orientation, it is understood that the radioactive waste system has to be dynamic, adaptive, and even experimental in its instruments, but not in its ultimate goal, *i. e.*, the passive protection of present and future generations and environments. The goal hierarchy is protection over control (see Section 9.6). The goal discussion has to be led in a broad and open manner, also because catchy but simplistic formulae (like the call for "reversibility of all decisions", see Section 12.7) have to be exposed and fundamental inconsistencies have to be dispelled. Impacts from unfounded decisions will likely be at the expense of future generations, inconsistencies are detrimental to the credibility of the entire system, and corrections made afterwards are at any rate expensive in view of the dimension of the programme if, at all, practical and efficient.

According to Strohl 1995, "experience has shown that the functioning of institutional mechanisms is generally efficient and permanent when their purpose is to protect society's vital interests; a well-informed public, together with other factors, can contribute to the maintenance of these" [G254:125-126].

Such a shared interest relies on some form of co-operation (Sections 9.9 and 12.4), which is driven by social trust based on common group membership. Social trust, in turn, affects performance-oriented confidence by conditioning judged performance. It is, therefore, obvious that only if a minimum relationship of trust is given, can discussions of technical aspects of risk management be usefully conducted [R15].

These relations have to be kept in mind when setting up institutional lines of defence in depth (*Table 13-2* on page 258). The choice of regulatory and other control is largely decided by non-scientific considerations. Institutional control is active (maintenance and monitoring/validation, reviewing), passive (zoning and land marking against human intrusion), and intermediate (documentation, know-how transfer, and training). As the IAEA puts it: "Control must, thus, contribute to making the storage site a social reality, *i. e.*, the control should be implemented in an 'active' way, allowing the stakeholders involvement in it" [G114:238].

As mentioned, besides the objective longevity of radioactive waste, we have to also deal with an institutional long-term dimension. Thus, managerial concepts and principles have to be integrated into a consistent and incremental decision-making path. The various major stakeholders actually involved in Switzerland are listed in *Table 13-2*. In the phased governance of radioactive waste, Level I denotes the physical input of material, Level II the implementation process, Level III the technical advice and review process, and, last but not least, Level IV, the political and societal backup. For each level and its stakeholders products, responsibilities, and functions/activities are set out. The overall system robustness strived at, across all technical and non-technical subsystems, is more than "engineering resilience", the speed of return to a stable steady state following some perturbation [M69]. It might amount to what Gunderson *et al.* 2002 call "ecological resilience", *i. e.*, the capacity with which a system may absorb disturbance before having to be restructured [M33:4*passim*][M38].

Table 13-2. Complex radioactive waste governance: stakeholders and their functions and responsibilities according to phases and institutional system levels. QA = quality assurance. Between proponent/implementer/operator there might be a continuous transition. Source: Flüeler & Scholz 2004 [G77:134-135].

Levels	Stakeholders	Products	Responsibilities	Functions/Activities							
				Concept; All phases	Site exploration	Construction	Operation	Monitoring	Sealing	Surveillance	Memory
I Input	Suppliers Producers Conditioners Storers	Quality material Quality waste Quality package Quality storage	Quality input, QA Funding, QA QA QA								
II Preparation and implementation	Proponents (Applicants)	Satisfactory safety case	Design	Performance assessment Modelling,	Safety case						
	Implementers	Demonstration of this safety case	Know-how transfer	Documentation, Training, Testing	Safety case	Safety case	Control Evaluation	Validation	Markers Sealing, Documentation		
	Operators	Safe operation	Operation	Operation Know-how maintenance			Maintenance Evaluation	Control Decision on sealing			
III Technical backup	Experts	Reports	Consultancy, reviewing	Consulting, reviewing							
	Safety authorities	Licences, reports, inspections, guidelines, standards	Licensing, inspecting, reviewing, independent evaluation, know-how maintenance	Conceptual framework, Criteria for inventory and break-off, design basis, procedure Reviewing Peer-reviewing	Model verification Decision on continuation	Verification	Independent control and evaluation, Validation	Independent monitoring Decision on sealing or break-off	Decommissioning Know-how transfer		
	Oversight bodies	Reports, audits									
	Advisory bodies International organisations	Reports Recommendations, Guidelines									
IV Political and societal backup/environment	Public administration Government Parliament NGOs Interveners Public: local, regional, national		Notification, zoning, public ownership Major decisions Problem identification See footnote 156.	National decision Reviewing Vote/ approval?	Approval of construction Vote/ approval?	Notification			Documentation Decision on closure Vote/ approval?	Land-use restriction, (monitoring) Approval of final assessment	Intrusion control

Integrated risk analysis: outline of an overall system robustness 259

The system calls for multiple technical barriers against the release of radioactivity, as well as for phased societal checks to achieve and sustain confidence in technical assessments and, hence, acceptance or at least tolerability in society (*Table 13-2* graphically converted to *Figure 13-6*).

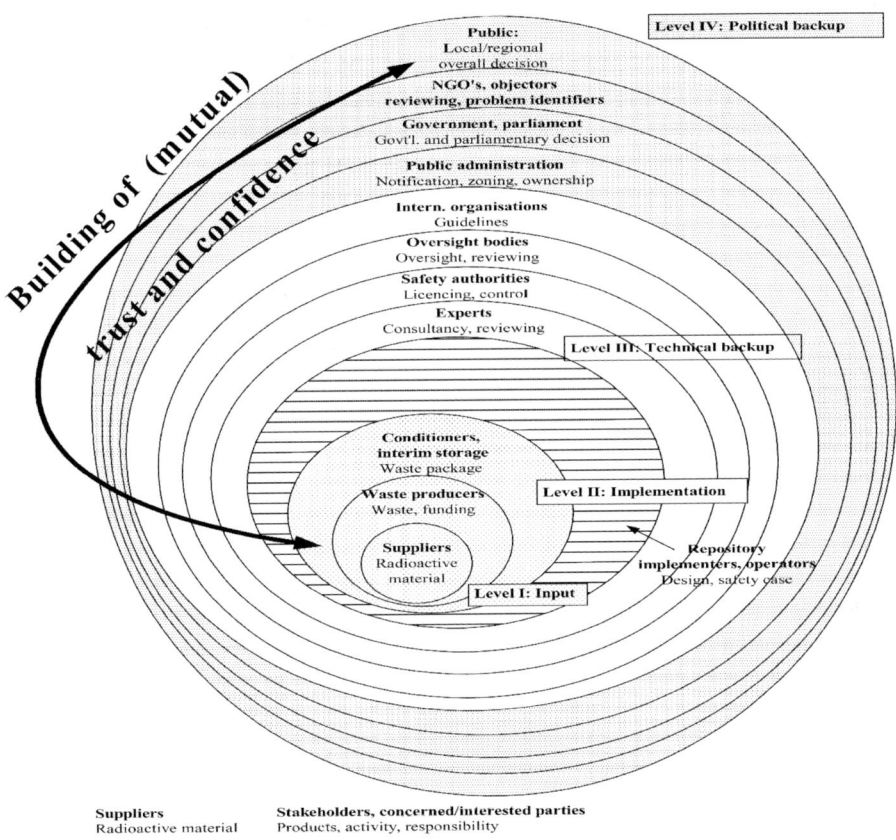

Figure 13-6. Societal/institutional robustness. Stakeholders are to act according to their respective responsibilities. Dependent on their mutual trust, their activities serve as institutional barriers and potentially lead to a consistent, *i. e.*, robust decision, backed up by incremental building of confidence in the overall disposal system. System levels are indicated. For details see *Table 13-2* and text. Source: Flüeler & Scholz 2004 [G77:136], after Flüeler 2002e [G72:218].

Institutional control goes far beyond administrative aspects even though these issues are not trivial in view of the time dimensions involved [G256], and the corresponding framework has not reached a degree of maturity [G124]. It has to include programme and financial control and auditing. This has, up to date, never been the case in Switzerland, as the position of AGNEB, the potential control board, shows (see Section 6.4.5). Out of the systemic approach, and according to the extended decision model, it is proposed to create such a conceptual – "guardian" – body (see below).

Trust and confidence of the public depend on the "degree of organised safety" [R91], which was supposedly erected by the institutions involved. Since expertise by external experts is necessary, the public has to gain trust in the scientific-technical community. Their judgement base, therefore, does not solely rest on expertise but is also, if not primarily, process-based. Consequently, not only is confidence in technical performance assessments needed, but also trust in the persons and institutions in charge and participating in the chosen procedure. In complex technical domains, trust (in experts and their work) is a key notion in the transfer of knowledge. Particularly, when dealing with radioactive waste, one cannot rely on known techniques (state of the art), but has to compensate ignorance (*i. e.*, the absence of knowledge) by trust in the specialised institutions (regulators, safety authorities, applicants, "independent" scientists). Their relevance is increased in authoritarian procedures (Decide–Announce–Defend strategies) where little active public participation exists and the public increasingly seeks trust in diverse information holders (authorities, applicants, experts, "counter experts", NGOs *etc.*).

(Sufficient) knowledge is a prerequisite to an informed judgement which itself is the basis for a decision. Goal-oriented knowledge (information) reduces uncertainty with regard to a decision. Since disposal of radioactive waste is a complex socio-technical problem, embedded in a highly politicised debate on energy options, several aspects of knowledge or expertise are important: the type and quality of knowledge, its origin (the suppliers) and the access to it. The following aspects are judged to be crucial:

Type and quality of expertise

Complex multidisciplinary (transdisciplinary) topics require broad-based approaches to problem solving (scientific, technical and societal), with interaction on various levels (technical, societal, ethical, conceptual, *etc.*). Diverse target groups have to be supplied with appropriate information. Since knowledge does not just exist "objectively" but is interest-bound, expertise, independent of the applicant, has to be built up in order to obtain a pluralistic perspective. Making one single option available is insufficient (of the type that a "choice of one is not a choice").

Origin/sender of expertise

Differences in perspective or focus are due to the distinct nature of the various stakeholders. On the one hand, a delegation of knowledge to experts occurs through the general division of labour, on the other hand the population, in part, lives more closely to a given reality in a given place. They are the most knowledgeable about their local affairs (as if to say "laymen are the experts of everyday life"[153]). Certain national and international NGOs are oriented towards a "global view" and federal stakeholders think "nationally", the citizens in the vicinity of a potential site, however, would normally maintain a local perspective. The "hidden agendas" of waste producers, some NGOs, and experts, have to be brought into the open and in this respect it should be realised that the "(in)dependence" of experts might be compromised.

General set-up: access to knowledge, resources

Decision makers depend on knowledge from diverse sources to reach an inclusive judgement considering all relevant aspects. Sufficient resources can be crucial. The "applicant's expertise" as well as "counter-expertise" have to be traded off prudently. Decision makers on the spot, *in situ*, should be given the opportunity to discuss controversial issues on a continuous basis and in a competent manner. Information must not be withheld deliberately.

Good practice, with regard to the general framework, is demonstrated by Sweden, where potential siting municipalities may build up expertise or consult experts on their own (see Section 11.3.1). Financing is secured through a state-administered fund and following unified rules. How the money is used is left up to the communities. According to the causality principle, the fund is accumulated by the waste producers (*i. e.*, the NPP operators). In Belgium the local initiatives (MONA at Mol and STOLA at Dessel), assisted by a secretariat as well as a technical and a social scientist, are directly reimbursed by the applicant (ONDRAF-NIRAS) [G186:131-127][G94]. In Switzerland the Government of a potential host canton (Nidwalden) appointed a special expert group, KFW, who was solely answerable to the Cantonal Government but whose expenditure was covered by the applicant, GNW. In both states, again according to the causality principle, the costs have to be borne by the waste producers or the electricity consumers. The issue of the critical mass of "local expertise" remains open, one suggestion might be to create an international expert pool [MA10:33].

---

[153] Otway 1987 called "the people whose lives are affected" "the true experts on questions of value regarding the risks of technology" [R63:125].

Schematically, the reflections made can be presented in an integrated expert system[154] as shown in *Figure 13-7*:

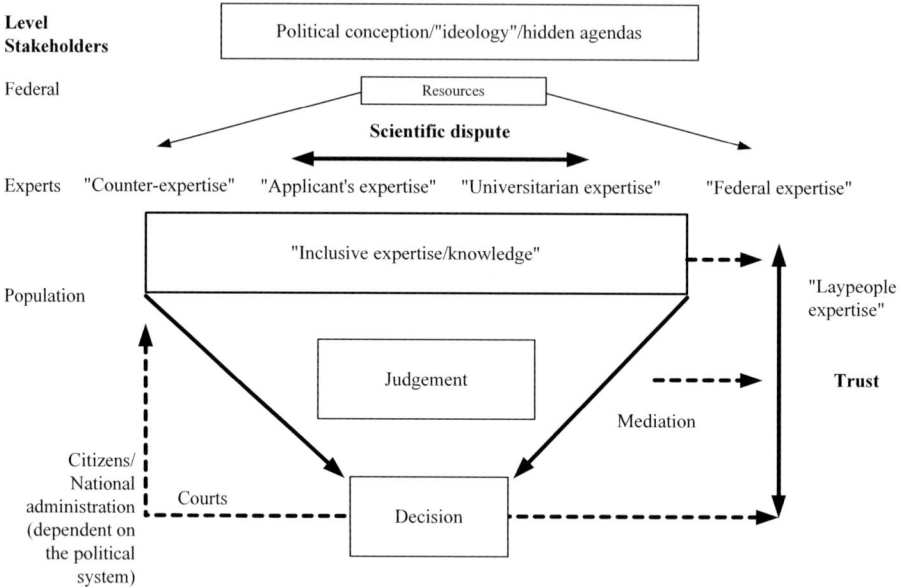

Figure 13-7. Transfer of knowledge with respect to decision (and implementation). The system is held together by a transparent, traceable scientific-technical (and societal) discourse, by trust in the stakeholders and a common understanding of the political conception: sustainability of waste disposition, passive safety combined with control and retrievability (or another set). A stepwise and recursive procedure should ensure the quality of expertise and decisions. For a short definition of the types of knowledge refer to Chapter 3.

In summary, knowledge is not equated with expert knowledge, but above all technical expertise. It is equally obvious, though, that basic knowledge for a repository rests with expert people however they are called. Following the differentiated division of labour as experienced, expert culture *per se* is undemocratic. This fact cannot be argued away, given all the support of involvement of concerned parties. Some aspects, however, have to be considered:

1. By definition, experts are only experts in their field of work. Since we deal with a complex issue in toxic wastes, interdisciplinarity is needed. It is already at the expert level that the technical debate, with different conceptual and model notions, takes place in a

---

[154] Here an "expert system" is deliberately defined by persons and not instruments (like computers, *etc.*).

meta-scientific domain. The frictional loss is minimised but existent.

2. Decisions (not only in democratic societies) are taken by non-experts, even if, ideally, on the basis of expert knowledge. In addition, impacts of decision are generally borne by non-experts as well. Radioactive waste governance is a trans-scientific and, for that matter, transdisciplinary issue. At any rate, experts have to deal with "laypeople" ... and laypeople with experts. It is true whether the process follows the recent paradigm of expanded participation or according to the linear decision model of "Decide–Announce–Defend". Additional stakeholders take part in the modern decision model, which has some bearing on the time frame but not necessarily on the quality of the decision by itself. Besides, a "good" decision as such does not ensure a "good" result [D12:7].

The decision rests with others, but the basic knowledge stems from specialists. In this sense, *Figure 10-2* deliberately is drawn in a dichotomic manner with experts/expertise "*vs.*" laypersons/trust. In turn this also signifies that a decision is only as good as the "worst"/"weakest" partial knowledge underlying it. After Kissling-Näf & Knoepfel 1996 "a policy ... might be the more capable of learning the better-developed societal feedback and the simpler the access to novel stocks of knowledge are organised" [D47:182]. Structuring information and re-examining, also by special bodies, create a prolific ground to facilitate "reflexivity", "which serves to enhance the knowledge about side effects in actions by stakeholders, in politics, economy, and society" [M61:143].

3. Finally, we have to recognise the special case of the waste field (the passages above are valid for many other complex areas as well): Virtually no one is genuinely interested in waste, except for the directly involved technical waste community. Most stakeholders instrumentalise the issue as was shown in Section 6.4.1. The circumstance of factual constraint (the waste "is there") leads to the difficulty that a "solution" has to be found.

The underlying assumption of "societal robustness" is the quest to overcome the binary positions of technical experts *vs.* laypeople, technical *vs.* political perspectives, objective *vs.* subjective approaches, rational *vs.* irrational views, *etc.* It is not a naïve appeal to harmonise society or to blend out controversies but to, nevertheless, achieve some sort of "co-production", the "simultaneous production of knowledge and social order" as Sheila

Jasanoff 1996 put it [M44:393]. At any rate, if the science and technology communities do not engage in this discourse, they will be held accountable by society anyway, as has often happened to the waste community in the past. The philosopher of science, Helga Nowotny, phrases it in a general way: "A 21st century view of science must embrace not only a wider societal context, but be prepared for the context to begin to talk back. Reliable knowledge will no longer suffice, at least in those cases, where the consensuality reached within the scientific community fails to impress those outside ... more will be demanded ..., namely a shift towards socially robust or context-sensitive knowledge" [M64:253].

4. APPROACH TO ROBUSTNESS OF THE OVERALL "RADIOACTIVE WASTE SYSTEM"

The primary goal in radioactive waste governance is the stability of the system: the permanent protection of human life and environment from release of (harmful) radioactivity. The complimentary goal is flexibility, defined as the potential to intervene.[155] A conclusive programme for control and monitoring has to be specified, including publication of the work, intensive reviewing, respective quality assurance, and a wide involvement of affected and concerned parties. In compliance with the International Waste Convention of 1997 [G110] the issue is considered a national task, which has to be carried out on the territory of Switzerland and on the basis of current knowledge [G142:4].

We are faced with a pronouncedly long-term project: "A repository is, by definition, a long term project, extending over centuries ... or even much longer periods ... involves a relatively long lead time (possibly more than 20 years for HLW or spent fuel) and is then anticipated to receive waste during several decades. After closing the repository, a surveillance and monitoring period will almost certainly be carried out even [*sic!*] for *shallow land burial* type repositories with LILW. This underlines once again the importance of the continuity factor not only from a contractual, but also from a technical, point of view (possibility/obligation to transfer/receive waste, waste acceptance criteria and quality of waste, control and monitoring, etc.)" [G112:9]. Consequently, it is of utmost relevance that the various "barriers" adequately

---

[155] One has to agree with Zuidema's respective tentativeness: "There is also broad agreement that retrievability should never be used as an excuse to make any compromises with respect to the level of scientific and technical soundness needed before starting emplacement of waste packages in the repository" [P287:[2]]. Positively interpreted, the tenacity of rejecting concept modifications in the course of time may be ascribed to the safety-oriented and evidence-based approach of the expert community. In this respect, "robustness" serves to meticulously scrutinise novel proposals.

Integrated risk analysis: outline of an overall system robustness 265

mesh. *Figure 13-8* below depicts the integration of technical and societal/institutional aspects into overall – or "integral" – robustness:

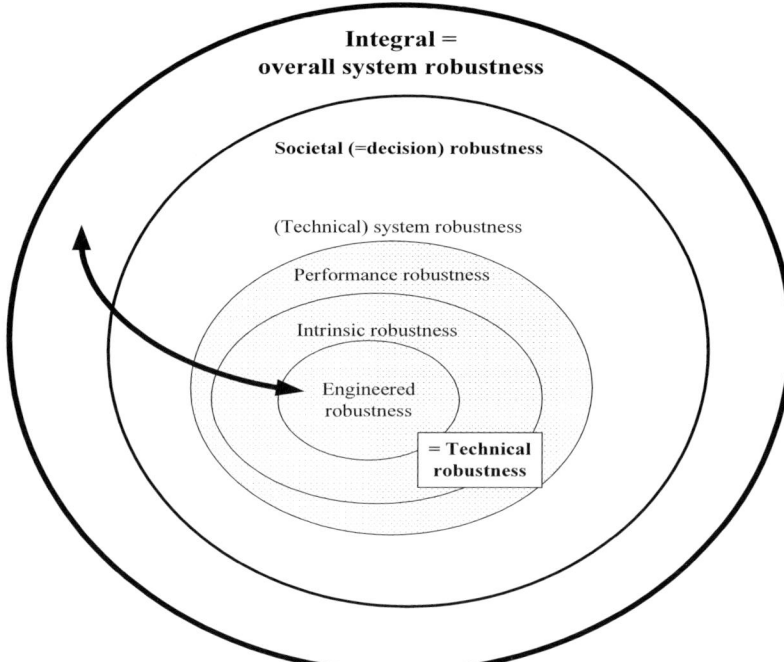

Figure 13-8. Integral robustness. Overall system robustness depends on (dotted) technical robustness and societal (=decision) robustness. The sequenced symbolic portrayal of the types of robustness shall not infer an objectivistic approach. The final and decisive validation of the concept is the implementation of a disposal concept with demonstrated long-term safety backed up by respective decisions and actions. This presupposes an effective coupling, *i. e.,* complementing/meshing, of the various types of robustness. Source: Flüeler 2001b [G67:319].

In order to achieve this aim, it is crucial that long-term comprehensive, but also stepwise, planning is set up, as was outlined by KNE 1998 for the HLW programme: "KNE ... proposes that phase-related programme flow chart and scheduling are drawn up ... where goal agreements and estimations of work load of the separate phases are laid down" [P167:5]. If the concerned public indeed is given an adequate share, the chances rise that the decisions taken will be accepted also in the future as being legitimate [G56:10]. See *Figure 13-9* overleaf.

Data acquisition	Phase	System component	Regulatory act	Responsibility (intensity of work)	Public involvement	Remarks
Integrated	Framework: criteria for inventory and break-off, design basis, procedure			AA**, G	Vote/approval of site and preparatory work	All period length depending on results
	Characterisation of site (siting)	Initial boreholes, *etc.*	Licence for preparatory work	II*	A, G	
		Exploratory gallery				
		Test facility			Vote/approval of Main facility	Stepwise modification of design
	Design		General licence	II	A, G	
	Construction	Main facility	Construction licence	II	A, G	Systematic and continuous evaluation of the system
		Pilot facility				
	Operation		Operating licence	III	A, G	
		Stepwise cavern backfilling				
	Surveillance (post-operational)			III	AA, G	Independent monitoring programme (for quality assurance and credibility reasons)
	Stepwise closure		Closure licence	II	AA, G	
					Vote/approval of final sealing	
	Sealing (=final closure including backfilling)	Main facility/ Test facility	Decommissioning licence (on the grounds of positive Pilot facility results)	II	AA, G	Decision on sealing or break-off
Additional	(Passive) monitoring	Main facility	State ownership		AA	
	Active monitoring	Pilot facility (access shaft closure undefined)			A, G	Vote/approval of "final" assessment
	Memory				A	

Figure 13-9. Proposed schematic overview of disposition phases, data gathering and responsibilities of the implementer (I*) and the safety authorities (A). It is meant to indicate the degree of intensity of work to be done by the parties (e. g., A<AA). Also indicated is the involvement of independent oversight groups (G) and of the public at major decision points. The concept of Test facility (a rock laboratory), Main facility (the actual final repository) and Pilot facility (with a controlling program to confirm the long-term safety analyses) is adopted from EKRA [P60]. Source: Flüeler 2001a [G66:793].

If appropriately set up, votes may be viewed as inclusive reviewing; as a much more representative instrument they are not replaceable by "more modern" modes of participation but may be complemented [P174][P259].[156]

From this it is apparent that the process of disposition will last over decades, from site characterisation via construction, emplacement, surveillance, closure, monitoring, and sealing. Such a situation helps to ensure that the principles of the "chain of obligation" and of the "rolling present" are not violated: The needs of the living and secondary generations are met, the secondary and third generations will take crucial decisions with respect to project management (see also *Figure 12-2*). Such a process contributes to the interest of the issue. It increases awareness and ownership of the problem, and it obviously has an effect on knowledge transfer, updating archives, and securing financing. There is a clear need for thoughts and action in this field, for there have been elaborated only a few studies on the topic [G222] [G134][G256][G136][G19][G244][G115].

Having said this, nevertheless, this is a rejection of perpetual surveillance, as recent analyses of institutional monitoring of radioactive legacies in the USA demonstrate in frustrating openness: "It is now becoming clear that relatively few ... DOE waste sites will be cleaned up to the point where they can be released for unrestricted use. 'Long-term stewardship' (activities to protect human health and the environment from hazards that may remain at its sites after cessation of the remediation) will be required for over 100 of the 144 waste sites under DOE control .... The details of long-term stewardship planning are yet to be specified, the adequacy of funding is not assured, and there is no convincing evidence that institutional controls and other stewardship measures are reliable over the long term" [G268:2]. Strohl was of the opinion in 1995 already: "... institutional instruments, although indispensable with regard to long-term safety, should only be considered as making a contribution of relative importance and of limited duration, and this must be made clear" [G254:127].

The system "radioactive waste" of Switzerland, *i. e.,* of Swiss origin, has to undergo a thorough overall evaluation. Partial systems must not be assessed in isolation, *e. g.,* favourable geoscientific conditions, a local population predominantly supportive, or a bulging fund. The entire picture has to be appraised: above all, competent and independent waste institutions, a realisable programme with realistic milestones, assured know-how, a strong oversight, extensive reviewing, ample funds, and sustained political support on the national and the regional levels. In this sense, the option "abroad"

---

[156] In my view, decisions on siting should be taken by the concerned region due to the inequal cost-benefit distribution (see *Figure 12-2*). Votes on a national scale will probably handicap the (regional) risk aspect with respect to the (national) benefit of the majority.

obtains low marks, not because it is *a priori* excluded but because a whole host of factors to date remain insecure and intangible [G54][G112]. Parker *et al.* concluded, thus, back in 1984: "... a 'solution' is to be found only in a national, not a scientific, context" [G217:108]. This goes with the requirement by the International Waste Convention of 1997 to consider the issue a national task as the signatory states are "[c]onvinced that radioactive waste should, as far as is compatible with the safety of the management of such material, be disposed of in the State in which it was generated" [G110]. Such a concept of (national) overall responsibility shall comply with the approach of overall robustness. A stipulation of international responsibility is deemed to fail in view of the (over-) complexity of the issues involved (though a role of relatively stable international bodies, such as the IAEA or even the UN, is desirable). The gap between benefit-risk distribution may be minimised and the bridge of the rolling generations maximised on a national scale.

Conceptional decisions are to be taken at an early stage and widely secured in society. With respect to retrieval, an IAEA working group concluded in 1996 that it is "important that possible post-closure actions are considered in advance of closure and are part of an overall plan to implement disposal" [G109:14]. For conceptional issues, a "National Council for the Safe Management of Radioactive Waste", or something alike, is appointed (see *Figure 13-10*), consisting of representatives of all relevant stakeholder groups (*Figure 12-3*) and supplied with adequate resources. It supervises goal compliance, programmes, planning, *etc.*, organises a national discourse, as sketched above, reports periodically to the Federal Government and is advisory to them.

Nagra and related organisations, with perseverance, have to execute the programmes once decided on a broad societal base. If the co-operative is not able to comply with its mandate within the current institutional framework (owned by the waste producers), one might envisage the establishment of an independent agency with a broad-based supervisory committee[157] (see *Figure 13-10*). The implementer's arguments have to be safety-oriented throughout, it works by national order but independently from the authorities. It declines financial compensation for host communities because it does not pay "risk premiums" even if they only seem to be such. Worth considering is an environmental compensation as Vatter 1995 [P280] proposed (see Section 10.4).

---

[157] Apart from an infringement of the causality principle, I am sceptical towards a "federal solution", particularly in the Swiss case. The small waste community virtually does not allow a strict separation of implementing and supervisory functions. The large attempts to install the state as implementer (USA, Germany, also France) do not speak for such an option, whereas Sweden stands out positively, with its clear separation of functions and strong independent actors. This, however, presupposes strong authorities (see just below).

The participants in the discourse treat each other with respect and take other people's opinions seriously. Total instrumentalisation and marginalisation of views is to be rejected, as Director Kiener of FOE did in 1992: "Not to solve the waste disposal issue would be equal to an escape from responsibility. If we are not able to overcome disposal, not just nuclear by the way, then doubts are justified whether we are at all capable of mastering problems of the future" [P161:9-10].

According to the Swedish model, a periodic review of the entire disposal programme takes place, *e. g.,* also every three years. Indeed, the disposal projects "present great technical-scientific challenges to HSK" as the safety authority realised in 2002 [P290:[1]]. This holds not only for the authority and not only for "technical-scientific" aspects (see below).

Just to sum up, there is no ideal method or solution to lead the way to a safe and sustainable radioactive waste governance, let alone to ensure that. The combination, however, of a widely discussed concept, state of the art, extensive reviewing and, finally, political support by the relevant decision makers, appears to be the prerequisite for a widely sustained disposition programme.

5. SPECIAL ROLE OF THE REGULATORY BODIES

Looking at the vast objective and institutional dimensions of radioactive waste management, it is reasonable, and necessary, to examine the central and challenging roles of the regulatory bodies on the way to a sustainable long-term governance. Increased stakeholder involvement and the charged regulators' role as trustees of the public not only pose novel demands on traditionally technocentric institutions but may conflict with long-established relations to implementers and proponents and, potentially, with long-term-safety responsibilities.

Regulators, political and safety authorities, are obliged to deal with several regulatory fields:

- Legislation: international conventions; laws, decrees, ordinances; guidelines;
- Licensing;
- Inspection;
- Assessment (and reviews, also by advisory bodies);
- Enforcement.

Having a leading role in licensing, the regulatory authorities are also entrusted with disseminating information to all stakeholders, *i. e.,* affected or interested actors, in the respective procedure, like licensees, interveners, or the general public, especially as for the latter the authorities may be considered trustees.

To keep radioactive waste management in accordance with regulations, the authorities have to see to the compliance with requirements, goals, and boundary conditions. In the case of Switzerland these are the following:

- Protection of man and environment, "at any time", *i. e.*, of present and future generations (according to the Nuclear Energy Act [P264], the Radiological Protection Act [P265][P270], and the Guideline R-21 on protection goals [P133]);
- "Safe and permanent [final] disposal" (Federal Decree of 1978 [P39]);
- Causality (Polluter pays) principle (Nuclear Energy Act [P264]), so no "state solution" should be envisaged except for the case below;
- Domestic solution, *i. e.*, to enforce such a programme even if, in the worst case, the waste producers were not willing or able to do so (Radiological Protection Act, IAEA Waste Safety Convention [G110]);
- Oversight and, if needed, correction of the disposal programme (revised Nuclear Energy Act, in force from 2005);
- Validation of safety analyses (Guideline R-21);
- Long-term monitored geological disposal (so-called "EKRA concept" [P60], to reconcile passive safety with limited confirmatory *in situ* monitoring) (revised Nuclear Energy Act);
- Effective separation of functions, independent regulatory body (Nuclear Safety and Waste Safety Conventions [G103][G110]);
- Adequate and competent staff and resources (Nuclear Safety and Waste Safety Conventions);
- Stakeholder involvement (revised Nuclear Energy Act, Aarhus Convention [G262]).

Due to the driving force of technology (radioactive waste, with distinct characteristics, "exists" and "seeks" a "solution" whatever that might be), the entire socio-technical system is heavily predetermined. Being stakeholders themselves, the regulators have to establish a platform for inclusive knowledge generation where a common set thereof has to be negotiated, based on a defined set of criteria (see Chapter 10 and Section 13.2 for concrete examples). This necessity to integrate different requirements, the step-by-step approach, the chance of "institutional constancy", and the special "national" task of the issue call special attention to the role of the regulatory authorities [G196]. We also remember the insight of risk perception that institutions and, above all, regulation [R68:52] play a pivotal role in accepting or non-accepting risks.

The regulatory authorities have to be active, strong, accountable and independent in providing a technical and an institutional monitoring program, as well as, on the whole, to ensure an open and transparent, yet reliable process (see *Figure 13-9*):

- An *early and broad goal discussion* has to be established to increase the legitimacy of the key conceptual decisions [R42:159], regarding decision impacts to be borne by future generations (termination of control, uncovering reasons for retrievability, *etc.*) and for an early detection of "system and decision weaknesses" (no technical or political, safety-compromising, "fixes"); thereby society's risk of enormous costs of corrective actions may be reduced.
- In the *concept phase*, the regulator has to specify the safety requirements and the "rules of the game": Since gaining an overview, let alone a comprehensive understanding, of the complex reasoning of safety in radioactive waste governance is virtually impossible for most stakeholders, proxies are called in: in the technical field (*e. g.,* exclusion criteria, safety indicators) as well as in procedural issues (*e. g.,* separation of promotion and oversight within the Administration, extensive publication of documents, stepwise and phased procedure, external reviewing).
- *During the whole project* the public authorities have to ensure that waste producers, applicants and operators carry out their work, until validation of the safety analyses, according to the state of the art (as is regularly done in the case of SKB's RD&D Programme for Sweden [G145]).
- After definitive *closure ("sealing")* the overall responsibility has to be taken over by an institution with a chance of surviving longer historical periods (presumably a national state); consequently, the authorities as trustees of the public(s) have to guarantee long-term safety by independently monitoring the repository over a sufficient but limited period of time.
- A sustained *knowledge management*, including a robust quality management, is to be established so that the reasoning developed over decades to substantiate and refine the "safety case" is transparent and traceable and may be shared by the operating and observing generations to come (know-how transfer).
- Traditionally regulators did not deal with *implementers' budgets*. In times of open markets and concurrent pressures on implementers the Federal officials need to be informed about financial corner stones of

disposal and research. Risks are mitigated if the fund paid by the waste producers is state administered.
- *Anticipatory regulatory research* has to be intensified because many issues cannot be resolved today (*e. g.,* on information closure). Relevant gaps (including uncertainties), goals, and milestones of a comprehensive research programme do have to be identified.

In the case of radioactive waste management, the disposal system has to have a conservative, passively stable design with adequately built-in control and intervention mechanisms. The approach suggested is dynamic, adaptive, even experimental [G43] in its instruments, but not in its ultimate goal, *i. e.,* the passive protection of present and future generations and environments. In 1999 IAEA peer discussions, senior regulators "agreed that regulatory bodies need to accept the value of being learning organizations" [G116:11].

One side of the coin is the fact that radioactive waste governance is faced with the need for filling in the "democratic deficit" [MA1:15] as having showed up in a lack of public participation in, and public legitimacy of, democratic institutions. Decision makers, above all in technical government administrations, have to recognise the dominance of policy. Alec Baer of the IAEA observed in 2000: "Whenever a technological approach to a given issue conflicts with a sociopolitical approach, the latter will prevail. Human societies are not organized according to scientific principles and they are not necessarily rational in their decisions. Even though technology is essential to society, it is only a small cog in a highly complicated organism" [G12: 19-20]. Modesty is a wanted characteristic.

The other side of appropriate governance is the need, as mentioned, for "institutional constancy" [M39][G143:26-26]. This aspect again illustrates the uselessness of statements such as "What is more stable, society or rocks?" Safety-related long-term radioactive waste governance could not afford unstable institutions [D85] and vulnerable regimes as are described in studies on "failing states" [D7]. According to Goldstone *et al.* 2000 there are two major characteristics that affect a state's capacity to resist political crisis: the "organizational effectiveness" of that state and the legitimacy of its authority [D29:15]. No other institution than a state is likely to maintain some basic stability.

This constellation, in turn, presupposes adequate resources, extensive reviewing, appropriate anticipatory (regulatory) research in diverse technical, non-technical and institutional fields as well as a continuous international and intergenerational knowledge transfer. These challenges cover way more than "technical-scientific" ones. How such an institutional set-up might be organised is outlined in *Figure 13-10*, whereby Switzerland just serves as an illustration.

Integrated risk analysis: outline of an overall system robustness 273

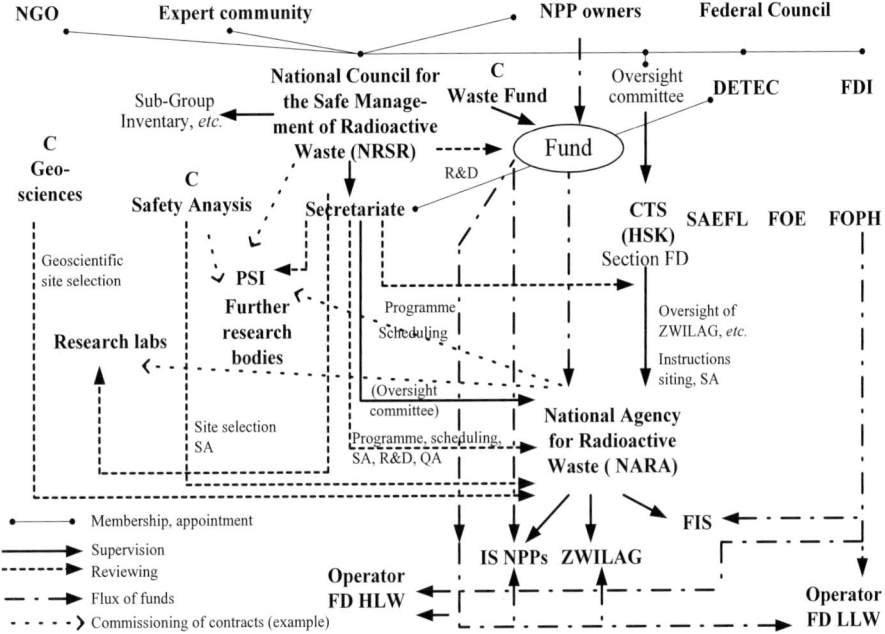

Figure 13-10. Proposal for re-structuring the institutional setting of radioactive waste governance in Switzerland. Decisive is the control of resources ("Fund") and process ("National Council" as its "guardian"). See text.

Abbreviations: C Committee, CTS Centre for Technical Safety (new, independent Agency, for current HSK), DETEC Department of the Environment, Transport, Energy, and Communications (Ministry of Environment), FD final disposal, FDI Federal Department of the Interior (Ministry of the Interior, responsible for collection MIR waste), Federal Council (=Swiss Federal Government), FIS Federal Interim Storage (for waste from medicine, industry, and research, MIR), FOE Federal Office of Energy (subject to DETEC), FOPH Federal Office of Public Health (subject to FDI), HLW high-level waste, HSK Swiss Federal Nuclear Safety Inspectorate (subject to DETEC, independent Agency, proposed CTS), ILW long-lived intermediate-level radioactive waste, IS interim storage, LLW low- and short-lived intermediate-level waste, MIR waste from medicine/industry/research, NARA National Agency for Radioactive Waste (new, previously Nagra), NGOs non-governmental organisations, NPPs nuclear power plants, NRSR National Council for the Safe Management of Radioactive Waste (new oversight and "guardian" of the process), PSI Paul Scherrer Institute (nuclear research centre), QA quality assurance, R&D Research & Development, SAEFL Swiss Agency for the Environment, Forests, and Landscape (subject to DETEC), SA safety analysis, ZWILAG Central Interim Storage (waste from NPPs). Source: Flüeler 2003a [G73:139].

In short, the regulators, assisted by conceptual (advisory and oversight) bodies, ensuring continuity have to take the lead over the *entire* programme (respecting the causality principle), including goal and programme control, concept and project management issues (*e. g.,* in case of delays), and key economic aspects. Consequently, the upper right of the "learning curve" (*Figure 11-2*) leads to the notion of stewardship and the decision model of "Propose–Learn–Share–Decide (PLSD)": Leadership has to be exerted in conjunction of an inclusive body of knowledge, broad societal support and strong regulatory bodies, *i. e.,* safety authorities and pluralistic oversight committees. In RISCOM terms, the proposed "National Council for the Safe Management of Radioactive Waste" would be the "guardian of process integrity" [MA1:11,106-108]. In my view, it should not be the government or the parliament – as RISCOM proposes [*ibid.*:56] – but a pluralistically composed body, independent of the "nuclear community" yet knowledgeable about the issue and not driven by daily politics.

6. REQUIREMENTS IN VIEW OF AN INCLUSIVE TECHNICAL-SCIENTIFIC AND SOCIETAL DISCOURSE

The two-fold approach (with criteria of risk perception "from below" and of decision science "from above"), and some generic insights, might be fruitful for the governance of *other* complex socio-technical systems. On general behalf of the dialogue in such a complex context, the elaborations in Chapter 13 can be transposed to basic propositions that have been addressed in radioactive waste governance over and over again (see *Figure 11-1*):

- Systemic risks call for systemic, and systematic, approaches: The systems approach has the potential to overcome misleading routes along the binary thinking of experts *vs.* laypersons, technical *vs.* political factors, *etc.* The legendary question "how safe is safe enough?" [R23] can *only* be answered politically but *only* on a sound technical basis. "Robustness" in this sense means to reconcile these two approaches into one, and thus into a more inclusive one. Only an over-arching body of knowledge, comprising both technical and non-technical perspectives, leads to satisfactory governance of complex socio-technical issues. And: A long-term complex socio-technical issue, in the end, will probably never be "closed" [M70].
- Transparency of decision making: Concept and programme formulation have to be examined in detail.

- Clear but open management structures are necessary to assure consistency and constancy of programming with regard to goal achievement; splitting up into many (partially) responsible bodies is not advisable.
- A stepwise procedure with interim decision points and public involvement under the motto "detours increase the local knowledge" secures the interim goals achieved and the process: "Speediness" is no quality feature by itself, narrow time limits may, in the end, lead to considerable delay.
- Formulation and discussion of alternative options: Time and again, one has to anticipate unexpected events, of technical or political nature, in complex systems. Accordingly, when setting relevant courses, back up scenarios have to be elaborated so that deadlocks and failures do not amount to a general stalemate of the programme (*cf.* the Wellenberg fiasco experienced twice, in 1995 and 2002).
- Traceable and plausible evaluation criteria are to be laid down and be open.
- Right of inspection of all data and publication, respectively: In the radioactive waste field it has not been state of the art to lead the discourse in peer-reviewed journals but in so-called "grey" literature; periodic publication in open sources is advisable.
- Reviewing process, "independent" second assessment: Dissent should be regarded as a chance of improvement not as an impediment right from the outset.
- Scientific attendance: For quality assurance reasons and for the sake of an open debate, it is essential and confidence-building to periodically and officially review the programme with appropriate resourcing.
- Inclusive participation of the public and concerned parties: Such a procedure helps, considering relevant aspects up front, and securing decisions broadly in society, for "[e]ach Swiss region will only give assistance on solving difficult issues if it is treated correctly" as the Cantonal Council of Uri stated in 1990 when called upon for hosting test drillings [P183:4].
- Utilisation of adequate dispute resolving techniques: Since "process is of paramount importance [i]n understanding what makes participation successful" [M5:74], it is vital to professionally cope with non-technical aspects and to continually consult respective experts.

7. CONCLUSION

The content analysis over a lapse of 40 years substantiates that the radioactive waste proponent's, partly also the safety authorities' course of action, was not transparent nor traceable for many years, even so for experts. The reasons for the initial over-optimism are manifold but three main aspects are paramount:

- Well-known structural engineering was prioritised, whereas the decisive issue of <u>long-term safety was underestimated.</u>
- Political considerations predominated the fact-based ones, whereby <u>the waste issue was instrumentalised</u> in energy politics by all parties.
- <u>Procedural issues</u> of such a complex disposal programme were <u>not duly followed,</u> and if they were, they were dealt with as "political" aspects under point 2.

The response to the hypotheses is as follows:

<u>Hypothesis 1</u> (page 13):

The asserted <u>dichotomy of "political" *vs.* "technical" perspectives</u> is an <u>unjustified simplification of the issue.</u>

The aspect of control, as a proxy in radioactive waste management, is an example of how in complex socio-technical fields, especially with respect to technological constraints, dimensions are often debated in reverse order: firstly, the technical and commercial aspects, followed by the political and economic, the social and, finally, the ethical aspects. Nevertheless, it ideally should be the other way around: First, one should have a broad debate and decision on political principles over ethical guidelines, this should in turn lead to the selection of the corresponding optimum technical variant, in consideration of ecology, economy and society.

An inclusive system understanding is crucial, particularly with waste. Waste forces us to consider and acknowledge the entire material flux. Complex issues may not be put down to a single cause: Neither a "technical fix" nor a "political fix", out of topicalities, are sustainable solutions. The programme is comprehensive, protracted, and prolonged. Wherever possible, it should be determinable, clearly structured, and manageable. An "international" solution or the expectation of technological leaps would increase imponderabilities, apart from the infringement of the causality principle or the violation of the International Waste Convention.

The (technical) decision basis is in close interconnection with the (political) policy and consequential decisions. In the "grey range" of this interrelationship there are aspects such as the definition of criteria, transparency, duty of publication, traceability of arguments, plausibility, reviewing process, and step-by-step procedure. Resilient political decisions in the socio-technical system called "long-term disposal of radioactive waste" are, consequently, based on solid technical grounds, which in turn can only be erected if corresponding pre-decisions leave them a sufficient framework (including resources). Conceptually, it is recognised that the waste producers, the applicants, and the authorities have integrated many demands over the years, even though under pressure and only with reservations. They exhibit adaptive learning so far, but during the implementation phase it will have to be proven how seriously the concept modifications are taken.

At any rate, as long as the "technical" and the "political" approaches do not converge, there is no chance for "closure" of the issue.

Hypothesis 2 (page 18):

Long-term safety in radioactive waste management intrinsically cannot be mathematically demonstrated. A diversified approach is needed which may be based on the concept of "defence in depth" with successive barriers known to the engineering community. A proposal towards robustness of the overall "radioactive waste system" shall serve as an answer to Hypothesis 2. It is an expansion of the multi-barrier principle by institutional and societal aspects in a stepwise decision-making process. It is a proposal to show components of a bridge between the hitherto irreconcilable camps of "political" vs. "technical" adherents. It is, hence, an attempt towards "closure" of the issue, knowing that such an endeavour is virtually impossible in view of its given characteristics [D92].

The "inner" circle of decision makers has been gradually expanded, strengthened, and professionalised; increasingly relevant arguments are being used in a well-informed discourse. Herewith the chances rise that an "extended final disposal" of radioactive waste may be achieved whose primary goal is passive long-term safety while allowing for control mechanisms to validate the performance assessments and to enhance confidence and securing a broad political backup (the specified notion of sustainability favours passive safety without abandoning technical and institutional control). In the long run, the generations benefiting from the utilisation of nuclear energy have to recognise ownership and responsibility of the concomitant problem, the waste.

The empirical content analysis and the literature study suggest that controllability, retrievability and procedural issues like transparency, traceability of arguments, and stakeholder involvement are key elements in the safety assessment of the system "radioactive waste". Because gaining an overview, let alone a comprehensive understanding, of the complex reasoning of safety in radioactive waste governance is virtually impossible for most stakeholders, proxies are called in: in the technical field (*e. g.,* exclusion criteria, safety indicators) as well as in procedural issues (*e. g.,* separation of promotion and oversight within the Administration, extensive publication of documents, stepwise and phased procedure, external reviewing).

A corresponding decision is well supported if it integrates relevant parts of both the problem and solution ranges of the main stakeholders (see *Table 9-2*). Therefore, the principle issues have to be thoroughly and broadly discussed. As the NEA stated in 1999: " ... an informed societal judgement is necessary" [G181:23]. A recourse to the semi-direct democratic structures as in Switzerland does not suffice; public and stakeholder involvement has to be actively, continuously and credibly sought for.

To reach this stage, an extended "control" by third parties has to be implemented – through strengthening the safety authorities, intensifying the review process and involving stakeholders hitherto excluded or not judged equivalent. As a counter to it, all partners have to proactively discuss eventual contradictions and inconsistencies and to duly consider time dimensions with respect to the construction of disposal facilities and system impacts.

Two crucial boundary conditions have changed since the Federal Decree to the Atomic Energy Act of 1978, according to which "the permanent and safe final disposition and disposal of the ... radioactive wastes" have to be "guaranteed" [P39]: The general research and concept phase is converting to an implementation phase and the procedure takes place in a deliberately process-based manner and with the involvement of the major stakeholders. The strategy of linear decision making (by the authorities and proponents) (*Figure 8-1*), eventually expanded to risk communication (*Figure 8-2*), is tentatively superseded by a dynamic and integrative, pluralistic, decision model (*Figure 13-5, Figure 11-2*). The Swedish advisory committee KASAM wrote in 1998 that radioactive waste governance was an issue "how the current generation is generally assuming responsibility for the long-term consequences of human impact on the environment" [G143:3].

In short, the following **insights** arise:

> **A.** The primary goal in radioactive waste governance is the stability of the disposition system as it performs: The protection against harmful emission of radioactivity is to be continuously ensured.
> **B.** The disposition of radioactive waste also is a long-term project: Long-term, comprehensive yet stepwise planning is needed.
> **C.** In compliance with the International Waste Convention, the issue has to be regarded as a national issue, which has to be carried out on the territory

of Switzerland, within the national institutional framework, and on the basis of current knowledge, albeit under international oversight.

D. Conceptual decisions have to be taken well in advance and widely supported in Swiss society.

E . A clear and unambiguous understanding of roles by the stakeholders is necessary.

From this, **propositions for specification and implementation** result:

ad **A.** The complementary goal is flexibility, defined as the potential to intervene. Controllability and retrievability have to be situated accordingly.

ad **B.** A representative monitoring and surveillance programme of the respective facility systems has to be specified. It comprises full publication of the work, an intensive reviewing, quality assurance, and an extensive participation of concerned and affected parties.

ad **B.** A broad-based periodical review process of the disposition programme has to be set up. It explicitly includes institutional and procedural aspects as well as research, development and demonstration (RD&D). The productive but pilot-like extended final disposal concept of "monitored long-term geological disposal" has to be specified, verified, and validated. It must undergo an extensive international review.

ad **C/D.** A National Council for the Safe Management of Radioactive Waste, as a "guardian" of the complex and long-term process, is to be established, with the involvement of all relevant stakeholder groups. It obtains the mandate to adequately accompany the programme outlined.

ad **D.** An extensive participation of the potential host regions is to be comprehended as an extended programme and project reviewing. Decision making and decisions, also political ones, may be interpreted as reviews in order to back up "good" projects.

ad **E.** According to the causality principle, the waste producers take care of the safe and long-term radioactive waste management, including sufficient funding. Accordingly, the Confederation has to clearly be positioned as the collector and producer of waste from medicine, industry, and research.

ad **E.** A rigorous leadership by the Confederation must be assured in the previously broadly decided programme. The Confederation has the oversight and control of programme and financing.

ad **E.** A persistent implementation of the programme and projects must be guaranteed by independent and well-resourced waste institutions.

Acknowledgements

So many people and institutions were involved in the thoughts developed here that I cannot single out any without drawing up a very long list indeed – from Córdoba to Östhammar and from Berkeley to Bratislava. Particular thanks are, however, extended to all discussion partners from the Swiss Federal Nuclear Safety Commission (KSA), the Cantonal Expert Group Wellenberg (KFW), within the EU project Community Waste Management (COWAM), the Swiss National Science Foundation project "risk-based regulation", the IAEA Senior Consultants meetings, and from the social science programme as laudably set up by the Belgian Nuclear Research Centre SCK•CEN. This study was supported by the Swiss Federal Nuclear Safety Inspectorate (HSK) and the Cogito Foundation. The empirical analysis was sponsored by the Swiss Federal Office of Energy (BFE), and the Gerold-und-Niklaus-Schnitter-Fonds für Technikgeschichte of ETH Zurich. I wish to thank Dr. Philip Tipping for reviewing and editing the text, Edwin Beschler for his considerate language-editing, and my counterpart at Springer's, Esther Verdries, for her enduring patience and support.

References

1. PRIMARY LITERATURE

[P1] Abstimmungskomitee STOP WELLENBERG ([1995]): Der Wellenberg: Ein Stück Heimat. Stans.
[P2] Aebersold, T. (1997): Kontrolliert – unkontrolliert – unkontrollierbar. Haftungsfragen bei der Endlagerung radioaktiver Abfälle. NZZ. 7.3.97:16.
[P3] Aebersold, T. (o.J.): Rechtsfragen zur unterirdischen Endlagerung radioaktiver Abfälle. Lizentiatsarbeit. Universität Bern.
[P4] AGNEB (1979): [1.] Tätigkeitsbericht über die Zeitdauer vom 15.2.78 bis zum 28.2.79. Arbeitsgruppe des Bundes für die nukleare Entsorgung AGNEB. Bundesamt für Energiewirtschaft BEW, Bern.
[P5] AGNEB (1980a): Bericht über die Aussprache mit Geologen am 30.8.1979. Februar 1980. BEW, Bern.
[P6] AGNEB (1980b): «Beseitigung der Abfälle durch den Bund». Grundlagen, heutige Tätigkeiten und mögliche neue Aufgaben des Bundes im Hinblick auf die nukleare Entsorgung, Vor- und Nachteile einer Bundeslösung. 1.12.1980. BEW, Bern.
[P7] AGNEB (1980c): 2. Tätigkeitsbericht. Berichtsperiode: 1.3.79–31.12.79. BEW, Bern.
[P8] AGNEB (1981): 3. Tätigkeitsbericht. Berichtsperiode: 1.1.80–31.12.80. BEW, Bern.
[P9] AGNEB (1982): 4. Tätigkeitsbericht. BEW, Bern.
[P10] AGNEB (1983): Netzplan «Gesamtprogramm für die nukleare Entsorgung». Stand: 31. Dezember 1982. Beilage I zum 5. Tätigkeitsbericht. BEW, Bern.
[P11] AGNEB (1990): 12. Tätigkeitsbericht. BEW, Bern.
[P12] AGNEB (1992): 14. Tätigkeitsbericht. BEW, Bern.
[P13] AGNEB (1995): 17. Tätigkeitsbericht. BEW, Bern.
[P14] AGNEB (1998): 20. Tätigkeitsbericht. Bundesamt für Energie, BFE, Bern.
[P15] AGNEB (1999): 21. Tätigkeitsbericht. BFE, Bern.
[P16] AGNEB (2000): 22. Tätigkeitsbericht. BFE, Bern.
[P17] AGNEB (2001): 23. Tätigkeitsbericht. BFE, Bern.
[P18] AGNEB (2002a): 24. Tätigkeitsbericht. BFE, Bern.
[P19] AGNEB (2002b):Arbeitsgruppe des Bundes für nukleare Entsorgung übernimmt Führungsrolle. Press release of 2002-6-28.
[P20] AGNEB (2003): 25. Tätigkeitsbericht. BFE, Bern.
[P21] Arbeitsgruppe Kristallin Nordschweiz (1996): Schlussbericht der Arbeitsgruppe. HSK, Villigen-HSK.
[P22] Arbeitsgruppe kritisches Wolfenschiessen AkW (1996): Chronik Wellenberg. 1934 bis Juni 1996. AkW, Wolfenschiessen.

[P23] ASK (1980): Sicherheit der Endlagerung radioaktiver Abfälle. Schwach- und mittelaktive Abfälle in der mittelländischen Molasse. Modellstudie. Basler&Hofmann, csd Colombi Schmutz Dorthe. Bericht ASK-E7. EAEW/ASK. Aarau.

[P24] Bauer (1977): Interpellation 77.379 vom 14.6.1977 – Radioaktive Abfälle. Amtliches Bulletin der Bundesversammlung. Nationalrat. Herbstsession:1365f., Wintersession:161-168.

[P25] Berner Zeitung (1984): Article on Nagra trip to Sweden, 1984-8-27.

[P26] BEW/Rechtsdienst (1985): Zur Philosophie in der Beseitigung der radioaktiven Abfälle. Diskussionspapier. 17. Juli. BEW, Bern.

[P27] BFE (1998, Hg.): Energie-Dialog Entsorgung. Schlussbericht des Vorsitzenden zu Handen des Eidg. Departements für Umwelt, Verkehr, Energie und Kommunikation. Bundesamt für Energie BFE, Bern.

[P28] BFE (1999): Verordnung über den Entsorgungsfonds für Kernkraftwerke. Vernehmlassungsentwurf und Erläuternder Bericht. BFE, Bern.

[P29] BFE (2000): Bericht über das Ergebnis der Vernehmlassung zum Vorentwurf Kernenergiegesetz. 3.10.2000. Bundesamt für Energie, BFE, Bern.

[P30] BFE (2003): Kernenergiegesetz: Referendumsfrist läuft ungenutzt ab [Period for referendum expires missed]. Press release of 2003-9-2.

[P31] BFE (2004): Entsorgungsnachweis für hochaktive Abfälle: Frühzeitige Information [Demonstration of feasibility of disposal: early information] Press release of 2004-1-13.

[P32] BEW (1993): Teilrevision des Atomgesetzes und des Bundesbeschlusses zum Atomgesetz. Auswertung der Vernehmlassung zum Vorentwurf vom Februar 1993. Sept. 1994. BEW, Bern.

[P33] BKW Energie AG (1995): Kontrollierbarkeit und Rückholbarkeit atomarer Abfälle. Brief vom 10.7.1995 an Grossrat Breitschmid. BKW, Bern.

[P34] Boos, S. (1999): Anti-AKW-Bewegung: Zwist wegen Atommüllfrage. Das strahlende Dilemma. Wochen-Zeitung. Nr. 27. 8.7.1999:1,3.

[P35] Breitschmid, F. u. a. (1987): Atommüll in der Schweiz. Keine Gewähr für die sichere Endlagerung. Energie+Umwelt 3/87:8-21.

[P36] Breitschmid, F. (2000): Wissenschaftliche, technische und gesellschaftliche Dilemmas bei der Lagerung radioaktiver Abfälle. 26.8.2000. Manuskript. (reprinted in natur+mensch 6/2000:14-17).

[P37] Brennstoffkommission der Überlandwerke (1994): Die Wiederaufarbeitung von bestrahltem Kernbrennstoff. Juni 1994.

[P38] Buclin, J.-P. (1972): Les déchets radioactifs. In: Association Suisse pour l'Energie Atomique ASPEA (ed.): Centrales nucléaires – source d'énergie propre. Novembre. ASPEA, Berne:33-35.

[P39] Bundesbeschluss [Federal Decree] zum Atomgesetz (BB AtG) vom 6.10.1978, amended in 1983 (prolongation by 10 years), amendment as of 1990-6-22 (prolongation until 2000-12-31), amendment as of 2000-10-4 (in force since 2001-1-1, prolongation until end of 2010).

[P40] Bundesverfassung [Swiss Federal Constitution] der Schweizerischen Eidgenossenschaft (1999): Bewährtes erhalten. Zukunft gestalten. Schweiz stärken. Von Volk und Ständen angenommen am 18.4.1999. Bundesbeschluss vom 18.12.1998. Bundesblatt 1997 I 1.

[P41] Bürgisser, H. et al. (1979): Geologische Aspekte der Endlagerung radioaktiver Abfälle in der Schweiz. SES-Report Nr. 6. Schweizerische Energie-Stiftung SES, Zürich.

References

[P42] Buser, M. (1979): Lagerung radioaktiver Abfälle – Aktenzeichen XY ungelöst. Tages-Anzeiger, 29.8.1979:47.

[P43] Buser, M. (1988): Mythos "Gewähr". Geschichte der Endlagerung radioaktiver Abfälle in der Schweiz. Schweizerische Energie-Stiftung, SES, Zürich.

[P44] Buser, M. (1994): Fallstudie Siblingen: Eine Analyse der Ereignisse. BEW, Bern.

[P45] Buser, M. (1997): Guardianship versus disposal: a modern-day conflict with implications for the future. nagra bulletin. No. 30. August 1997:24-32.

[P46] Buser, M. (1998): «Hüte»-Konzept versus Endlagerung radioaktiver Abfälle: Argumente, Diskurse und Ausblick. Expertenbericht. Hauptabteilung für die Sicherheit der Kernanlagen HSK.

[P47] Buser, M. & W. Wildi (1981): Wege aus der Entsorgungsfalle. SES-Report Nr. 12. SES, Zürich.

[P48] Buser, M. & W. Wildi (1984): Das «Gewähr»-Fiasko. Materialien zum gescheiterten Projekt «Gewähr» der Nagra. SES, Zürich.

[P49] Colombi Schmutz Dorthe (1984): Beurteilung der Sondiergesuche für Endlager von schwach-mittelaktiven Abfällen [sic!]. Hauptabteilung für die Sicherheit der Kernanlagen, HSK. 6.4.1984. csd, Aarau.

[P50] CORE, Eidg. Energieforschungskommission (1999): Konzept der Energieforschung des Bundes 2000-2003. 31.5.1999. Bundesamt für Energie, Bern. www.energieschweiz.ch/internet/forschung/index.html?lang=en

[P51] CORE, Eidg. Energieforschungskommission (2004): Konzept der Energieforschung des Bundes 2004-2007. 1.1.2004. Bundesamt für Energie, Bern. www.energieschweiz.ch/internet/forschung/index.html?lang=en

[P52] COWAM Network (2003): Nuclear waste management from a local perspective. Reflections for a better governance. Final report. Community Waste Management (COWAM). Nov. 2003. Mutadis, Paris. www.cowam.com/zBoard.asp.

[P53] Crevoisier (1981): Einfache Anfrage 81.713 vom 23.9.1981 – Behandlung von Nuklearabfällen. Amtliches Bulletin der Bundesversammlung. Nationalrat. Wintersession:1785.

[P54] Damveld, H. & R. J. van den Berg (1998): Atomabfall und Atomethik. Gesellschaftliche und ethische Aspekte rücknehmbarer Lagerung von Atomabfall. Ein Untersuchungsbericht im Überblick. Dezember 1998. (translation of the report "Waste and nuclear ethics")

[P55] Damveld, H. & R.J. van den Berg (1999): Social and ethical aspects of retrievable disposal. METRA. Abridged version. 23.6.1999.

[P56] Dichter-Institut (1992): Schweizer Stimmbürgerinnen und Stimmbürger erwarten von Politikern und Parlamentariern tatkräftiges Vorwärtsmachen bei der Entsorgung radioaktiver Abfälle. Dichter-Institut, Zürich.

[P57] Eckhardt, A. et al. (1995): Risiko – Dialog zwischen Theorie und Praxis. Leitfaden zur Information und Kommunikation über Risiken des Kantons Zürich. Koordinationsstelle für Störfallvorsorge, Zürich.

[P58] Econcept (1998): Auswirkungen der Stromliberalisierung. Forschungsprogramm Energiewirtschaftliche Grundlagen. BEW/BFE, Bern.

[P59] EGES, Expertengruppe Energieszenarien (1988): Energieszenarien. Möglichkeiten, Voraussetzungen und Konsequenzen eines Ausstiegs aus der Kernenergie. EVED. EDMZ, Bern.

[P60] EKRA, Expert Group on Disposal Concepts for Radioactive Waste (2000): Disposal concepts for radioactive waste. Final report. On behalf of the Swiss Federal Department for the Environment, Transport, Energy and Communication. Federal Office of

Energy, Bern.www.energie-chweiz.ch/imperia/md/content/informationenlinks/broschren/2.pdf.

[P61] EKRA (2002): Beitrag zur Entsorgungsstrategie für die radioaktiven Abfälle in der Schweiz. Im Auftrag des Departements für Umwelt, Verkehr, Energie und Kommunikation (UVEK). Bundesamt für Energie, Bern.

[P62] Enderlin Cavigelli, R. (1993): Auswertungsbericht über die Arbeit der ersten Arbeitsgruppe der KORA zum Thema Wiederaufarbeitung. Manuskript.

[P63] Enderlin Cavigelli, R. (1995): Meditationsverfahren in der schweizerischen Kernenergiepolitik. Auswertungsbericht zur Arbeit der ersten Arbeitsgruppe der Konfliktlösungsgruppe radioaktive Abfälle (KORA) zum Thema Wiederaufarbeitung. Schriften zum Mediationsverfahren im Umweltschutz Nr. 12. Wissenschaftszentrum Berlin für Sozialforschung (WZB), Berlin.

[P64] Enderlin Cavigelli, R. & P. Schmid (1998): PUBLIFORUM "Strom und Gesellschaft". Evaluationsbericht der Stiftung Risiko-Dialog. Document de travail/Arbeitsdokument TA/DT 21/1998. Schweizerischer Wissenschaftsrat, Programm TA, Bern.

[P65] Energieforum (1994): Energie-Nachrichten. Nr. 6.

[P66] EVED (later on UVEK, DETEC) (1978): Die Entsorgung der schweizerischen Kernkraftwerke. EVED, Bern.

[P67] EVED (1985): Vorentwurf und Erläuternder Bericht zu einem Kernenergiegesetz vom 1. Oktober 1985. EVED, Bern.

[P68] EVED (1993): Vorentwurf einer Änderung des Atomgesetzes und des Bundesbeschlusses zum Atomgesetz. Eröffnung des Vernehmlassungsverfahrens. Beilagen Vernehmlassungsentwurf, Erläuternder Bericht. 17.2.1993. EVED, Bern.

[P69] EVED (1997): Endlager Wellenberg: Einsetzen einer technischen Arbeitsgruppe. Pressemitteilung vom 6.3.97. EVED, Bern.

[P70] Facts (1995): AKW/Endlager für Atomträume. Uranhandel/Das Ende eines strahlenden Geschäfts. Nr. 26. 29.6.1995:18-22, 23-25.

[P71] Favez, J.-C. & L. Mysyrowicz (1987): Le nucléaire en Suisse. Jalons pour une histoire difficile. L'Age d'Homme, Lausanne.

[P72] Federal Council, Schweizerischer Bundesrat (1968): Bewilligung eines Objektkredites für die Erstellung eines Lagerhauses zur Einlagerung schwach radioaktiver Abfälle in Lossy/Passafou. Botschaft 9872 vom 21.2.1968 und Entwurf Bundesbeschluss. Bundesblatt 1968 f:441-449.

[P73] Federal Council (1974): Elektrizitätsversorgung. Botschaft und Beschlussentwurf vom 11.9.1974. Amtliches Bulletin der Bundesversammlung. Nationalrat. Wintersession:1658,1667. (resultierender Beschluss vom 13.12.1974)

[P74] Federal Council (1978): Volksabstimmung vom 18. Februar 1979. Erläuterungen zur Atominitiative:9-11,15f.

[P75] Federal Council (1985): Entscheid über das Gesuch der Nagra um die Erteilung der Bewilligung von vorbereitenden Handlungen im Hinblick auf die Errichtung eines Endlagers für schwach- und mittelradioaktive Abfälle in Bauen (UR) [in analogy: Mesocco und Rosso/Ollon]. Entscheid vom 30.9.1985. Bern.

[P76] Federal Council (1988): Beschluss Nukleare Entsorgung: Projekt Gewähr, Materielle Beurteilung. 3.6.1988. Bern.

[P77] Federal Council (1989): Volksinitiativen «Stopp dem Atomkraftwerkbau (Moratorium)» und «für den Ausstieg aus der Atomenergie». Botschaft 89.032 vom 12.4.1989. Bundesblatt II:1.

[P78] Federal Council (2001): Kernenergiegesetz. Entwurf und Botschaft 01.022. Bundesblatt 2001:2754-2805,2829-2867;2717f.,2740-2753.

[P79] Federal Council/House of Representatives, Nationalrat (1980/1981): Verlängerung des Bundesbeschlusses über die Elektrizitätsversorgung. Botschaft 80.086 und Beschlussentwurf vom 26.11.1980. Bundesblatt 1981:223-231. Amtliches Bulletin der Bundesversammlung. Nationalrat. Frühjahrssession:361-364.

[P80] Federal Council/Parliament (1994) (1996): Botschaft 94.008 über die Teilrevision des Atomgesetzes und des Bundesbeschlusses zum Atomgesetz. Änderungen. Bundesblatt:1361-1393. Amtliches Bulletin der Bundesversammlung. Ständerat, Herbstsession 1994:956-960; Nationalrat. Frühjahrssession 1995:274f. Frühjahrssession 1996:58f.

[P81] Federal Department of the Interior (EDI) (1993): Bericht über die Ergebnisse des Vernehmlassungsverfahrens zur Strahlenschutzverordnung. Oktober. EDI, Bern.

[P82] Federal House of Representatives/Nationalrat (1959): 7587. Atomenergie und Strahlenschutz. Bundesgesetz. Amtliches Bulletin der Bundesversammlung. Nationalrat. Herbstsession:598-641.

[P83] Federal House of Representatives /Nationalrat (1990): Atomenergie. Volksinitiativen. 89.032. Amtliches Bulletin der Bundesversammlung. Nationalrat. Frühjahrssession:92-122.

[P84] Federal Parliament (2003): Nuclear Energy Act. Final decision. 2003-3-21.

[P85] Flüeler, T. (1991): Atommüll – von der Sorge um die Nachsorge des Energiefriedens. Energie+Umwelt 2/91:4-5.

[P86] Flüeler, T. (1993): Wege aus dem Atommüll-Labyrinth? Energie+Umwelt 4/93:4-5.

[P87] Flüeler, T. (1994): Radioaktive Abfälle: Forderungen an die Langzeitlagerung. Energie+Umwelt 1/94:8-11.

[P88] Flüeler, T. (1998): «Hüten oder endlagern?» – Ist das die Frage? Referat an der Veranstaltung zur «Problematik der radioaktiven Abfälle: Hüten? Endlagern? Kontrollieren?», 2.3.98. Parlamentarische Gruppe für Bildung, Wissenschaft, Forschung und Technologie, Bern.

[P89] Flüeler, T. (2000a): Mutual learning [in] radioactive waste management: Transcontinental gridlock – transdisciplinary solution? In: R. W. Scholz et al. (eds.): Transdisciplinarity: Joint problem-solving among science, technology and society. Proc. of the International Transdisciplinarity 2000 Conference, Zurich, Feb 27–Mar 1, 2000. Workbook II: Mutual learning sessions. Vol. 2. Haffmanns Sachbuch, Zürich:304-307.

[P90] Flüeler, T. (2002b): KFW Cantonal Expert Group Wellenberg. An advisory body to support regional decision makers in LLW siting. EU Concerted Action Project Community Waste Management COWAM. 3rd Seminar. Fürigen, Switzerland. Sep 12–15. www.cowam.com/Wellenberg.htm.

[P91] Flüeler, T., C. Küppers & M. Sailer (1997): Die Wiederaufarbeitung von abgebrannten Brennelementen aus schweizerischen Atomkraftwerken. «Recycling» von atomarem Material aus der Schweiz im Ausland. Analysen der Konsequenzen für Umwelt und Energiepolitik. CAN Anti Atom Koalition, Zürich.

[P92] Forum für verantwortbare Anwendung der Wissenschaft (1976): Radioaktive Abfälle. Hokus Pokus Verschwindibus. Forum für verantwortbare Anwendung der Wissenschaft, Basel.

[P93] Forum vera (1994): Statement of Charles McCombie, Nagra. Bulletin. No.3.

[P94] Forum vera (1997): Bulletin. news release. No.4.

[P95] Forum vera (1998a): Bulletin. No.3.

[P96] Forum vera (1998b): Bulletin. No.4.

[P97] Forum vera (no yr.): Die Entsorgung radioaktiver Abfälle: Eine Umweltaufgabe, die uns alle angeht. Forum vera, Zürich.

[P98] GAB/AMüs (1993): Atommüll: AbSOLUTION ins Reduit. Geiseln der Atomwirtschaft usw. Mühleberg stillegen! Nr. 30. Nov. 1993:6-13.

[P99] Gerwig (1982): Einfache Anfrage 82.724 vom 6.10.1982 – Argentinien. Export einer Schwerwasseranlage. Amtliches Bulletin der Bundesversammlung. Nationalrat. Herbstsession:1846f.

[P100] Geschäftsprüfungskommission des Nationalrats/Schweizerischer Bundesrat (1980): Sicherheit der Kernkraftwerke. Bericht 80.072 der Geschäftsprüfungskommission an den Nationalrat vom 14.11.1980 und Stellungnahme des Bundesrates vom 7.1.1981. Bundesblatt 1981 II 467-490.

[P101] GfS-Forschungsinstitut (1991): Analyse der eidgenössischen Abstimmung vom 23. September 1990 ("Volksinitiative für den Ausstieg aus der Atomenergie", Volksinitiative "Stop dem Atomkraftwerkbau (Moratorium)", Energieartikel). VOX-Publikation Nr. 40. gfs, Zürich.

[P102] GfS-Forschungsinstitut (1997): UNIVOX 6-97. gfs, Zürich.

[P103] gfs, Gesellschaft für praktische Sozialforschung (1979): Analyse der eidgenössischen Abstimmung vom 20. Mai 1979 (Revision des Atomgesetzes). VOX-Publikation Nr. 10. gfs, Zürich.

[P104] gfs, Gesellschaft für praktische Sozialforschung (1984): Analyse der eidgenössischen Abstimmung vom 23. September 1984 (Atom-Initiative, Energie-Initiative). VOX-Publikation Nr. 23. gfs, Zürich.

[P105] gfs, Gesellschaft für praktische Sozialforschung (1997/2002): UNIVOX-Staat 1997, UNIVOX-Kunden-Zufriedenheitsbarometer, Mehrjahresvergleich 1996-2001. gfs, Zürich/Bern.

[P106] GNW (1994): Technischer Bericht zum Gesuch um die Rahmenbewilligung für ein Endlager schwach- und mittelaktiver Abfälle am Wellenberg, Gemeinde Wolfenschiessen, NW. GNW TB 94-01.

[P107] GNW (1995): Das Endlager im Wellenberg: Antworten auf Ihre Fragen. GNW c/o NAGRA, Wettingen.

[P108] GNW (2000a): Zeit reif für einen Sondierstollen. Medienmitteilung vom 7.2.2000 zum EKRA-Bericht.

[P109] GNW (2000b): Lager für schwach- und mittelaktive Abfälle am Wellenberg, Gemeinde Wolfenschiessen NW. Bericht zu Handen der Kantonalen Fachgruppe Wellenberg. Nov. GNW TB 00-01.

[P110] GNW (2001): Gesuch um die Erteilung einer Verleihung zur Benutzung des Untergrunds ... für untertägige Sondieranlagen (Sondierstollen) am Wellenberg, Gemeinde Wolfenschiessen. 30.1.2001. GNW, Wolfenschiessen.

[P111] Graf, P. & H. Zünd (1969): Die Behandlung und Beseitigung radioaktiver Abfälle. In: SVA (ed.): Aktuelle Probleme der Atomenergie und Kerntechnik in der Schweiz. Luzern, 29.-30.5.1969. Separatdruck aus Neue Technik, Ausgabe B. Zürich:16-31.

[P112] Greenpeace ([1992]): Trittst im Morgenrot daher, she' ich dich im Strahlenmeer.... Faktenmappe. August. Greenpeace Schweiz, Zürich.

[P113] Greenpeace (1993): Wellenberg-Entscheid: Augenwischerei der Atomlobby. Medienmitteilung [press release of 1993-6-29].

[P114] Greenpeace (1995a): Atommüll: aus den Augen, aus dem Sinn [Atomic waste: out of sight, out of mind]. Zwar-Magazin 1/95:6-14.

[P115] Greenpeace (1995b): Kein atomarer Friedhof am Wellenberg! Brief an Greenpeace-Freundinnen und -Freunde. Beiblatt: Wohin mit den radioaktiven Abfällen? Das Kon-

References 289

zept der Nagra und der AKW-Betreiber/Das Greenpeace Konzept. 18.5.1995. Greenpeace, Zürich.

[P116] Gruppe Ökologie (1994): Stellungnahme zum Nachweis der grundsätzlichen Eignung des Standortes Wellenberg (Gemeinde Wolfenschiessen, NW) für die Errichtung eines Endlagers für schwach- und mittelradioaktive Abfälle. Schweizerische Anti-Atom-Koalition. Hannover.

[P117] Hählen, P. (1990): Kernenergie in der Schweiz – wie weiter? Finanz-Revue, 22.6.1990. Abdruck in: INFEL herausgegriffen. Informationsstelle für Elektrizitätsanwendung, INFEL, Zürich

[P118] Heierli, W. (1979): Forschungs- und Entwicklungsbedürfnisse im Zusammenhang mit dem Bau und Betrieb eines tiefliegenden Kavernen-Endlagers für hochaktive verglaste Abfälle in der Schweiz. Eidg. Institut für Reaktorforschung, EIR. Bericht Nr. 1005.

[P119] House of Representatives (Nationalrat) (1959): 7587. Atomenergie und Strahlenschutz. Bundesgesetz. Amtliches Bulletin der Bundesversammlung. Nationalrat. Herbstsession:598-641.

[P120] House of Representatives (Nationalrat) (1975): Kaiseraugst und Kernkraftwerke. Amtliches Bulletin der Bundesversammlung. Nationalrat. Sommersession:857-904.

[P121] HSK (1983): Kommentare zum Zwischenbericht der NAGRA über das Projekt «Gewähr». HSK 23/14. 15.12.1983.

[P122] HSK (1986): Gutachten zum Projekt Gewähr 1985 der Nationalen Genossenschaft für die Lagerung radioaktiver Abfälle (Nagra). HSK 23/28.

[P123] HSK (1987): Technischer Bericht zum Gutachten über das Projekt Gewähr 1985. HSK 23/29.

[P124] HSK (1997): Nukleare Entsorgung in der Schweiz. Übersicht und Standpunkt der Aufsichtsbehörde. 15.10.1997. HSK 21/61.

[P125] HSK (1999): Entsorgungsnachweis für HAA/LMA – Option Endlager im Opalinuston. Beurteilungskonzept für den Standortnachweis. 25.1.1999. HSK 23/57.

[P126] HSK (2000): Anforderungen der HSK an das Projekt eines Lagers für schwach- und mittelaktive Abfälle (SMA) am Wellenberg. Nov. 2000. HSK 30/15.

[P127] HSK (2001a): Wasserflusskriterien für den Sondierstollen am Wellenberg. 25.1.2001. HSK 30/16.

[P128] HSK (2001b): Kurzer Überblick über das Auswahlverfahren eines Sedimentstandortes für die Endlagerung hochradioaktiver Abfälle. 10.4.2001. HSK 23/62.

[P129] HSK (2001c): Annual Report 2000. HSK, Würenlingen.

[P130] HSK (2002a): Annual Report 2001. HSK, Würenlingen.

[P131] HSK (2002b): Unternehmensplan der HSK für die erste Leistungsauftragsperiode (2004–2007) der FLAG-Einführung. Anhang. HSK, Würenlingen.

[P132] HSK (2004): Electronic mail of G. Schwarz, 2004-4-23. HSK, Würenlingen.

[P133] HSK, KSA (1980, rev. 1993): Protection objectives for the disposal of radioactive waste. HSK-R-21/e. November 1993. HSK, Villigen-HSK.

[P134] Hubacher (1982): Einfache Anfrage 81.762 vom 8.12.1981 – Atommüll. Endlagerung. Amtliches Bulletin der Bundesversammlung. Nationalrat. Frühjahrssession:1027.

[P135] Hufschmid, P. *et al.* (2002) "Monitored long-term geological disposal". A new approach to the disposal of radioactive waste in Switzerland. Transactions, European Nuclear Conference (ENC) 2002, Lille, France, 7–9 Oct. ENC, Brussels (unpubl.).

[P136] Hug, P. (1987): Geschichte der Atomtechnologie-Entwicklung in der Schweiz. Lizentiatsarbeit. Universität Bern.

[P137] Hug, P. (1991): Die Anfänge der Atomtechnologieentwicklung in der Schweiz. La genèse de la technologie nucléaire en Suisse. Relations internationales. No. 68. Hiver 1991:325-344.

[P138] Hug, P. (1994): Elektrizitätswirtschaft und Atomkraft. Das vergebliche Werben der Schweizer Reaktorbauer um die Gunst der Elektrizitätswirtschaft 1945-1964. In: D. Gugerli (Hg.): Allmächtige Zauberin unserer Zeit. Zur Geschichte der elektrischen Energie in der Schweiz. Chronos, Zürich:167-184.

[P139] Hug, P. (1998): Atomtechnologie in der Schweiz zwischen militärischen Interessen und privatwirtschaftlicher Skepsis. In: B. Heintz & B. Nievergelt (Hg.): Wissenschafts- und Technikforschung in der Schweiz. Sondierungen einer neuen Disziplin. Seismo, Zürich:225-242.

[P140] IAEA (1991b): Inventory of radioactive material entering the marine environment: Sea disposal of radioactive waste. TECDOC-588.

[P141] IAEA (1999): Report of the International Regulatory Review Team (IRRT) to Switzerland. 30 Nov - 11 Dec 1998. IAEA/NSNI/IRRT/99/01. January 1999. Division of Nuclear Installation Safety. (recommendation #33)

[P142] IHA.GfM (1995): Telefonische Meinungsumfrage im Auftrag der Nagra. Juli 1995 (interpretation in nagra report 1/96:2-3).

[P143] Imboden, M. (1964): Helvetisches Malaise. Polis 20. Evangelische Zeitbuchreihe. EVZ-Verlag, Zürich.

[P144] Interdepartementaler Ausschuss Rio IDARio (1996): Nachhaltige Entwicklung in der Schweiz. Bericht. BUWAL, Bern.

[P145] Interdepartementaler Ausschuss Rio IDARio (1997): Nachhaltige Entwicklung in der Schweiz. Stand der Realisierung. BUWAL, Bern.

[P146] Issler, H. (1980): Der gesetzliche Auftrag und das Forschungsprogramm der Nagra. Nagra-Pressekonferenz, 8.1.1980. Medienunterlagen.

[P147] Issler, H. (1994): Kein Ausstieg aus der Entsorgungs-Verantwortung. In: Öffentliches Hearing «Verantwortung für eine ferne Zukunft. Vom Umgang mit radioaktiven Abfällen am Beispiel Wellenberg». 14.12.1994. CAN, Zürich.

[P148] Iten Joseph (1993): Interpellation 92.3461 vom 30.11.1992 - Standortwahl für Nagra-Bohrungen. Amtliches Bulletin der Bundesversammlung. Nationalrat. Frühjahrssession:631f.

[P149] Jäckli, H. (1975): Opposition gegen die Endlagerung radioaktiver Abfälle. NZZ. 17.5.1975. Nr. 88:35.

[P150] Jäckli, H. (1976): Konzept für die Einlagerung radioaktiver Abfälle in geologischen Formationen in der Schweiz. Zürich-Höngg.

[P151] Jäckli, H. (1977): Die Endlagerung radioaktiver Abfälle in geologischen Formationen. Bull. ASE/UCS (VSE). Nr. 1. 8.1.1977.

[P152] Joss, S. & A. Brownlea (1998): Verfahrensgerechtigkeit in der partizipativen Technikfolgenabschätzung am Beispiel des PubliForum "Strom und Gesellschaft". Konzepterarbeitung und Evaluation. TA/DT 22/1998. Dezember 1998. Schweizerischer Wissenschaftsrat, Bern.

[P153] KARA (1975): Koordinations-Ausschuss des Amtes für Energiewirtschaft auf dem Gebiet der radioaktiven Abfälle KARA. Zusammenfassendes Protokoll der Sitzung vom 28. November 1975 in Bern/auch: ASK-internes Protokoll vom 4.12.75. Abteilung für die Sicherheit der Kernanlagen, ASK, Würenlingen.

[P154] KARA (1976): Protokoll der 2.Sitzung des KARA vom 24.2.1976. KARA-08. ASK, Würenlingen.

[P155] Kasser, U., A. von Däniken & G. Furrer (1984): Die Endlagerung schwach- und mittelradioaktiver Abfälle in der Schweiz: Schwachstellen in der Risikoberechnung. ökos Beratergemeinschaft für angewandte Ökologie, Zürich.

[P156] KFW, Kantonale Fachgruppe Wellenberg (2001a): Zweiter Zwischenbericht an den Koordinationsausschuss. Grundlagen für die Beurteilung eines Sondiergesuchs. 23.2.2001. KFW, Stans.

[P157] KFW (2001b): Bericht zum Konzessionsgesuch Sondierstollen der GNW. 10.4.2001. KFW, Stans.

[P158] KFW (2002a): Bericht zur Standortwahl Wellenberg. Januar 2002. KFW, Stans.

[P159] KFW (2002b): Abfallinventar SMA Wellenberg. Juni 2002. KFW, Stans.

[P160] Kiener, E. (1985): Was Sicherheit ist, muss die Politik entscheiden. Das «Projekt Gewähr» der Nagra liegt jetzt zwar vor, doch ob es hält, was es verspricht, zeigt sich erst in einem Jahr. Interview. Weltwoche. 6.6.1985.

[P161] Kiener, E. (1992): Grusswort. In: Energieforum (Hg.): Nukleare Entsorgung – eine Bestandesaufnahme. Tagung, Bern, 19.5. EF-Dokumentation. Bd. 59. Energieforum Schweiz, Bern: 7-17.

[P162] Kiener, E. (1996): Internationale Lösung notwendig. Forum vera. Bulletin 2/96:2-4.

[P163] Klügel, J.-U., R. Gilli & J. Vigfusson (1999): Ergebnisse einer ersten sicherheitstechnischen Beurteilung des Konzeptes eines bleigekühlten Energy Amplifiers aus behördlicher Sicht. atomwirtschaft atw. Heft 11. November 1999:664-669.

[P164] KNE (1990): Stellungnahme zur Sedimentstudie der Nagra (NTB 88-25). Kommission Nukleare Entsorgung KNE. Bericht zuhanden des Bundesamtes für Energiewirtschaft.

[P165] KNE (1994): Stellungnahme der Kommission Nukleare Entsorgung (KNE) zur Standortwahl der Nagra (NTB 93-02). Bericht zuhanden der Hauptabteilung für die Sicherheit der Kernanlagen, BEW.

[P166] KNE (1995): Stellungnahme zu den Sondiergesuchen NSG 19 und NSG 20 der Nagra. Mai 1995. Bericht zuhanden der Hauptabteilung für die Sicherheit der Kernanlagen des Bundesamtes für Energiewirtschaft.

[P167] KNE (1998): Entwicklung der Arbeiten zur Endlagerung der radioaktiven Abfälle in der Schweiz. Positionspapier der KNE. 18.11.1998. KNE, Bern (reprinted in Appendix IV of the AGNEB 1998 Annual Report [P15]).

[P168] Knoepfel, P., R. Enderlin Cavigelli, F. Varone, S. Wälti & H. Weidner (1997): Energie 2000: Evaluation der Konfliktlösungsgruppen. Bericht zuhanden des Bundesamtes für Energiewirtschaft. Nov. EDMZ, Bern.

[P169] Konfliktlösungsgruppe radioaktive Abfälle KORA (1992): Bericht der Arbeitsgruppe Wiederaufarbeitung. Schlussentwurf vom Dezember 1992. (not published)

[P170] Kowalski, E. (1986): Betrachtungen über Abfälle und Demokratie – sowie darüber, dass es nicht genügt, das Recht zu haben (Editorial). Nagra informiert 4/86:3f.

[P171] Kowalski, E. (1990a): Kontrolle und Endlagerung – ein Widerspruch? NZZ. 16.2.90.

[P172] Kowalski, E. (1990b): Endlager für kurzlebige Abfälle. In: Energieforum (Hg.): Nukleare Entsorgung – eine Bestandesaufnahme. Tagung, Bern, 19.5. EF-Dokumentation. Bd. 59. Energieforum Schweiz, Bern: 97-110.

[P173] Kowalski, E. (1996): Waste as an argument against nuclear energy. nagra bulletin. Nr. 28. October:3-9 (transl. reprint of NZZ 1996-2-19).

[P174] Kowalski, E. & H. Issler (1994): Dialog bei der Entsorgung radioaktiver Abfälle. Suche nach einvernehmlicher Lösung am Wellenberg. NZZ. 6.4.1994:22.

[P175] Kreuzer, K. (1990): Hütet unser Strahlen-Erbe! anstelle des Märchens «Gewähr». nux. Nrn. 66-67. Okt. Flüh SO:1-11.

[P176] Kreuzer, K. (1997): Forum, Brief an W. Wildi. Energie+Umwelt. Nr. 4. SES, Zürich.
[P177] KSA (1992): Sicherheitsprinzipien für die Entsorgung radioaktiver Abfälle. Klausurtagung 1992 des Ausschusses «Strahlenschutz und Entsorgung». Tagungsbericht. KSA 21/75.
[P178] KSA, Kommission für die Sicherheit von Kernanlagen (1998): Aktuelle Fragen zur Endlagerung radioaktiver Abfälle in der Schweiz – Position der KSA. 22.9./30.11.1998. KSA 21/124 (reprinted in Appendix V of the AGNEB 1998 Annual Report [P15]).
[P179] KSA (1999): Kommentare zum Bericht: Konzept der Energieforschung des Bundes 2000-2003; Teil Kernspaltung. Schreiben vom 3.12.1999. KSA-AN-2088.
[P180] KSA (2000): Kernauslegung für Kernkraftwerke: Erhöhter Abbrand. Verlängerte Betriebszyklen. MOX-Brennelemente. KSA-Klausurtagung vom 28.-30. Oktober 1998 in Meiringen. KSA-AN-2027.
[P181] Küffer, K. (1994): Wiederaufarbeitung von Kernbrennstoff: Ökonomische, ökologische und technische Aspekte. Referat des Präsidenten der Zwilag Zwischenlager Würenlingen AG anlässlich der GV vom 6.4.1994. SVA-Bulletin Nr. 7-8/1994:22-24.
[P182] Küffer, K. (1998), SVA-Vertiefungskurs Kostenoptimierung in Kernkraftwerken, 22. – 24.4.98. Schweizerische Vereinigung für Atomenergie (SVA), Bern:1.1-1–1.1-13.
[P183] Landammann und Regierungsrat des Kantons Uri (1990): [Schreiben vom 9.4.1990]. Landammann und Regierungsrat, Altdorf.
[P184] Laubscher, H. P. (1985): Struktur des Grundgebirges und des Paläozoikums in der Nordschweiz. HSK-Expertenbericht. Geologisch-paläontologisches Institut, Univ. Basel.
[P185] Lüthi, H. R (1981): Radioaktive Abfälle. In: BEW (Hg.): Die schweizerische Energiewirtschaft 1930 – 80:50-53.
[P186] Lutz, H. R. (1970): Gefahren der Kernkraftwerke IV. Radioaktive Abfälle und ihre Lagerung. Basler Nachrichten. 2.7.70:3.
[P187] Lutz, H. R. (1997): Das achte Weltwunder [The eighth Wonder of the World]. Aargauer Zeitung, 8.11.97.
[P188] Mauch Ursula (1979): Frage 43 vom 10.12.1979. Energieversorgung. Amtliches Bulletin der Bundesversammlung. Nationalrat. Wintersession:1570.
[P189] Mauch Ursula (1986): Einfache Anfrage 86.647 vom 9.6.1986 – Nagra-Bericht «Gewähr 1985». Veröffentlichung der internen Dokumente. Amtliches Bulletin der Bundesversammlung. Nationalrat. Herbstsession:1526.
[P190] McCombie, C. (1994): Interfaces and interactions in the Swiss waste disposal programme. 5th ICHLRWM, Las Vegas. Vol. 1. ANS, La Grange Park, IL:238-248.
[P191] McCombie, C. (1997a): Radioactive waste disposal and the environment. In: IAEA, Radiation and society: comprehending radiation risk. IAEA Conference, 24–28 Oct. 1994, Paris. Vol. 3:153-157. IAEA, Vienna.
[P192] McCombie, C. (1997b): In the eye of the beholder: different perceptions of the problem of waste disposal. No. 30. Nagra, Wettingen:15-23. (based on a presentation given at the 7th International Conference on HLRWM, Las Vegas, 1996)
[P193] McCombie, Ch. (1998): R&D in support of repository implementation – Do we need any more? In: I. G. McKinley & Ch. McCombie (eds.): Scientific Basis for Nuclear Waste Management XXI. Sep 28–Oct 3, 1997. Davos. Symposium Proc. Vol. 506. Materials Research Society, Warrendale, PA:33-36.
[P194] Morf (1980/1981) Interpellation 80.446 vom 11.6.1980. – Entsorgung. Oberexpertise. Votum vom 12.3.1980. Amtliches Bulletin der Bundesversammlung. Nationalrat. Frühjahrssession 1980:199ff. Sommersession 1981:909ff.

References 293

[P195] Naegelin, R. (1996): Vielfältige Aufsicht über Kernanlagen. Die Aufgaben der HSK sind immer komplexer geworden. NZZ. 10.4.1996. Nr. 83.
[P196] Nagra (1972): Statuten der Nationalen Genossenschaft für die Lagerung radioaktiver Abfälle NAGRA vom 4.12.72. Nagra, Baden.
[P197] Nagra (1976): Geologische Studien. Stüblenen, Gemeinden Lenk und Lauenen/BE. Technischer Bericht. 30.4.1976. Nagra, Baden.
[P198] Nagra (1980): Projektstudie P1. Endlagertyp B für schwach- und mittelaktive Abfälle. Motor Columbus. Mai 1980. Interner Bericht 52.
[P199] Nagra (1983a): Nukleare Entsorgung Schweiz – Konzept und Stand der Arbeiten 1982. NTB 83-02.
[P200] Nagra (1983b): Die Endlagerung schwach- und mittelradioaktiver Abfälle in der Schweiz: Evaluation der potentiellen Standortgebiete. Band 1 (Grundlagen, Bewertungskriterien und Ergebnisse), Band 2 (Standortbezogene Anhänge und Beilagen). NTB 83-15.
[P201] Nagra (1985a): Nukleare Entsorgung Schweiz: Konzept und Übersicht über das Projekt Gewähr 1985. NGB 85-01.
[P202] Nagra (1985b). Projekt Gewähr 1985. Endlager für hochaktive Abfälle: Das System der Sicherheitsbarrieren. NGB 85-04.
[P203] Nagra (o.J.): Projekt Gewähr 1985. Realisierbarkeit und Langzeitsicherheit der Endlager für radioaktive Abfälle.
[P204] Nagra (1987): Entscheiderhebliche Gutachten und Stellungnahmen zum Projekt Gewähr 1985. 3 Beilagen. 23.2.1987. Nagra, Baden. (reference: Annex 3)
[P205] Nagra (1988a): Übersicht über die Untersuchungen der Phase I an den potentiellen Standorten Bois de la Glaive, Oberbauenstock und Piz Pian Grand. NTB 88-20.
[P206] Nagra (1988b): Sedimentstudie – Zwischenbericht 1988. Möglichkeiten zur Endlagerung langlebiger radioaktiver Abfälle in den Sedimenten der Schweiz. Textband/Beilagenband. NTB 88-25.
[P207] Nagra (1988c): Antrag zur Bewilligung der Fortsetzung der Untersuchungsprogramme ... durch den Vortrieb von Sondierstollen [Ollon VD, Bauen UR, Mesocco und Rossa GR]. 23.11.88.
[P208] Nagra (1992): Nukleare Entsorgung Schweiz – Konzept und Realisierungsplan. NTB 92-02.
[P209] Nagra (1993a): ?Fragen und Antworten! zur Entsorgung radioaktiver Abfälle. März. Nagra, Wettingen.
[P210] Nagra (1993b): Vergleichende Beurteilung der Standorte Bois de la Glaive, Oberbauenstock, Piz Pian Grand und Wellenberg. Endlager für kurzlebige schwach- und mittelaktive Abfälle (Endlager SMA). Textband/Beilagenband. NTB 93-02.
[P211] Nagra (1993c): Endlager für kurzlebige Abfälle. Vorbericht zur Standortwahl [Preliminary report on site selection]. NTB 93-15.
[P212] Nagra (1993d): Geologische Grundlagen und Datensatz zur Beurteilung der Langzeitsicherheit des Endlagers für schwach- und mittelaktive Abfälle am Standort Wellenberg (Gemeinde Wolfenschiessen, NW). NTB 93-28.
[P213] Nagra (1994a): Model radioactive waste inventory for Swiss waste disposal projects. Vol. 1: Main report. NTB 93-21.
[P214] Nagra (1994b): Kristallin-I. Safety assessment report. NTB 93-22.
[P215] Nagra (1994c): Bericht zur Langzeitsicherheit des Endlagers SMA am Standort Wellenberg. NTB 94-06.
[P216] Nagra (1995): nagra report. No. 4. Juli/August.
[P217] Nagra (1996): nagra report. No. 1. Januar/März.

[P218] Nagra (1997): Which is more stable: a rock formation or a social structure? nagra bulletin. No. 30. August.
[P219] Nagra (1998a): Endlager für schwach- und mittelaktive Abfälle am Standort Wellenberg. Etappen auf dem Wege zum Verschluss; präzisierende Darstellung der Kontrollierbarkeit und Rückholbarkeit. NTB 98-04.
[P220] Nagra (1998b): Entsorgungskonzept. FOCUS 01. Themenheft zur nuklearen Entsorgung. Dezember.
[P221] Nagra (1999): Abfälle lagern. FOCUS 02. Themenheft zur nuklearen Entsorgung. Dezember. Reprint 2001.
[P222] Nagra (2002): Project Opalinus Clay. Safety report. Demonstration of disposal feasibility of spent fuel, vitrified high-level waste and long-lived intermediate-level waste (Entsorgungsnachweis). NTB 02-05.
[P223] Nagra (1980*passim*): nagra aktuell. Various issues.
[P224] Nagra (1980*passim*): nagra bulletin/nagra informiert/report. Various issues.
[P225] Nagra (1991*passim*): Nagra. Geschäftsbericht [company report]. Nagra, Wettingen.
[P226] Nagra/Cédra (1981): Le stockage final des déchets faiblement et moyennement radio-actifs en Suisse. Vol. 1 (Principes de base et procédures pour la sélection d'une région), Band 2 (Standortgebiete der engeren Wahl). NTB 81-04.
[P227] Naturfreunde Schweiz *et al.* (1993): NAGRA: Politischer, nicht geologischer Widerstand entschied. Medienmitteilung von Umwelt- und Energieverbänden vom 29.6.1993.
[P228] Nidecker, A. (1995): Das Wellenbergprojekt aus der Sicht der ÄrztInnen für Soziale Verantwortung (PSR/IPPNW) Schweiz. Stans. 28.3. PSR, Basel.
[P229] NEA, Nuclear Energy Agency (2004): Safety of disposal of spent fuel, HLW and long-lived ILW in Switzerland. An international peer review of the post-closure radiological safety assessment for disposal in the Opalinus Clay of the Zürcher Weinland. OECD, Paris.
[P230] Nowotny, H. & R. Eisikovic (1990): Entstehung, Wahrnehmung und Umgang mit Risiken [Generation, perception, and management of risks]. Forschungspolitische Früherkennung (FER) B/34. Schweizerischer Wissenschaftsrat [Swiss Science Council] SWR, Bern.
[P231] Nuclear Engineering International (2002): Canton rejects waste repository. November 2002:6.
[P232] nux (2000): Rückholbar lagern – – endlich doch! – – ? Kontrolliertes geologisches Langzeitlager, ... Nr. 107. März.
[P233] Oberholzer-Gee, F. (1998): Die Ökonomik des St. Florianprinzips. Warum wir keine Standorte für nukleare Endlager finden. Helbing & Lichtenhahn, Basel.
[P234] Oberholzer-Gee, F., B. S. Frey, A. Hart & W. W. Pommerehne (1995): Panik, Protest und Paralyse. Eine empirische Untersuchung über nukleare Endlager in der Schweiz. Schweiz. Zeitschrift für Volkswirtschaft und Statistik. Vol. 131. Nr. 2:147-177.
[P235] Patak, H.N. & R.H. Stapper (1998): Grundzüge: Kostenfaktor Brennstoffzyklus. SVA-Vertiefungskurs Kostenoptimierung in Kernkraftwerken, 22.-24.4.98. Schweizerische Vereinigung für Atomenergie, SVA, Bern:4.1.-1 bis 12.
[P236] Petitpierre (1990): Interpellation 89.754 vom 11.12.1989 – Radioaktive Abfälle. Lagerung in Kernkraftwerken. Amtliches Bulletin der Bundesversammlung. Nationalrat. Frühjahrssession:754.
[P237] Prêtre, S., W. Jeschki, J. Nöggerath & U. Schmocker (1998): Kosten und Sicherheit. SVA-Vertiefungskurs Kostenoptimierung in Kernkraftwerken, 22.-24.4.98. SVA, Bern.

References

[P238] PSR/IPPNW Ärzte und Ärztinnen für soziale Verantwortung *et al.* (1993): Positionspapier zur Energiepolitik der Schweiz.
[P239] Rausch, H. (1977): Rechtliche Probleme der Lagerung radioaktiver Abfälle aus Kernkraftwerken. Schweizerische Juristen-Zeitung. Heft 3. Feb.:33-36.
[P240] Rausch, H. (1980): Schweizerisches Atomenergierecht. Schulthess Polygraphischer Verlag, Zürich.
[P241] Rebeaud (1989): Einfache Anfrage 89.1089 vom 18.9.1989 – Radioaktive Abfälle und nukleare Zusammenarbeit mit China. Amtliches Bulletin der Bundesversammlung. Nationalrat. Wintersession:2288.
[P242] Rometsch, R. (1980): Auf dem steinigen Weg zu Lösung einer nationalen Aufgabe. Vortrag anlässlich der Sessionsveranstaltung vom 18.3.1980 in Bern. Energieforum Schweiz, Bern.
[P243] Rometsch, R. (1982): Nagra-Präsident Rudolf Rometsch: Die Bevölkerung unterstützt uns. Gelbes Heft. 12.10.1982:62f.
[P244] Rometsch, R. (1984): Wie zuverlässig verhält sich die Natur? Endlagerung radioaktiver Abfälle: Nagra-Präsident Rudolf Rometsch im Wettlauf mit der Zeit. Interview. Weltwoche. 9.8.1984.
[P245] Rometsch, R. (1991): Kann Endlagersicherheit vorausberechnet werden? NZZ. 7.3.1991.
[P246] Sager (1989): Interpellation 89.379 vom 13.3.1989 – Entsorgung radioaktiver Abfälle. Amtliches Bulletin der Bundesversammlung. Nationalrat. Sommersession:1219f.
[P247] Sarasin, Ph. (1984): Die kommerzielle Nutzung der Atomenergie in der Schweiz. In: S. Füglister (Hg.): Darum werden wir Kaiseraugst verhindern. Texte und Dokumente zum Widerstand gegen das geplante AKW. orte-Verlag, Zürich.
[P248] Schindler, C. (1986): Beurteilung der geologischen Aspekte des Gewähr-Projektes der Nagra für ein Endlager Typ B (schwach- und mittelaktive Abfälle). Expertenberichte zum Projekt Gewähr. BEW. Hauptabteilung für die Sicherheit der Kernanlagen HSK. ETH Zürich.
[P249] Schweizerischer Bundesrat [Federal Council] (1957): Botschaft über die Ergänzung der Bundesverfassung durch einen Artikel betreffend Atomenergie und Strahlenschutz. Bundesblatt No. 19:1137-1158.
[P250] Schweizerischer Bundesrat (1974): Elektrizitätsversorgung. Botschaft und Beschlussentwurf vom 11.9.1974. Amtliches Bulletin der Bundesversammlung. Nationalrat. Wintersession:1658,1667.
[P251] Schweizerischer Nationalfonds, SNF (1997): Immer weniger Vertrauen in die Behörden. Pressemitteilung vom 27.10.97. SNF, Bern.
[P252] Schweizerischer Wissenschaftsrat (1998): Konzept Umwelt- und Nachhaltigkeitsforschung. Vorschläge der Kommission "Strategie Umweltforschung und Nachhaltige Entwicklung". Forschungspolitik FOP 52/1998. SWR, Bern.
[P253] Seiler, H. (1986): Das Recht der nuklearen Entsorgung in der Schweiz. Stämpfli, Bern.
[P254] SES (1988): SES fordert lückenlose Stoffkontrolle von radioaktivem Material. Pressemitteilung vom 23.2.88. SES, Zürich.
[P255] SES (1999): Brief vom 30.6.1999 an UVEK, H. Werder, Generalsekretär. SES, Zürich.
[P256] SES, Greenpeace, MNA (1999a): Kernaussagen Umweltorganisationen (vertreten durch SES, Greenpeace, MNA [Komitee für die Mitsprache des Nidwaldner Volkes bei Atomanlagen]. Hearing Expertengruppe Entsorgungskonzepte. 6.9.1999.

[P257] SES, Greenpeace, MNA (1999b): Brief an die Expertenkommission Radioaktive Abfälle, W. Wildi, Präsident, M. Aebersold, BFE. 15.12.1999.
[P258] Ständerat (Senate) (1990): Volksinitiativen. Atomenergie. 89.032. Amtliches Bulletin der Bundesversammlung. Ständerat. Frühjahrssession:23-35.
[P259] Steiner, P. (2002): Of local public consultation and decision making on nuclear issues: the Nidwalden case. In: PSR/IPPNW Switzerland (ed.): Rethinking nuclear energy and democracy after 09/11. Basel, April 26–27, 2002.
[P260] Steinmann, W. (2002): Rolle und Erwartungen des Bundes zur nuklearen Entsorgung. Referat anlässlich der Tagung Nagra/UAK vom 9. Januar 2002. Bundesamt für Energie, Bern.
[P261] Stüssi-Lauterburg, J. (1995): Historischer Abriss zur Frage einer Schweizer Nuklearbewaffnung. Silvester.
[P262] SVA, Swiss Association of Atomic Energy (1998): Bulletin.
[P263] SVA (2001): Bulletin. Vol. 21. No. 1, other issues.
[P264] Swiss Federal Nuclear Energy Act of 2003-3-21. Presumably put into force on 2005-1-1 (according to [P30]).
[P265] Swiss Federal Radiation Protection Act of 1991-3-22. Put into force on 1994-10-1. SR 814.50.
[P266] Swiss Federal Radiation Protection Ordinance of 1963-4-19.
[P267] Swiss Federal Ordinance Concerning Protection Against Major Accidents (Major Hazards Ordinance, OPAM) of 1991-2-27. SR 814.512.
[P268] Swiss Federal Ordinance on Preparatory Measures of 1979-11-15 (status as of 1989-11-27). SR 732.012
[P269] Swiss Federal Radioactive Waste Disposal Fund Ordinance of 2000-3-6. SR 732.014
[P270] Swiss Federal Radiation Protection Ordinance of 1994-6-22. SR 814.501 (art. 37).
[P271] TAG, Technische Arbeitsgruppe Wellenberg (1998): Endlager von schwach- und mittelradioaktiven Abfällen: Stellungnahme zu technischen Aspekten des Projekts Wellenberg. Bundesamt für Energie, Bern.
[P272] Thut, M. (1974): Untertagspeicherung radioaktiver Abfälle in der Schweiz. Beilage zum SVA-Bulletin. Nr. 6. Schweizerische Vereinigung für Atomenergie (SVA), Bern.
[P273] Untergruppe (Subgroup) Geologie der AGNEB (1983): Meinungsäusserung zu den potentiellen Standortgebieten für Endlager vom Typ B. 22.2.1983. Beilage IV zum 5. Tätigkeitsbericht der AGNEB. April 1983. BEW, Bern.
[P274] Untergruppe Geologie der AGNEB (1986): Stellungnahme zum Projekt «Gewähr». 27.3.1986. Basel.
[P275] UVEK, DETEC (1998): Energie-Dialog zur Entsorgung der nuklearen Abfälle. Pressemitteilung vom 10.2.98. Eidg. Departement für Umwelt, Verkehr, Energie und Kommunikation UVEK, Bern.
[P276] UVEK (1999a): Energiepolitische Weichenstellungen: Elektrizitätsmarktöffnung und Kernenergie. Medienmitteilung vom 10.6.1999. UVEK, Bern.
[P277] UVEK (1999b): Kernenergiegesetz. Vernehmlassungsentwurf und Erläuternder Bericht. [5.10. 1999]. UVEK, Bern.
[P278] UVEK (2000a): Departementsstrategie UVEK. UVEK, Bern.
[P279] UVEK (2000b): Kernenergiegesetz. Eröffnung des Vernehmlassungsverfahren. Vernehmlassungsentwurf und Erläuternder Bericht. 6./10.3.2000. UVEK, Bern.
[P280] Vatter, A. (1995): Qualifiziertes Referendumsrecht für Betroffene. Zum Vorgehen bei der Standortsuche für ein Atommüllager. NZZ. 21.12.95:15.
[P281] Vorsteher des Eidg. Verkehrs- und Energiewirtschaftsdepartement, EVED (1995): Entsorgung und Lagerung radioaktiver Abfälle. Schreiben von Bundesrat A. Ogi an

References 297

D. Schaer-Born, Bau-, Verkehrs- und Energiedirektorin des Kantons Bern. 29.9.1995. Bern.
[P282] VSE, Verband Schweizerischer Elektrizitätswerke, Gruppe der Kernkraftwerkbetreiber und -projektanten, GKBP & Konferenz der Überlandwerke, UeW (1977): Schweizerische Elektrizitätswirtschaft. Konzept für die nukleare Entsorgung in der Schweiz. Zielsetzungen, Grundsätze und Aufgabenteilung. 6.4.1977. No loc.
[P283] VSE, GKBP, UeW & Nationale Genossenschaft für die Lagerung radioaktiver Abfälle, NAGRA (1978): Die nukleare Entsorgung in der Schweiz. 9.2.1978. [VSE, Zürich]
[P284] Wälti, S. (1993): Neue Problemlösungsstrategien in der nuklearen Entsorgung. In: Vollzugsprobleme. Problèmes de la mise en œuvre des politiques publiques. Schweizerisches Jahrbuch für Politische Wissenschaft. Bd. 33:205-224.
[P285] Weber Agnes (1997): Motion 96.3644 vom 12.12.1996 – Auflösung der NAGRA in ihrer heutigen Form. Amtliches Bulletin der Bundesversammlung. Nationalrat. Frühjahrssession.
[P286] Weber, R. (1976): Entsorgung der Kerntechnik. Beilage zum SVA-Bulletin Nr. 7. Schweizerische Vereinigung für Atomenergie SVA, Bern.
[P287] Zuidema, P. (2000): Comments on the international situation on waste management. International Symposium on Radioactive Waste Management – Sustainable Disposal or Tentative Solutions? March 30, Bern. Forum vera, Zurich.
[P288] Zuidema, P. & H. Issler (2001): Kosten der Endlagerung radioaktiver Abfälle. SVA-Informationstagung «Die Kernenergie im offenen Strommarkt». 12. – 13.11.2001. Zürich. SVA, Bern.
[P289] Zünd, M. (2003): Regulatorische Sicherheitsforschung. Überblicksbericht zum Forschungsprogramm 2002. HSK. BFE, Bern.
[P290] Zurkinden, A. (2002): Aufgaben der HSK bei der nuklearen Entsorgung. 18.2.2002. HSK 21/72. HSK, Villigen-HSK.

2. SECONDARY LITERATURE

2.1 Methods (M)

[M1] Allen, P. T. (1998): Public participation in resolving environmental disputes and the problem of representativeness. Risk: Health, Safety & Environment. No. 9. Fall: 297-308.
[M2] Ascher, W. (1999): Resolving the hidden differences among perspectives on sustainable development. Policy Sciences. Vol. 32:251-377.
[M3] Bijker, W. E. (1995): Of bicycles, bakelites, and bulbs: Toward a theory of sociotechnical change. MIT Press, Cambridge, MA.
[M4] Bijker, W. E., T. P. Hughes & T. Pinch (1989): The social construction of technological systems. New directions in the sociology and history of technology. MIT Press, Cambridge, MA.
[M5] Beierle, T. C. & J. Cayford (2002): Democracy in practice. Public participation in environmental decisions. Resources for the Future, Washington, DC.
[M6] Beierle, T. C. (2004): Democracy in practice. Public participation in environmental decisions. In: NEA (*ed.*): Stakeholder participation in decision making involving radiation: Exploring processes and implications. [3rd] Workshop Proc., Villigen, Switzerland, 21–23 October 2003. OECD, Paris.

[M7] Bohnenblust, H. & P. Slovic (1998): Integrating technical analysis and public values in risk-based decision making. Reliability Engineering & System Safety. Vol. 59: 151-159.

[M8] Borowsky, P. *et al.* (1989): Einführung in die Geschichtswissenschaft I. Grundprobleme, Arbeitsorganisation, Hilfsmittel. Westdeutscher Verlag, Opladen.

[M9] Böschen, S. (2000): Risikogenese. Prozesse gesellschaftlicher Gefahrenwahrnehmung: FCKW, DDT, Dioxin und Ökologische Chemie. Leske + Budrich, Opladen.

[M10] CEC, Commission of the European Communities (1999): New Approach and Global Approach & Legislation. Guide to the Implementation of Directives based on New Approach and Global Approach. www.europa.eu.int/comm/enterprise/newapproach/-standardization/document/1999_1982_en.pdf.

[M11] CEC (2000): Communication from the Commission on the precautionary principle. www.europa.eu.int/comm/off/com/health_consumer/precaution_en.pdf.

[M12] CEC (2001): European Governance. A White Paper. COM(2001) 428 final. 2001-7-25. See also www.europa.eu.int/comm/governance/governance.

[M13] Collins, H. M. (1981): Stages in the empirical programme of relativism. Social Studies of Science. Vol. 11:3-10.

[M14] Collins, H. M. (1992): Changing order: replication and induction in scientific practice. The Univ. of Chicago Press, Chicago.

[M15] Cornelis, G.C. (2002): Transgenerational ethics. Protecting future generations against nuclear waste hazards. In: NEA (*ed.*: Better integration of radiation protection in modern society. [2nd] Workshop Proc. Villigen, 23–25 January 2001. OECD, Paris: 93-102. (transparency 12)

[M16] Covello, V. T & J. Mumpower (1985): Risk analysis and risk management. An historical perspective. Risk analysis. Vol. 5. No. 2:103-120.

[M17] Dobson, A. (1996): Representative democracy and the environment. Political Studies. Vol. 44. No. 3. December 1996.

[M18] Director-General for Research (2000): The TRUSTNET framework: A new perspective on risk governance. European Commission Nuclear Science and Technology. Brussels. European Communities, Luxembourg.

[M19] Enderlin Cavigelli, R. (1993): Auswertungsbericht über die Arbeit der ersten Arbeitsgruppe der KORA zum Thema Wiederaufarbeitung. Manuskript.

[M20] Enderlin Cavigelli, R. (1995): Mediationsverfahren in der schweizerischen Kernenergiepolitik. Auswertungsbericht zur Arbeit der ersten Arbeitsgruppe der Konfliktlösungsgruppe radioaktive Abfälle (KORA) zum Thema Wiederaufarbeitung. Schriften zum Mediationsverfahren im Umweltschutz Nr. 12. Wissenschaftszentrum Berlin für Sozialforschung (WZB), Berlin.

[M21] Ericsson, K. .A. & H. A. Simon (^{2}1993): Protocol analysis. Verbal reports as data. Revised edition. Bradford Book, MIT Press, Cambridge, MA.

[M22] Erdmann, G. (1995): Energieökonomik. Theorie und Anwendungen. Verlag der Fachvereine vdf, Zürich.

[M23] European Environment Agency (2001): Late lessons from early warnings: the precautionary principle 1896–2000. Environmental issue report No. 22. Office for Official Publications of the European Communities, Luxembourg.

[M24] Flüeler, T. (2000a): Mutual learning [in] radioactive waste management: Transcontinental gridlock – transdisciplinary solution? In: R. W. Scholz *et al.* (*eds.*): Transdisciplinarity: Joint problem-solving among science, technology and society. Proc. of the International Transdisciplinarity 2000 Conference, Zurich, Feb 27–Mar 1, 2000.

Workbook II: Mutual learning sessions. Vol. 2. Haffmanns Sachbuch, Zürich: 304-307.

[M25] Flüeler, T. (2000b) (ed.): Mutual learning session 15. Radioactive waste management: Transcontinental gridlock – transdisciplinary solution? A process document of the Mutual Learning Session 15. Working paper. Swiss Federal Institute of Technology ETH, Chair of Environmental Sciences: Natural and Social Science Interface UNS, Zurich. (unpublished)

[M26] Flüeler, T. & H. Seiler (1999): Risk-based regulation: A suitable concept to legislate and regulate technical risks? Evaluation of various case studies in Switzerland. In: L. H. J. Goossens (ed.): Proc. 9th Annual Conference [on] Risk Analysis: Facing the new millennium. Rotterdam, Oct 10–13, 1999. Society for Risk Analysis Europe. Delft Univ. Press, Delft:593-597.

[M27] Flyvberg, B., N. Bruzelius & W. Rothengatter (2003): Megaprojects and risk. An anatomy for ambition. Cambridge Univ. Press, Cambridge.

[M28] Frey, R. L. & Chr. A. Schaltegger (2000): Abgeltungen bei Strukturanlagen? Grossprojekte zwischen Widerstand und Akzeptanz. Forschungsprogramm Energiewirtschaftliche Grundlagen. BFE, Bern.

[M29] Früh, W. (1998): Inhaltsanalyse. Theorie und Praxis. UVK Medien, Konstanz.

[M30] Galison, P. (1997): Image and logic: A material culture of microphysics. Univ. of Chicago Press, Chicago.

[M31] Gibbons, M. et al. (1994): The new production of knowledge. The dynamics of science and research in contemporary societies. Sage, London.

[M32] Goedevert, D. (1996): Wie ein Vogel im Aquarium. Aus dem Leben eines Managers. Rowohlt, Reinbek bei Hamburg.

[M33] Gunderson, L. H., Prichard Jr., L. (2002, eds.): Resilience and the behavior of large-scale systems. Scope 60. Island Press, Washington, DC.

[M34] Haimes, Y. Y. (1998): Risk modeling, assessment and management. Wiley, New York.

[M35] Hatfield, A. J. & K. W. Hipel (2002): Risk and systems theory. Risk Analysis. Vol. 22. No. 6:1043.1057.

[M36] Health Council of the Netherlands, Committee on Risk Measures and Risk Assessment (1996): Risk is more than just a number. Reflections on the development of the environmental risk management approach. To the Minister of Health, Welfare and Sports, the Minister of Housing, Spatial Planning and the Environment. No. 1996/03E, 31 March. The Hague, NL.

[M37] Heinze, T. (1995): Qualitative Sozialforschung. Erfahrungen, Probleme und Perspektiven. Westdeutscher Verlag, Opladen.

[M38] Holling, C. S. (1996): Engineering resilience versus ecological resilience. In: P. C. Schulze (ed.): Engineering within ecological constraints. National Academy Press, Washington, DC:31-43.

[M39] Inglestam, L. (ed., 1996): Complex technical systems. Swedish Council for Planning and Coordination of Research. Affärs Litteratur, Stockholm.

[M40] IRGC, International Risk and Governance Council (2002): Proposal. Supplement Round Table, 24/25 January. January 2002. Paul Scherrer Institut, Villigen PSI.

[M41] ISO, International Organization for Standardization (2000): Quality management and quality assurance. Quality management systems – Fundamentals and vocabulary (ISO 9000:2000), – Requirements (ISO 9001:2000), Guidelines for performance (ISO 9004:2000).

[M42] ISO (2002): Risk management – Vocabulary – Guidelines for use in standards. ISO/IEC Guide 73.
[M43] Jasanoff, S. (1990): The fifth branch: Science advisors as policy makers. Harvard Univ. Press, Cambridge, MA.
[M44] Jasanoff, S. (1995): Beyond epistemology: Relativism and engagement in the politics of science. Social Studies of Science. Vol. 26:393-418.
[M45] Jonas, H. (1984, orig. 1977): The imperative of responsibility. In search of an ethics for the technological age. Univ. of Chicago, Chicago.
[M46] Jonas, H. (1987): Warum die Technik ein Gegenstand der Ethik ist. In: H. Lenk & G. Ropohl (Hg.): Technik und Ethik. Reclam, Stuttgart:81-91.
[M47] Joint Research Centre (2000) Draft consensus report: Proposal towards a project on the development of a Compass for Risk Analysis. Based on the results of the EC-JRC International Workshop on Promotion of Technical Harmonisation on Risk-Based Decision Making, Stresa/Ispra, Italy, 22–25 May. European Communities-Joint Research Centre, Ispra (workshop under www.mahbsrv.jrc.it/stresa/Risk-Workshop-2000.html).
[M48] Kaiser, F. G. & J. Frick (2002): Entwicklung eines Messinstrumentes zur Erfassung von Umweltwissen auf der Basis des MRCML-Modells. Diagnostica. Bd. 48:181-189.
[M49] Kaiser, F. G. & U. Fuhrer (2000): Wissen für ökologisches Verhalten. In: H. Mandl & J. Gerstenmaier (eds.): Die Kluft zwischen Wissen und Handeln: Empirische und theoretische Lösungsansätze. Hogrefe, Göttingen:51-71.
[M50] Kaplan, S. & B. J. Garrick (1981): On the quantitative definition of risk. Risk Analysis. Vol.1. No. 1:11-27.
[M51] Kim, T. C. & J. A. Dator (1999): Co-creating. Public philosophy for future generations. Praeger, Westport, CT.
[M52] Krippendorff, K. (1980): Content analysis. An introduction to its methodology. Sage, Newbury Park.
[M53] Kröger, W. (2001): Dealing with risks and safety of technical systems – need for a new commission?. Risk Research. Vol. 4. Issue 4. Oct. 2001:405-410.
[M54] Kuhn, T. S. (1962, 1970, ^{3}1996): The structure of scientific revolutions. Univ. of Chicago Press, Chicago.
[M55] Latour, B. (1987): Science in action. How to follow scientists and engineers through society. Harvard Univ. Press, Cambridge, MA.
[M56] Linnerooth-Bayer, J. & B. Wahlström (1991): Applications of probabilistic risk assessments: the selection of appropriate tools. Risk Analysis. Vol. 11. No. 2:239-248.
[M57] Majone, G. (1989): Evidence, argument, and persuasion in the policy process. Yale Univ. Press, New Haven.
[M58] Mayring, Ph. (1997): Qualitative Inhaltsanalyse. Grundlagen und Techniken. Beltz, Deutscher Studienverlag, Weinheim.
[M59] Menkes, J. (1977): Is there a future in history: The applicability of historical analysis to policy research. In: J. A. Tarr (ed.): Retrospective technology assessment – 1976. San Francisco Press, San Francisco:321-324.
[M60] Merten, K. (1995): Inhaltsanalyse. Einführung in Theorie, Methode und Praxis. Westdeutscher Verlag, Opladen.
[M61] Minsch, J. et al. (1998): Institutionelle Reformen für eine Politik der Nachhaltigkeit. Springer, Berlin.
[M62] Neuendorf, K. A. (2002): The content analysis guidebook. Sage, Thousand Oaks, CA.
[M63] Noble, B. F. (2000): Strengthening EIA through adaptive management: a systems perspective. Environmental Impact Assessment Review. Vol. 20:97-111.

[M64] Nowotny, H. (1999): The place of people in our knowledge. European Review. Vol. 7. No. 2:247-262.
[M65] Nowotny, H., P. Scott & M. Gibbons (2001): Re-thinking science. Knowledge and the people in an Age of Uncertainty. Polity Press, Cambridge, UK.
[M66] OECD (2003): Emerging risks in the 21st century: An agenda for action. Organisation for Economic Co-operation and Development (OECD), Paris.
[M67] Panel of the National Academy of Public Administration (1997): Deciding for the future: Balancing risks, costs, and benefits fairly across generations. Report for the U. S. Department of Energy. National Academy of Public Administration, Washington, DC.
[M68] Pidgeon, N. F. (1991): Safety culture and risk management in organizations. Journal of Cross-Cultural Psychology. Vol. 22. No. 1:129-140.
[M69] Pimm, S. L. (1984): The complexity and stability of ecosystems. Nature. Vol. 307:321-326.
[M70] Pinch, T. J. & W. E. Bijker (1984): The social construction of facts and artefacts, or how the sociology of science and the sociology of technology might benefit from each other. Social Studies of Science. Vol. 14:399-441.
[M71] Rasmussen, J. (1997): Risk management in a dynamic society: a modelling problem. Safety Science. Vol. 27:183-213.
[M72] Rasmussen, J., A. M. Peijtersen & L. P. Goodstein (1994): Cognitive systems engineering. Wiley, New York.
[M73] Rawls, J. (1971): A theory of justice. Harvard Univ. Press, Harvard.
[M74] Reason, J. (1987): The Chernobyl errors. Bulletin of the British Psychological Society. Vol. 40:201-226.
[M75] Renn, O. (1993): Wie lässt sich der «Energiefriede» fortführen? Ein Vorschlag für einen kooperativen Diskurs. NZZ. Nr. 156. 9.7.1993.
[M76] Ropohl, G. (1978, ^{2}1999): Allgemeine Technologie. Eine Systemtheorie der Technik. Carl Hanser Verlag, München.
[M77] Rowe, W. D. (1977): An anatomy of risk. John Wiley & Sons, New York.
[M78] Rowe, W. D. (1994): Understanding uncertainty. Risk Analysis. Vol. 14. No. 5: 743-750.
[M79] Scholz, R.W. (2000): Mutual learning as a basic principle of transdisciplinarity. In: R. W. Scholz et al. (eds.): Transdisciplinarity: Joint problem-solving among science, technology and society. Proc. of the International Transdisciplinarity 2000 Conference, Zurich, Feb 27–Mar 1, 2000. Workbook II: Mutual learning sessions. Vol. 2. Haffmanns Sachbuch, Zürich:13-17.
[M80] Scholz, R. W. & O. Tietje (2002): Embedded case study methods. Integrating quantitative and qualitative knowledge. Sage, Thousand Oaks, CA.
[M81] Shrader-Frechette, K. (1991): Risk and rationality. Univ. of California Press, Berkeley.
[M82] Solow, R. M. (1974): The economics of resources and the resources of economics. The Review of Economic Studies. Vol. 41:29-45.
[M83] Weidner, H. (1993): Der verhandelnde Staat. Minderung von Vollzugskonflikten durch Mediationsverfahren. In: Vollzugsprobleme. Problèmes de la mise en œuvre des politiques publiques. Schweizerisches Jahrbuch für Politische Wissenschaft. Bd. 33:225-244.
[M84] Weinberg, A. M. (1972): Science and trans-science. Minerva. Vol. 10:209-222.
[M85] Wene, C.-O. & R. Espejo (1999): A meaning for transparency in decision processes. In: K. Andersson (ed.): VALDOR 1999 – VALues in Decisions On Risk. Proceed-

ings. Stockholm, June 13-17, 1999. EC DGXI, SKI, SSI, US NWTRB, KASAM, Stockholm:404-421.
[M86] Weinmann, A. (1991): Uncertain models and robust control. Springer, Vienna.
[M87] Wissenschaftlicher Beirat der Bundesregierung Globale Umweltveränderungen, WBGU (1998): Welt im Wandel. Strategien zur Bewältigung globaler Umweltrisiken. Jahresgutachten 1998. Kurzfassung. WBGU, Bremerhaven (complete with Springer, Berlin, 1999).
[M88] Zio, E. (1999): Decision making in environmental management: a game theoretic approach. Radioactive Waste Management and Environmental Restoration. Vol. 21: 211-233.

2.2 Decision science, learning organisations, institutional aspects (D)

[D1] Abelson, R. P. & A. Levi (1985): Decision making and decision theory. In: G. Lindsey & E. Aronson (*eds.*): Handbook of social psychology. Vol. 1. Theory and Method. Lawrence Erlbaum, New York:231-309.
[D2] Alexander, E. R. (1970): The limits of uncertainty: A note. Theory and Decision. Vol. 6:363-370.
[D3] Argyris, C. (1976): Single-loop and double-loop models in research on decision making. Administrative Science Quarterly. Vol. 21:363-375.
[D4] Argyris, C. (1982): Reasoning, learning and action. Jossey-Bass Publishers, San Francisco.
[D5] Argyris, C. & D. Schön (1978): Organizational learning: a theory of action perspective. Addison-Wesley Publisher, Reading, MA.
[D6] Bussmann, W., U. Klöti & P. Knoepfel (*eds.*, 1997): Einführung in die Politikevaluation. Helbing & Lichtenhahn, Basel.
[D7] Esty, D. C. *et al.* (1998): The State Failure project: early warning research for US foreign policy planning. In: J. L. Davies & T. R. Gurr (1998, *eds.*): Preventive measures: building risk assessment and crisis early warning systems. Rowman and Littlefield, Boulder, CO.
[D8] Cyert, R. M. & J. G. March (1995): Eine verhaltenswissenschaftliche Theorie der Unternehmung. 2. Auflage. Deutsche Ausgabe. Herausgegeben vom Carnegie Bosch Institut. Schäffer-Poeschel, Stuttgart.
[D9] Dörner, D. (1989): Die Logik des Misslingens. Strategisches Denken in komplexen Situationen. Rowohlt, Reinbek bei Hamburg.
[D10] Derrickson, W. B. (1992): Managing large complex projects. 3rd ICHLRWM, Las Vegas. Vol. 2. ANS, La Grange Park, IL:1751-1757.
[D11] Dutke, S. (1994): Mentale Modelle: Konstrukte des Wissens und Verstehens. Kognitionspsychologische Grundlagen für die Software-Ergonomie. Verlag für Angewandte Psychologie, Göttingen.
[D12] Edwards, W. *et al.* (1984): What constitutes "a good decision"? Acta Psychologica. No. 56:5-27.
[D13] Elster, J. (1993, ^{2}1997): Risiko, Ungewissheit und Kernkraft. In: G. Bechmann (*ed.*): Risiko und Gesellschaft. Grundlagen und Ergebnisse interdisziplinärer Risikoforschung. Westdeutscher Verlag, Opladen:59-87.

[D14] Esser, J. K. & J. S. Lindoerfer (1988): Groupthink and the space shuttle Challenger accident: toward a quantitative case analysis. Behavioral Decision Making. Vol. 2:167-177.
[D15] Evers, A. & H. Nowotny (1987): Über den Umgang mit Unsicherheit. Die Entdeckung der Gestaltbarkeit von Gesellschaft. Suhrkamp, Frankfurt a. M.
[D16] Feick J. & W. Jann (1988): Nations matter – Vom Eklektizismus zur Integration in der vergleichenden Policy-Forschung. Sonderheft der Politischen Vierteljahresschrift. Nr. 19:197-221.
[D17] Fernandes, R. & H. A. Simon (1999): A study of how individuals solve complex and ill-structured problems. Policy Sciences. Vol. 32:225-245.
[D18] Fietkau, H.-J. (1984): Bedingungen ökologischen Handelns. Beltz, Weinheim.
[D19] Finger, M., S. Bürgin & U. Haldimann (1996): Ansätze zur Förderung organisationaler Lernprozesse im Umweltbereich. In: M. Roux & S. Bürgin (Hg.): Förderung umweltbezogener Lernprozesse in Schulen, Unternehmen und Branchen. Birkhäuser, Basel:43-70.
[D20] Finkel, A. M. (1990): Confronting uncertainty in risk management. A guide for decision-makers. Center for Risk Management. Resources for the Future, Washington, DC.
[D21] Fischhoff, B. (1985): Cognitive and institutional barriers to "informed consent". In: M. Gibson (ed.): To breathe freely. Risk, consent, and air. Rowman & Allanheld Publishers, Totowa, NJ:169-232.
[D22] Flüeler, T. (1998): Entscheidungsprozesse in der Entsorgung von Energieabfällen: Eindrücke eines Beobachters. Weiterbildungskurs ETH Zürich/Forum VERA, 10.-12.9.98. (Manuskript)
[D23] Franzen, A. (1998): Persönliches Umweltverhalten zwischen kollektivem Nutzen und individuellem Interesse. In: G. Schneider (Hg.): Zwischen allen Stühlen. Entscheidfindung in Konfliktsituationen. Rüegger, Chur.
[D24] Freiburghaus, D. & W. Zimmermann (1985): Wie wird Forschung politisch relevant? Erfahrungen in und mit den schweizerischen nationalen Forschungsprogrammen. Paul Haupt, Bern.
[D25] Frensch, P. A. & J. Funke (1995, eds.): Complex problem solving. the European perspective. Lawrence Erlbaum Assoc., Hillsdale, NJ:185-222.
[D26] Frey, B. S. & I. Bohnet (1996): Tragik der Allmende. Einsicht, Perversion und Überwindung. In: A. Diekmann & C. Jäger (Hg.): Umweltsoziologie. Sonderheft der (Special issue of) Kölner Zeitschrift für Soziologie und Sozialpsychologie. Nr. 36.
[D27] Funke, J. (1991): Solving complex problems: Exploration and control of complex systems. In: R. Sternberg & F. Frensch (eds.): Complex problem solving – Principles and mechanisms. Lawrence Erlbaum Assoc., Hillsdale, NJ:185-222.
[D28] Gigerenzer, G. & R. Selten (eds., 2002): Bounded rationality. The adaptive toolbox. MIT Press, Cambridge, MA.
[D29] Goldstone, J. A. (2000): State Failure Task Force Report: phase III findings. Sept. 2000. Science Applications International Corporation (SAIC). McLean, VA.
[D30] Gomez, P. & G. Probst (^{2}1997): Die Praxis des ganzheitlichen Problemlösens. Vernetzt denken. Unternehmerisch handeln. Persönlich überzeugen. Haupt, Bern.
[D31] Guba, E. G. & Y. S. Lincoln (1989): Fourth generation evaluation. Sage, Newbury Park.
[D32] Haberfellner, R. et al. (1976, 1992, ^{10}1999): Systems Engineering. Methodik und Praxis. Hg. W. F. Daenzer & F. Huber. Verlag Industrielle Organisation/Orell Füssli, Zürich.

[D33] Habermas, J. (1971) Toward a rational society. Heineman, London.
[D34] Habermas, J. (1981, 1995): Theorie des kommunikativen Handelns. Band I. Handlungsrationalität und gesellschaftliche Rationalisierung. Suhrkamp, Frankfurt a. M.
[D35] Hanson, S.O. (1991): An overview of decision theory. SKN Report 41. National Board for Spent Nuclear Fuel, SKN, Stockholm.
[D36] Hellstern, G.-M. (1988): Können Institutionen lernen? Wissensstrukturen, Informationsprozesse und Transfermechanismen in der Kommunalverwaltung. Teil 1. Europäische Hochschulschriften. Reihe V. Volks- und Betriebswirtschaft. Bd. 895. Peter Lang, Frankfurt a. M.
[D37] Helton, J. C. & D. E. Burmaster (1996): Guest editorial: treatment of aleatory and epistemic uncertainty in performance assessments for complex systems. Reliability Engineering and System Safety. Vol. 54:91-94.
[D38] Huber, O. (1998): Unsicherheit – Risiko – Chance. In: G. Schneider (Hg.): Zwischen allen Stühlen. Entscheidfindung in Konfliktsituationen. Rüegger, Chur.
[D39] Hussy, W. (1993): Denken und Problemlösen. Grundriss der Psychologie, Bd. 8. W. Kohlhammer, Stuttgart.
[D40] Japp, K. P. (1992): Selbstverstärkungseffekte riskanter Entscheidungen. Zur Unterscheidung von Rationalität und Risiko. Zeitschrift für Soziologie. Jg. 21. Heft 1:31-48.
[D41] Janis, I. L. (1972): Victims of groupthink. Houghton Mifflin, Boston, MA.
[D42] Janis, I. L. & L. Mann (1977): Decision making: a psychological analysis of conflict, choice, and commitment. Free Press, New York.
[D43] Jasanoff, S. (1985): Peer review in the regulatory process. Science, Technology & Human Values. Vol. 10. Issue 3:20-32.
[D44] Jungermann, H., H.-R. Pfister & K. Fischer (1998): Die Psychologie der Entscheidung. Eine Einführung. Spektrum Akademischer Verlag, Heidelberg.
[D45] Kahneman, D. & A. Tversky (1979): Prospect theory: an analysis of decision under risk. Econometrica. Vol. 47:263-291.
[D46] Kahneman, D. & A. Tversky (1981): The framing of decisions and the psychology of choice. Science. Vol. 211. Jan 30:453-458.
[D47] Kissling-Näf, I. & P. Knoepfel (1996): Lernfiguren in der schweizerischen Umweltpolitik. In: M. Roux & S. Bürgin (Hg.): Förderung umweltbezogener Lernprozesse in Schulen, Unternehmen und Branchen. Birkhäuser, Basel:159-184.
[D48] Kleindorfer, P. R., H. C. Kunreuther & P. J. H. Schoemaker (1993, [3]1998): Decision sciences. An integrative perspective. Cambridge Univ. Press, Cambridge, UK.
[D49] Klose, W. (1994): Ökonomische Analyse von Entscheidungsanomalien. European Univ. Studies. Series V. Economics and Management. Vol. 1533. Peter Lang, Frankfurt a. M./New York.
[D50] Krücken, G. (1989): Gesellschaft/Technik/Risiko: Analytische Perspektiven und rationale Strategien unter Gewissheit. Wissenschaftsforschung, Bd. 36. Kleine, Bielefeld.
[D51] LaPorte, T. R. (1982): On the design and management of nearly error-free organizational control systems. In: D. L. Sills, C. P. Wolf & V. B. Shelanski (*eds.*): Accident at Three Mile Island: the human dimensions. Westview Press, Boulder, CO:185-200.
[D52] Leeuw, F. L. R. C. Rist & R. C. Sonnichsen (1994): Can governments learn? Comparative perspectives on evaluation and organisational learning. Transaction Publishers, New Brunswick, USA.
[D53] Luce, R. D. & H. Raiffa (1957): Games and decisions. Wiley, New York.
[D54] Luhmann, N. (1968, 1973): Zweckbegriff und Systemrationalität. Suhrkamp, Frankfurt a. M.

References 305

[D55] MacCrimmon, K. R. & R. N. Taylor (1976): Decision making and problem solving. In: M. D. Dunnette (*eds.*): Handbook of Industrial and organizational psychology. Rand McNally, Chicago.
[D56] Mag, W. (1990): Grundzüge der Entscheidungstheorie. Franz Vahlen, München.
[D57] Majone, G. (1989): Evidence, argument, and persuasion in the policy process. Yale Univ. Press, New Haven, CO.
[D58] March, J. G. (*ed.*, 1991): Decisions and organisations. Basil Blackwell, Oxford (from R. M. Cyert & J. G. March (^{2}1995): Eine verhaltenswissenschaftliche Theorie der Unternehmung. Schäffer-Poeschel, Stuttgart)
[D59] Mazmanian, D. A. & P. A. Sabatier (1983): Implementation and public policy. Scott, Foresman and Company, Glenview, IL.
[D60] Mérö, L. (1998): Optimal entscheiden? Spieltheorie und die Logik unseres Handelns. Birkhäuser, Basel.
[D61] Mintzberg, H. (1994): The rise and fall of strategic planning. Prentice-Hall, New York.
[D62] Mintzberg, H., D. Raisinghani & A. Théorêt (1976): The structure of "unstructured" decision processes. Administrative Science Quarterly. Vol. 21, June:246-275.
[D63] Morone, J. G. & E. J. Woodhouse (1986): Averting catastrophe. Strategies for regulating risky technologies. Univ. of California Press, Berkeley.
[D64] Ninck, A. *et al.* (1997): Systemik. Integrales Denken, Konzipieren und Realisieren. Verlag Industrielle Organisation/Orell Füssli, Zürich.
[D65] Perrow, C. (1982): The President's Commission and the normal accident. In: D. L. Sills, C. P. Wolf & V. B. Shelanski (*eds.*): Accident at Three Mile Island: the human dimensions. Westview Press, Boulder, CO:173-184.
[D66] Perrow, C. (1983): The organizational context of human factors engineering. Administrative Science Quarterly. Vol. 28:521-541.
[D67] Perrow, C. (1984): Normal accidents. Living with high-risk technologies. Basic Books, New York.
[D68] Perrow, C. (1986): Lernen wir etwas aus den jüngsten Katastrophen? Soziale Welt. Bd. 37:390-401.
[D69] Rapoport, A. (1967): Escape from paradox. Scientific American. No. 217:50-56.
[D70] Reitman, W. R. (1964): Heuristic decision procedures, open constraints, and the structure of ill-defined problems. In: W. M. Shelly II & G. L. Bryan (*eds.*): Human judgments and optimality. Wiley, New York.
[D71] Rip, A. (1987): Controversies as informal technology assessment. Knowledge: Creation, Diffusion, Utilization. Vol. 8. No. 2:349-371.
[D72] Rohrmann, B. (1991): Akteure der Risiko-Kommunikation. In: H. Jungermann, B. Rohrmann & P. M. Wiedemann (*eds.*): Risikokontroversen. Konzepte, Konflikte, Kommunikation. Springer, Heidelberg:355-370.
[D73] Sabatier, P. (1987): Knowledge, policy-oriented learning, and policy change. Knowledge: Creation, Diffusion, Utilization. Vol. 8. No. 4:649-692.
[D74] Samuelson, W. & R. Zeckhauser (1988): Status quo bias in decision making. Risk and Uncertainty. Vol. 1:7-59.
[D75] Schein, E. H. (1985): Organizational culture and leadership. Jossey-Bass, San Francisco.
[D76] Simon, H. (1955): A behavioral model of rational choice. Quarterly Journal of Economics. Vol. 69. February:99-118.
[D77] Simon, H. (1973): The structure of ill structured problems. Artificial Intelligence. Vol. 4:181-201.

[D78] Sowden, L. (1984): The inadequacy of Bayesian decision theory. Philosophical Studies. Vol. 45:293-313.
[D79] Spillmann, W. (1996): Nachhaltigkeit – vom Grundsatz zur Umsetzung. Unbestreitbarer und umfassender Handlungsbedarf. NZZ. 15.3.96.
[D80] Tillman, A.-M. (2000): EIA procedure. Significance of decision making for LCA methodology. Environmental Impact Assessment Review. Vol. 20:113-123.
[D81] Travis, G. D. L. & H. M. Collins (1991): New light on old boys: cognitive and institutional particularism in the peer review system. Science, Technology & Human Values. Vol. 16. No. 3:322-341.
[D82] Tversky, A. & D. Kahneman (1981): The framing of decisions and the psychology of choice. Science. Vol. 211:453-458.
[D83] von Weizsäcker, E. U. & C. von Weizsäcker (1984): Fehlerfreundlichkeit. In: K. Kornwachs (ed.): Offenheit – Zeitlichkeit – Komplexität. Zur Theorie der Offenen Systeme. Campus, Frankfurt a. M.
[D84] Weber, M. (^{5}1963): Gesammelte Aufsätze zur Religionssoziologie. Band I. Tübingen.
[D85] Weick, K. E. (1976): Educational organizations as loosely coupled systems. Administrative Science Quarterly. Vol. 21. No. 1:1-19.
[D86] Weick, K. E. (1987): Organizational culture as a source of high reliability. California Management Review. Vol. 29. No. 2:112-127.
[D87] Wiedemann, P. (1991): Strategien der Risiko-Kommunikation und ihre Probleme. In: H. Jungermann, B. Rohrmann & P. M. Wiedemann (eds.): Risikokontroversen. Konzepte, Konflikte, Kommunikation. Springer, Heidelberg:371-394.
[D88] Windhoff-Héritier, A. (1980): Politikimplementation, Ergebnisse der Sozialwissenschaften, Ziel und Wirklichkeit politischer Entscheidungen. Anton Hain, Königstein/Ts.
[D89] Windhoff-Héritier, A. (1987): Policy-Analyse. Eine Einführung. Campus, Frankfurt a. M.
[D90] World Commission on Environment and Development (Brundtland Commission) (1987): Our common future. Oxford Univ. Press, Oxford.
[D91] Wynne, B. (1975): The rhetoric of consensus politics: a critical review of technology assessment. Research Policy. Vol. 4:108-158.
[D92] Wynne, B. (1988): Unruly technology: practical rules, impractical discourses and public understanding. Social Studies of Science. Vol. 4:108-158.
[D93] Wynne, B. (1991): Knowledges in context. Science, Technology & Human Values. Vol. 16. No. 1:111-121.
[D94] Zillessen, H., P. C. Dienel & W. Strubelt (Hg., 1993): Die Modernisierung der Demokratie. Internationale Ansätze. Westdeutscher Verlag, Opladen.

2.3 Risk perception (R)

[R1] Barben, D. & M. Dierkes (1993): Un-Sicherheiten im Streit um Sicherheit. Zur Relevanz der Kontroversen um die Regulierung technischer Risiken. In: M. Dierkes (ed.): Die Technisierung und ihre Folgen: zur Biographie eines Forschungsfeldes. Edition sigma, Rainer Bohn, Berlin.
[R2] Barnes, D. G. (1994): Times are tough – Brother, can you paradigm? Risk Analysis. Vol. 14. No. 3:219-223.
[R3] Baird, J. C. (1985): Prediction of environmental risk over very long time scales. Environmental Psychology. Vol. 6. No. 3:233-244.

[R4] Bechmann, F. (1995): Im Zeitalter des Risikos. In the Age of Risk. In: J. Brauns & G. G. Eigenwillig (*eds.*): Entsorgung: Wiederverwertung – Beseitigung. Band I. 27. Jahrestagung. Wolfenbüttel, 25. – 29. September 1995. Fachverband für Strahlenschutz e. V. TÜV Rheinland, Köln:3-18.

[R5] Beck, U. (1995, 1986 orig.): Risk society. Towards a new modernity. Sage, London.

[R6] Beck, U. (1995, 1988 orig.): Ecological politics in an Age of Risk. Polity Press, Cambridge, UK.

[R7] Beratende Kommission für Umweltforschung BKUF/CCRE des BUWAL (2002): Erklärung von Gerzensee. Visionen zur Umweltforschung in der Schweiz. 21.3.2002.

[R8] Bibliographisches Institut Mannheim/Wien/Zürich (1983): Meyers grosses Taschenlexikon in 24 Bänden. Band 24: Wau – Zz. Bibliographisches Institut, Mannheim.

[R9] Bosshardt, Chr. (2001): Homo Confidens. Eine Untersuchung des Vertrauensphänomens aus soziologischer und ökonomischer Perspektive. Social Strategies. Vol. 33. Peter Lang, Bern.

[R10] Brehmer, B. (1987): The psychology of risk. In: W. T. Singleton and J. Hovden (*eds.*): Risks and decisions. John Wiley & Sons, Chichester:25-39.

[R11] CASS & ProClim (1997): Research on sustainability and global change. Visions in science policy by Swiss researchers [orig. Visionen der Forschenden. Forschung zu Nachhaltigkeit und Globalem Wandel – Wissenschaftspolitische Visionen der Schweizer Forschenden. Konferenz der Schweizerischen Wissenschaftlichen Akademien CASS, Forum für Klima und Global Change ProClim–, Schweizerische Akademie der Naturwissenschaften SANW. Juni 1997. ProClim–, SANW, Bern]. http://www.proclim.ch/Reports/Visions97/Visions_E.html.

[R12] Combs, B. & P. Slovic (1979): Causes of death: biased newspaper coverage and biased judgments. Journalism Quarterly. Vol. 56:837-843.

[R13] Daamen, D. D. L., B. Verplanken & C. J. H. Midden (1986): Accuracy and consistency of lay estimates of annual fatality rates. In: B. Brehmer, H. Jungermann, P. Lourens & G. Sévon (*eds.*): New directions in research on decision making. North-Holland, Amsterdam.

[R14] DeLuca, D. R., J. A. J. Stolwijk and W. Horowitz (1986): Public perceptions of technological risks. A methodological study. In: V. Covello, J. Menkes & J. Mumpower (*eds.*): Risk evaluation and management. Plenum Press, New York:25-67.

[R15] Earle, T. C. (2002): From risk perception to social trust: an outline of recent contributions of psychology to risk management. In: NEA (*ed.*): Better integration of radiation protection in modern society. [2nd] Workshop Proc. Villigen, 23–25 January 2001. OECD, Paris:27-33.

[R16] Edwards, W. & D. von Winterfeldt (1986): Public disputes about risky technologies. Stakeholders and arenas. In: V. Covello, J. Menkes and J. Mumpower (*eds.*): Risk evaluation and management. Plenum Press, New York:69-92.

[R17] Elster, J. (1987): Subversion der Rationalität. Campus, Frankfurt a. M.

[R18] Fischhoff, B. (1977): Cost benefit analysis and the art of motorcycle maintenance. Policy Sciences. Vol. 8:177-202.

[R19] Fischhoff, B. (1985): Managing risk perceptions. Issues in Science and Technology. Vol. 2 (Fall). No. 1:83-96.

[R20] Fischhoff, B. (1990): Understanding long-term environmental risks. Risk and Uncertainty. Vol. 3:315-330.

[R21] Fischhoff, B. (1995): Risk perception and communication unplugged: twenty years of process. Risk Analysis. Vol. 15. No. 2:137-145.

[R22] Fischhoff, B., P. Slovic & S. Lichtenstein (1977): Knowing with certainty: The appropriateness of extreme confidence. Experimental Psychology: Human Perception and Performance. Vol. 3:552-564.

[R23] Fischhoff, B., P. Slovic, S. Lichtenstein, S. Read & B. Combs (1978): How safe is safe enough? A psychometric study of attitudes towards technological risks and benefits. Policy Sciences. Vol. 9:127-152.

[R24] Frewer, L. J. (2003): Trust, transparency and social context: implications for social amplification of risk. In: N. Pidgeon, R. E. Kasperson & P. Slovic (eds.): The social amplification of risk. Cambridge Univ. Press, Cambridge, UK:123-137.

[R25] Freudenberg, W. R. & S. K. Pastor (1992): NIMBYs and LULUs: stalking the syndromes. Social Issues. Vol. 48. No. 4:39-61.

[R26] Gadomska, M. (1994): Risk communication. In: IAEA: Radiation and society: comprehending radiation risk. IAEA Conference, 24–28 Oct. 1994, Paris. Vol. 1: Comprehending radiation risks. A report to the IAEA. IAEA, Vienna:147-166.

[R27] Gardner, G. T. & L. C. Gould (1989): Public perceptions of the risks and benefits of technology. Risk Analysis. Vol. 9. No. 2:225-242.

[R28] Hansson, S. O. (1989): Dimensions of risk. Risk Analysis. Vol. 9. No. 1:107-112.

[R29] Hohenemser, C., R. W. Kates & P. Slovic (1983): The nature of technological hazard. Science. Vol. 220:378-384.

[R30] Hornborg, A. (1996): Ecology as semiotics. In: P. Descola & G. Pàlsson (eds.): Nature and society. Anthropological perspectives. Routledge, London:45-62.

[R31] Hunt, S., L. J. Frewer, & R. Shepard (1999): Public trust in sources of information about radiation risks in the UK. Journal of Risk Research. Vol. 2. No. 2:167-180.

[R32] Hynes, M. & E. VanMarcke (1976): Reliability of embankment performance prediction. ASCE Engineering Mechanics Division Specialty Conference. Univ. of Waterloo Press, Waterloo, Ontario.

[R33] Jungermann, H. & P. Slovic (1993a): Charakteristika individueller Risikowahrnehmung. In: W. Krohn & G. Krücken (eds.): Riskante Technologien: Reflexion und Regulation. Einführung in die sozialwissenschaftliche Risikoforschung. Suhrkamp, Frankfurt a. M.:79-100.

[R34] Jungermann, H. & P. Slovic (1993b, ^{2}1997): Die Psychologie der Kognition und Evaluation von Risiken. In: G. Bechmann (ed.): Risiko und Gesellschaft. Grundlagen und Ergebnisse interdisziplinärer Risikoforschung. Westdeutscher Verlag, Opladen:167-207.

[R35] Kahneman, D., P. Slovic & A. Tversky (eds., 1982): Judgement under uncertainty: Heuristics and biases. Cambridge Univ. Press, Cambridge, MA.

[R36] Kasperson J. X., R. E. Kasperson, N. Pidgeon & P. Slovic (2003): The social amplification of risk: assessing fifteen years of research and theory. In: N. Pidgeon, R. E. Kasperson & P. Slovic (eds.): The social amplification of risk. Cambridge Univ. Press, Cambridge, UK:13-46.

[R37] Kasperson, R. E. (1986): Six propositions on public participation and their relevance for risk communication. Risk Analysis. Vol. 6. No. 3:275-281.

[R38] Kasperson, R. E., D. Golding & S. Tuler (1992): Social distrust as a factor in siting hazardous facilities and communicating risks. Social Issues. Vol. 48. No. 4:161-187.

[R39] Kasperson, R. E., O. Renn, P. Slovic, H. S. Brown, J. Emel, R. Goble, J. X. Kasperson & S. Ratick (1988): The social amplification of risk: a conceptual framework. Risk Analysis. Vol. 8. No. 2:177-187.

[R40] Keeney, R. L. (1980): Equity and public risk. Operations Research. Vol. 28. No. 3. Part I:527-534.

[R41] Keeney, R. L. & D. von Winterfeldt (1986): Improving risk communication. Risk Analysis. Vol. 6. No. 4:417-424.
[R42] Klinke, A. and O. Renn (2001): Precautionary principle and discursive strategies: classifying and managing risks. Risk Research. Vol. 4. No. 2:159-173.
[R43] Krohn, W. & G. Krücken (1993): Risiko als Konstruktion und Wirklichkeit. Eine Einführung in die sozialwissenschaftliche Risikoforschung. In: *Id.* (*eds.*): Riskante Technologien: Reflexion und Regulation. Einführung in die sozialwissenschaftliche Risikoforschung. Suhrkamp, Frankfurt a. M.:9-44.
[R44] Krugman, P. (2003): Delusions of power. New York Times, 2003-3-28.
[R45] Lichtenstein, S., B. Fischhoff & L. D. Phillips (1977): Calibration of probabilities: the state of the art. In: H. Jungermann & G. de Zeeuw (*eds.*): Decision making and change in human affairs. D. Reidel, Dordrecht, NL.
[R46] Lichtenstein, S., P. Slovic, B. Fischhoff, M. Layman & B. Combs (1978): Judged frequency of lethal events. Experimental Psychology: Human Learning and Memory. Vol. 4:551-578.
[R47] Lidskog, R. & I. Elander (1992): Reinterpreting locational conflicts: NIMBY and nuclear waste in Sweden. Policy and Politics. Vol. 20. No. 4:249-264.
[R48] Luhmann, N. ([1969], 1975,1983, ^{2}1997): Legitimation durch Verfahren. Suhrkamp, Frankfurt a. M.
[R49] Luhmann, N. (1988): Familiarity, confidence, trust: Problems and alternatives. In: D. Gambetta (*ed.*): Trust: Making and Breaking cooperative relations. Basil Blackwell, Oxford:94-107.
[R50] Luhmann, N. (1990, ^{2}1993): Soziologische Aufklärung 5. Konstruktivistische Perspektiven. Westdeutscher Verlag, Opladen.
[R51] Luhmann, N. (1991): Soziologie des Risikos. Walter de Gruyter, Berlin.
[R52] Marks, G. & D. von Winterfeldt (1984): "Not in my back yard": influence of motivational concerns on judgments about a risky technology. Applied Psychology. Vol. 69. No. 3:408-415.
[R53] Marris, C. & I. Langford (1996): No cause for alarm. New Scientist. Vol. 151. Issue 2049:36.
[R54] Mazmanian, D. & D. Morell (1990): The "NIMBY" syndrome: Facility siting and the failure of democratic discourse. In: N. J. Vig & M. E. Kraft (*eds.*): Environmental policy in the 1990s. Toward a new agenda. CQ-Press, Washington, DC:125-143.
[R55] Nelkin, D. (1988): Risk reporting and the management of industrial crises. Management Studies. Vol. 25:341-351.
[R56] Nelkin, D. & M. Pollak (1979): Public participation in technological decisions: reality of grand illusion? Technology Review. Vol. 81. No. 8:55-64.
[R57] Nelkin, D. & M. Pollak (1980): Consensus and conflict resolution. In: M. Dierkes, S. Edwards & R. Coppock (*eds.*): Technological risk. Its perception and handling in the European Community. Oelgeschlager, Gunn & Hain, Cambridge/Hain, Königstein:65-75.
[R58] Nowitzky, K.-D. (1993, ^{2}1997): Konzepte zur Risiko-Abschätzung und -Bewertung. In: G. Bechmann (*ed.*): Risiko und Gesellschaft. Grundlagen und Ergebnisse interdisziplinärer Risikoforschung. Westdeutscher Verlag, Opladen:125-144.
[R59] Nowotny, H. & R. Eisikovic (1990): Entstehung, Wahrnehmung und Umgang mit Risiken. Forschungspolitische Früherkennung (FER) B/34. Schweizerischer Wissenschaftsrat SWR, Bern.
[R60] O'Riordan, T. (1983): The cognitive and political dimensions of risk analysis. Environmental Psychology. Vol. 3:345-354.

[R61] O'Riordan, T. (1995): Introduction: risk management in its social and political context. In: T. O'Riordan (ed.): Perceiving environmental risks. Readings in environmental psychology. Academic Press, London:1-10.

[R62] Otway, H. (1980): The perception of technological risks: a psychological perspective. In: M. Dierkes, S. Edwards & R. Coppock (eds.): Technological risk. Its perception and handling in the European Community. Oelgeschlager, Gunn & Hain, Cambridge/Hain, Königstein:35-44.

[R63] Otway, H. (1987): Experts, risk communication, and democracy. Risk Analysis. Vol. 7. No. 2:125-129.

[R64] Otway, H. & K. Thomas (1982): Reflections on risk perception and policy. Risk Analysis. Vol. 2. No. 2:69-82.

[R65] Otway, H. & D. von Winterfeldt (1992): Expert judgment in risk analysis and management: process, context, and pitfalls. Risk Analysis. Vol. 12. No. 1:83-93.

[R66] Otway, H. & B. Wynne (1989): Risk communication: paradigm and paradox. Risk Analysis. Vol. 9. No.2:141-145.

[R67] Piller, C. (1991): The fail-safe society. Community defiance and the end of American technological optimism. Basic Books, New York.

[R68] Ravetz, J. R. (1980): Public perceptions of acceptable risks as evidence for their cognitive, technical and social structure. In: Dierkes, M., S. Edward & R. Coppock (eds.): Technological risk. Its perception and handling in the European Community. Oelgeschlager, Gunn & Hain, Königstein:45-54.

[R69] RISKPERCOM (1998): Midterm report. International CEU project. CIEMAT (Spain), CRR (Sweden), IFR (UK), IPSN (France), NRPA (Norway). Center for Risk Research CRR, Stockholm.

[R70] Sandman, P. M. (1987): The public's role in risk communication. Speech at the Workshop on the Role of Government in Health Risk Communication and Public Education. Alexandria, VA, 1987-1-21.

[R71] Schweizerischer Nationalfonds (SNF) (2002): Wissen – und doch nicht handeln. Studie [der ETH Zürich und der Techn. Univ. Eindhoven] über umweltbezogenes Wissen und Handeln. Medienmitteilung vom 10.4.2002. SNF, Bern.

[R72] Siegrist, M., T. C. Earle & H. Gutscher (2003): Test of a trust and confidence model in the applied context of electromagnetic field (EMF) risks. Risk Analysis. Vol. 23. No. 4:705-716.

[R73] Sjöberg, L. (1980): The risks of risk analysis. Acta Psychologica. Vol. 45:301-321.

[R74] Sjöberg, L. (1998): Worry and risk perception. Risk Analysis. Vol. 18. No. 1:85-93.

[R75] Sjöberg, L. (2001): Limits of knowledge and the limited importance of trust. Risk Analysis. Vol. 21. No. 1:189-198.

[R76] Sjöberg, L. & B.-M. Drottz-Sjöberg (1994): Risk perception. In: IAEA (ed.): Radiation and society: comprehending radiation risk. IAEA Conference, 24–28 Oct 1994, Paris. Vol. 1: Comprehending radiation risks. A report to the IAEA. IAEA, Vienna: 29-59.

[R77] Slovic, P. (1987): Perception of risk. Science. Vol. 236:280-285.

[R78] Slovic, P. (1993): Perceived risk, trust, and democracy. Risk Analysis. Vol. 13. No. 6:675-682.

[R79] Slovic, P. (1999a): Perceived risk, trust, and democracy. In: Cvetkovich & R. E. Löfstedt (1999, eds.): Social trust and the management of risk. Earthscan, London: 42-52.

[R80] Slovic, P. (1999b): Trust, emotion, sex, politics, and science: Surveying the risk-assessment battlefield. Risk Analysis. Vol. 19. No. 4:689-701.

[R81]	Slovic, P., B. Fischhoff & S. Lichtenstein (1980): Facts and fears: understanding perceived risk. In: R. C. Schwing & W. A. Albers (*eds.*): Societal risk assessment. How safe is safe enough? Plenum Press, New York: 181-214.
[R82]	Slovic, P., B. Fischhoff & S. Lichtenstein (1985): Characterizing perceived risk. In: R. W. Kates, C. Hohenemser & J. X. Kasperson (*eds.*): Perilous progress: managing the hazards of technology. Westview, Boulder, CO:92-125.
[R83]	Slovic, P., B. Fischhoff and S. Lichtenstein (1986): The psychometric study of risk perception. In: V. Covello, J. Menkes and J. Mumpower (*eds.*): Risk evaluation and management. Plenum Press, New York:3-24.
[R84]	Slovic, P., S. Lichtenstein & B. Fischhoff (1984): Modeling the societal impact of fatal accidents. Management Review. Vol. 30, No. 4:464-474.
[R85]	Starr, C. (1969): Social benefit versus technological risk. What is our society willing to pay for safety? Science. Vol. 165:1232-1238.
[R86]	Svenson, O. & G. Karlsson (1989): Decision making, time horizons, and risk in the very long-term perspective. Risk Analysis. Vol. 9. No. 3:385-399.
[R87]	Tesh, S. N. (1999): Citizen experts in environmental risks. Policy Studies. Vol. 32: 39-58.
[R88]	Thomas, K., E. Swaton, M. Fishbein & H. J. Otway (1980): Nuclear energy: the accuracy of policy makers' perceptions of public beliefs. Behavioral Science. Vol. 25:332-344.
[R89]	Tversky, A. & D. Kahneman (1973): Availability: a heuristic for judging frequency and probability. Cognitive Psychology. Vol. 4:207-232.
[R90]	Tversky, A. & D. Kahneman (1974): Judgment under uncertainty: heuristics and biases. Science. Vol. 185:1124-1131.
[R91]	Vlek, C. & P.-J. Stallén (1981): Judging risks and benefits in the small and in the large. Organizational Behavior and Human Performance. Vol. 28:235-271.
[R92]	von Winterfeldt, D. & W. Edwards (1984): Patterns of conflict about risky technologies. Risk Analysis. Vol. 4. No. 1:55-68.
[R93]	von Winterfeldt, D., R. S. John & K. Borcherding (1981): Cognitive components of risk ratings. Risk Analysis. Vol. 1. No. 4:277-287.
[R94]	Weinstein, N. D. (1980): Unrealistic optimism about future life events. Personality and Social Psychology. Vol. 39:806-820.
[R95]	Wynne, B. (1983a): Redefining the issues of risk and public acceptance. The social viability of technology. Futures. Vol. 15. No. 1:13-32.
[R96]	Wynne, B. (1983b): Technologie, Risiko und Partizipation: Zum gesellschaftlichen Umgang mit Unsicherheit. In: J. Conrad (*ed.*): Gesellschaft, Technik und Risikopolitik. Springer, Berlin:156-187.
[R97]	Wynne, B. (1995): Public understanding of science. In: S. Jasanoff *et al.* (*eds.*): Handbook of science and technology. Sage, Thousand Oaks, London:361-388.

2.4 Radioactive Waste: General, international (G)

[G1]	Acland, A. (2001): Principles and characteristics of stakeholder dialogue processes. Consolidated draft for comment. Ref: CD2/10/5/01. May 2001. Dialogue by Design Ltd., Wotton-under-Edge, UK.
[G2]	Åhäll, K.-I. *et al.* (1988): Nuclear waste in Sweden – The problem is not solved! Lindbergs Grafiska, Uppsala.

[G3] Agreement between the Federal Government of Germany and the utility companies dated 14 June 2000. www.bmu.de.
[G4] AkEnd (2002): Selection procedure for repository sites. Recommendations of the AkEnd – Committee on a Selection Procedure for Repository Sites, AkEnd. Dec 2002. Federal Office for Radiation Protection, BfS, Salzgitter. www.akend.de.
[G5] Andersson, J. et al. (1999): Transparency and public participation in complex decision processes. Pre-study for a Decision Research Institute in Oskarshamn. Research Report. Division of Land and Water Resources, Dept. of Civil and Environmental Engineering, Royal Institute of Technology KTH, Stockholm.
[G6] Andersson, K. et al. (2004): Transparency and public participation in radioactive waste management. RISCOM II final report. Oct./Dec. 2003. SKI Report 2004:08. www.karinta-konsult.se/RISCOM.htm, www.valdoc.org, www.ski.se.
[G7] ANDRA, Agence nationale pour la gestion des déchets radioactifs (no yr.): Centre de l'Aube. ANDRA, Fontenay-aux-Roses.
[G8] ANDRA (2003): Thematic Network on the role of monitoring in a phased approach to geological disposal. Methods and techniques. Contribution to final report. Chapter 5. Presentation of draft report. EC/Euratom 5th Framework Programme. 2003-1-21. Brussels.
[G9] AVT (1993): The position of the Dutch Government on deep burial. The National Environmental Policy Plan, action point 62. 8 pp.
[G10] AECL, Atomic Energy of Canada Ltd. (1994): Environmental impact statement on the concept for disposal of Canada's nuclear fuel waste. AECL-10711, COG-93-1. AECL, Pinawa, Manitoba.
[G11] AECL (1996): The disposal of Canada's nuclear fuel waste: A study of postclosure safety of in-room emplacement of used CANDU fuel in copper containers in permeable plutonic rock. Vol. 1: Summary. AECL-11494-1, COG-95-552-1. AECL, Chalk River, Ontario.
[G12] Baer, A. J. (2000): Issues & answers towards improved management of radioactive waste. IAEA-Bulletin. Vol. 42. No. 3:19-20.
[G13] Baer, A. J. (2002): Summary by the Conference President. IAEA International Conference on Issues and Trends in Radioactive Waste Management. IAEA, Vienna, Dec 9–13, 2002. www-rasanet.iaea.org/downloads/meetings/waste-trends02.pdf.
[G14] Barrett, L. (1998): Deep geologic disposal: international perspectives. 8th ICHLRWM, Las Vegas. ANS, La Grange Park, IL. Manuscript (introductory speech).
[G15] Basset, G.-W., H. Jenkins-Smith & C. Silva (1996): On-site storage of high-level nuclear waste: attitudes and perceptions of local residents. Risk Analysis. Vol. 16. No. 3:309-319.
[G16] Baumann, W. (1995): Verfassungsrechtliche Fragen der Endlagerung atomarer Abfälle. In: IPPNW (ed.): Die Endlagerung radioaktiver Abfälle. Hirzel, Stuttgart.
[G17] Bell, T. (2002): The Department of Energy's environmental monitoring support for Rongelap resettlement in the Marshall Islands: A partnership for the future. In: NEA (ed.): Better integration of radiation protection in modern society. [2nd] Workshop Proc., Villigen, Switzerland, 23–25 January 2001. OECD, Paris:183-186.
[G18] Benson, A., W. Morgan, P. C. Reyes & P. Van Nelson (1993): A public involvement planning model for the Civilian Radioactive Waste Management Program. 4th ICHLRWM, Las Vegas. Vol. 1. ANS, La Grange Park, IL:1036-1039.
[G19] Berndes, S. & K. Kornwachs (1996): Transferring knowledge about high-level waste repositories: an ethical consideration. 7th ICHLRWM, Las Vegas. ANS, La Grange Park, IL:494-498.

References

[G20] Bjarnadóttir, H. & T. Hilding-Rydevik (2001): Final disposal of spent nuclear fuel in Sweden. Some unresolved issues and challenges in the design and implementation of the forthcoming planning and EIA processes. SKI Report 01:24. SKi, Stockholm.

[G21] Board on Radioactive Waste Management/National Research Council (2001): Disposition of high-level waste and spent nuclear fuel. The continuing societal and technical challenges. National Academy Press, Washington, DC.

[G22] Bradley, D. J., C. W. Franck & Y. Mikerin (1997): Contamination nucléaire en Sibérie. La Recherche. Vol. 304:56-59.

[G23] Brasser, T. (1996): Disposal of Radioactive Waste and Chemical Toxic Waste in Underground Repositories (part of a programme in the field of management and storage of radioactive waste (1990-1994) of the Commission of the European Communities). GTDC Workshop 96: Toxic Waste. 4–6 Sep., Vienna.

[G24] Brasser, T., I. Simón, R. Little & A. Van Dalen (1997): Comparison of disposal and safety assessment methods for toxic and radioactive wastes. In: T. McMenamin (ed.): Fourth European Conference on Management and Disposal of Radioactive Waste. European Communities, Luxembourg:708-722.

[G25] Brown, P. A. (2000): The Canadian experience with public interveners on the long-term management of nuclear fuel. In: NEA: Stakeholder confidence and radioactive waste disposal. Workshop Proc. Paris, 28–31 Aug 2000. OECD, Paris:53-57.

[G26] Carnes, S.A., E. Peelle, E. D. Copenhaver, J. H. Sorensen, E. J. Soderstrom, J. H. Reed, & D. J. Bjornstad (1983): Incentives and nuclear waste siting: prospects and constraints. Energy Systems and Policy. Vol. 7. No. 4:323-351.

[G27] Bullard, C. W. (1992): Low level radioactive waste. Regaining public confidence. Energy Policy. August: 712-720.

[G28] Bullock Flynn, C. (1982): Reactions of local residents to the accident at Three Mile Island. In: D. L. Sills, C. P. Wolf & V. B. Shelanski (eds.): Accident at Three Mile Island: the human dimensions. Westview Press, Boulder, CO:49-63.

[G29] Burdge, R. J. (1992): Mandated public participation in siting of hazardous and conventional waste facilities: the Illinois experience. 3rd ICHLRWM, Las Vegas. Vol. 2. ANS, La Grange Park, IL:1909-1915.

[G30] Carlsson, T. et al. (2001): The Oskarshamn model for public involvement in the siting of nuclear facilities. In: K. Andersson (ed.): VALDOR 2001 – Proceedings. Stockholm, June 10–14, 2001. SKI, SSI, EA UK, UK Nirex, EC DG-E, Stockholm: 334-343.

[G31] Centre for Economic and Environmental Development, UK CEED (2000/2001): Workshops on the monitoring and retrievability of radioactive waste, Manchester Town Hall, 2 Dec 2000/12 Feb 2001. Reports for Nirex prepared by UK CEED. www.nirex.co.uk/news/na10212.htm.

[G32] Canty, M. J. et al. (1990): Internationale Kernmaterialüberwachung und physischer Schutz. Status und Perspektiven. Jül-Spez-553. Forschungszentrum Jülich, Jülich.

[G33] Chapman, N. & Ch. McCombie (2003): Principles and standards for the disposal of long-lived radioactive wastes. Waste Management Series, No. 3. Pergamon/Elsevier, Amsterdam.

[G34] Civilian Radioactive Waste Management System, Management & Operating Contractor (1997): Reference design description for a geologic repository. Revision 01. Sept.. Yucca Mountain Project, Las Vegas, NV.

[G35] Clarke, R. H. (2003): the evolution of the system of radiological protection: the justification for new ICRP Recommendations. In: K. Andersson (ed.): VALDOR 2003 –

Proceedings. Stockholm, June 9–13, 2003. SCK•CEN, SKI, SSI, NKS, OECD/NEA, UK Nirex, Stockholm:1-11. (in oral presentation)

[G36] Coates, R. (2002): BNFL experience of public engagement: Expectations for risk policies. In: NEA (*ed.*): Better integration of radiation protection in modern society. [1st] Workshop Proc., Villigen, Switzerland, 23–25 January 2001. OECD, Paris: 151-156.

[G37] Commission of the European Communities (2001): Communication from the Commission ... On the sixth environment action programme of the European Community "Environment 2010: Our future, Our choice". COM (2001)31 final. 2001/0029 (COD). Brussels, 24.1.2001.

[G38] Commission of the European Communities (2002a): Amended proposals for Council Decisions concerning the specific programmes implementing the Sixth Framework Programme of the European ... Atomic Energy Community for research and training activities (2002–2006). COM(2002) 43 final. 30.1.2002. Brussels.

[G39] Commission of the European Communities (2002b): Draft proposal for a Council Directive (Euratom) on the management of spent nuclear fuel and radioactive waste. COM(2003) 32 final. 2003/0022(CNS). 2003-1-30. Brussels. www.europa.eu.int/-comm/energy/nuclear/new_package.htm.

[G40] Committee of the Regions (1998): Resolution of the Committee of the Regions on "Nuclear Safety and Local/Regional Democracy" (98/C 251/06). 10.8.1998. Official Journal of the European Communities. C 251/34-36.

[G41] Cordis RTD-projects (2002a): RISCOM II. Enhancing transparency and public participation in nuclear waste management (2000–2003, 36 months). Record Control Number: 53141. European Communities, Brussels.

[G42] Cordis RTD-projects (2002b): Thematic network on the role of monitoring in a phased approach to disposal (2001–2003, 24 months). Record Control Number: 58152. European Communities, Brussels.

[G43] Cook, B. J., J. L. Emel & R. E. Kasperson (1990): Organizing and managing radioactive waste disposal as an experiment. Policy Analysis and Management. Vol. 9. No. 3:339-366.

[G44] Cook, N. G. W., W. E. Kastenberg, T. LaPorte, M. O'Hare (1996): Institutional and technical issues regarding nuclear stewardship: radioactive waste and nuclear materials management. CNTWM 96-1. Center for Nuclear and Toxic Waste Management, Univ. of California, Berkeley.

[G45] Cook, P. (1997): The role of the earth sciences in sustaining our life-support system. British Geological Survey Technical Report WQ/97/1. Natural Environment Research Council, Swindon, Wiltshire, UK.

[G46] Cropper, M. & P. Portney (1992): Discounting human lives. Resources for the Future. No. 108. Summer 1992:4.

[G47] Cunningham, R. E. (1998): Summary and conclusions. In: NEA (*ed.*): The societal aspects of decisionmaking in complex radiological situations. [1st] Workshop Proc., Villigen, Switzerland, 13–15 January 1998. OECD, Paris:139-144.

[G48] de Marsily, G., E. Ledoux, A. Barbreau & J. Margat (1977): Nuclear waste disposal: Can the geologist guarantee isolation? Science. Vol. 197:519-527.

[G49] Directorate General for Environmental Protection at the Ministry of Housing, Physical Planning and Environment (1989): Premises for risk management. Risk limits in the context of environmental policy. Annex to the Dutch Environmental Policy Plan "Kiezen of Verliezen" (to Choose or to Lose). Second Chamber of the States General. 1988-89 session, 21137. No. 1-2 (no. 5).

[G50] Drottz-Sjöberg, B.-M. & L. Sjöberg (1990): Risk perception and worries after the Chernobyl accident. Environmental Psychology. Vol. 10:135-149.
[G51] Dutton, L. M. C. (2001): The role of monitoring in a phased approach to disposal. IBC's 3rd International Two-day Conference on Managing Radioactive Waste. London, 11–12 Dec, 2001. IBC Global Conferences, London.
[G52] Easterling, D. (1997): The vulnerability of the Nevada visitor economy to a repository at Yucca Mountain. Risk Analysis. Vol. 17. No. 5:635-647.
[G53] Easterling, J. B. & M. Peck (1991): Case studies in cooperation with external groups. 2nd ICHLRWM, Las Vegas. Vol. 2. ANS, La Grange Park, IL:1343-1346.
[G54] Eigenwillig, G. G. (1997): Ist eine Internationalisierung der Endlagerung problemlösend? In: Schwerpunktthema «Entsorgung: Das Reizthema des Atomzeitalters?». Strahlenschutz-Praxis. Heft 3/97. Organ des Fachverbandes für Strahlenschutz e.V.:44.
[G55] Eiser J. R. et al. (1990): Nuclear attitudes after Chernobyl: a cross-national study. Environmental Psychology. Vol. 10:101-110.
[G56] Espejo, R. & A. Gill (1998): The systemic roles of SKI and SSI in the Swedish nuclear waste management system. Synchro's report for project RISCOM. SKI Report 98:4/SSI-report 98-2. SKi, SSI, Stockholm.
[G57] European Council (2002): Council Decision of 30 September 2002 Adopting a Specific Programme (Euratom) for Research and Training on Nuclear Energy (2002–2006). Official Journal of the European Communities. L294/74-85. 2002-10-29.
[G58] EU, European Union (2002a): Research and Technology Development beyond 2002. Sixth Framework Programme. Priority thematic area of research in FP6. 1.7 Citizens and Governance in a Knowledge-based society. www.cordis.lu/fp6/eoi-instruments/-citizens.htm.
[G59] EU (2002b) Research and Technology Development beyond 2002. Sixth Framework Programme. Priority Thematic Area of Research in FP6. 2.2 Management of Radioactive Waste. www.cordis.lu/fp6/eoi-analysis.htm.
[G60] EU (2002c): Research and Technology Development beyond 2002. Sixth Framework Programme. Priority thematic area of research in FP6. 2.3 Radiation protection. www.cordis.lu/fp6/eoi-analysis.htm.
[G61] Fell, A. & T. Fell (no yr.): Radioactive waste and democracy. The rights of future generations:151-157. Centrum för regionalvetenskap (CERUM), Umeå Univ., Umeå.
[G62] Fleming, P. (1991): Expert judgment and high-level nuclear waste management. Policy Studies Review. Vol. 10 (Winter 1991/92), No. 4:114-126.
[G63] Flüeler, T. (1998a): Decision anomalies and institutional learning in radioactive waste management. 8th ICHLRWM. Las Vegas, 11–14 May 1998. ANS, La Grange Park, IL:796-799.
[G64] Flüeler, T. (1998b): Vier Jahrzehnte Umgang mit radioaktiven Abfällen in der Schweiz – Perspektiven an Hand der Risikowahrnehmungsforschung [Four decades of radioactive waste management in Switzerland – Perspectives supported by risk perception research]. Unveröffentlichter Bericht. Bundesamt für Energie, Bern. (Unpublished)
[G65] Flüeler, T. (1999): Pursuit of sustainability: Guardianship or final disposal? Input paper for the Workshop on Disposition of High-Level Radioactive Waste. National Research Council/Board on Radioactive Waste Management. Irvine, CA, Nov 4-5, 1999. (unpublished). See [G21].
[G66] Flüeler, T. (2001a): Options in radioactive waste management revisited: a proposed framework for robust decision making. Risk Analysis, Vol. 21. No. 4. Aug.:787-799.

[G67] Flüeler, T. (2001b): Robustness in radioactive waste management. A contribution to decision making in complex socio-technical systems. In: E. Zio, M. Demichela & N. Piccinini (*eds.*): Safety & Reliability. Towards a Safer world. Proc. of the European Conference on Safety and Reliability. ESREL 2001. Torino (I), 16-20 Sep. Vol. 1. Politecnico di Torino, Torino, Italy:317-325.

[G68] Flüeler, T. (2002a): Passive safety and controllability in radioactive waste management. A Swiss attempt on the borderline. Topical Day 2002: Ethics on Radioactive Waste. How radiant is our future? Belgian Nuclear Research Centre SCK•CEN, Mol (B), Jan 22.

[G69] Flüeler, T. (2002b): Robust radioactive waste management: Decision making in complex socio-technical systems. Part 1: Principles. Part 2: Implementation. UNS-Working Paper 31. April 2002. Umweltnatur- und Umweltsozialwissenschaften UNS, ETH Zürich. (reprints of Flüeler (2001a) and (2001b)).

[G70] Flüeler, T. (2002c): KFW Cantonal Expert Group Wellenberg. An advisory body to support regional decision makers in LLW siting. EU Concerted Action Project Community Waste Management COWAM. 3rd Seminar. Fürigen, Switzerland. Sep 12–15.

[G71] Flüeler, T. (2002d): Long-term radioactive waste management: challenges and approaches to regulatory decision making. HSK/IAEA/NEA Workshop on Regulatory Decision Making Processes. Grandhotel Giessbach, Brienz, Switzerland. Oct 15–18.

[G72] Flüeler, T. (2002e): Radioaktive Abfälle in der Schweiz. Muster der Entscheidungsfindung in komplexen soziotechnischen Systemen [Radioactive waste management in Switzerland. Patterns of decision making in complex socio-technical systems]. Vols. I and II. Doctoral dissertation Nr. 14645. Swiss Federal Institute of Technology ETH, Zurich. Published with dissertation.de, Berlin. www.dissertation.de, www.e-collection.ethbib.ethz.ch/show?type=diss&nr=14645

[G73] Flüeler, T. (2003a): Die Einbettung der Arbeit des AkEnd in den internationalen Kontext. Kommentar aus der Sicht eines Beobachters. In: A. Dally (Hg.): Atommüll und sozialer Friede. Strategien der Standortsuche für nukleare Endlager. Dokumentation einer Tagung der Evangelischen Akademie Loccum (D) vom 7. bis 9. Februar 2003. Loccumer Protokolle 05/03. Evangelische Akademie, Rehburg-Loccum:121-147.

[G74] Flüeler, T. (2003b): Robust decision making in radioactive waste management is process-oriented. In: K. Andersson (*ed.*): VALDOR 2003 – Proceedings. Stockholm, June 9–13, 2003. SCK•CEN, SKI, SSI, NKS, OECD/NEA, UK Nirex, Stockholm: 79-87.

[G75] Flüeler, T. (2004): Long-term radioactive waste management: Challenges and approaches to regulatory decision making. In: C. Spitzer, U. Schmocker & V. N. Dang (*eds.*): Probabilistic safety assessment and management 2004. PSAM 7 – ESREL '04. Berlin, June 14–18. Vol. 5. Springer, London:2591-2596.

[G76] Flüeler, T., B. Kastenberg, W. Wildi, P. Hufschmid (2002): Proposal for anticipatory radioactive waste research: Towards robust and long-term radioactive waste management and governance. In collaboration with the Swiss Federal Nuclear Safety Inspectorate, HSK. 2002-05-30.

[G77] Flüeler, T. & R. W. Scholz (2004): Socio-technical knowledge for robust decision making in radioactive waste governance. Risk, Decision and Policy. Vol. 9, Issue 2:129-159.

[G78] Flüeler, T. & F. van Dorp (2000): Risikobasiertes Recht. Fallstudie Abfälle. [inside: Gemeinsamkeiten und Unterschiede der Bereiche radioaktiv – nichtradioaktiv]

Schweizerischer Nationalfonds, Projekt Nr. 1113-52163.97: Risk Based Regulation – ein taugliches Konzept für das Sicherheitsrecht? Stämpfli, Bern:7-18.

[G79] Flynn, J., W. Burns, C. K. Mertz & P. Slovic (1992): Trust as a determinant of opposition to a high-level radioactive waste repository: analysis of a structural model. Risk Analysis. Vol. 12. No. 3:417-429.

[G80] Frischknecht, R., A. Braunschweig, P. Hofstetter & P. Suter (2000): Human health damages due to ionising radiation in life cycle impact assessment. Environmental Impact Assessment Review. Vol. 20:159-189.

[G81] Ginsburg, S. (1995): Nuclear waste disposal. Gambling on Yucca Mountain. Aegean Park Press, Laguna Hills, CA.

[G82] Greber, M. A. (1995): Addressing ethical considerations about nuclear fuel waste management. In: NEA (1995): Informing the public about radioactive waste management: Proc. of an NEA International Seminar, Rauma, Finland, 13–15 June 1995. OECD, Paris:135-146.

[G83] Grupa, J. B. *et al.* (2000): Concerted action on the retrievability of long-lived radioactive waste in deep underground repositories. Final report. EURATOM. EUR 19145. European Communities, Brussels.

[G84] Goodwin, B. W. *et al.* (1994): Scenario analysis for the postclosure assessment of the Canadian concept for nuclear fuel waste disposal. AECL-10969, COG-94-247. AECL, Chalk River, Ontario.

[G85] Gouvernement Français (1998): Décisions du 9 décembre 1998 concernant les déchets nucléaires et la mise sur pied d'une nouvelle Autorité de Sûreté Nucléaire indépendante du Gouvernement disposant d'une compétence réglementaire et d'un pouvoir de sanction administrative.

[G86] Hadjilambrinos, C. (2000): An egalitarian response to utilitarian analysis of long-lived pollution: The case of high-level radioactive waste. Environmental Ethics. Vol. 22: 43-62.

[G87] Hammond, R. P. (1979): Nuclear wastes and public acceptance. American Scientist. Vol. 67:146.

[G88] Hériard Dubreuil, G. *et al.* (2002): A report of TRUSTNET on risk governance – lessons learned. Risk Research. Vol. 5. No. 1:83-95.

[G89] Hériard Dubreuil, G. *et al.* (2004): Local communities in nuclear waste management: the COWAM European Project. Euradwaste '04 – Community policy and research initiatives. Sixth European Commission Conference on the Management and Disposal of Radioactive Waste. Luxembourg, Mar 29–Apr 1.

[G90] Hériard Dubreuil, G. & T. Schneider (1998): The decision-making process in dealing with populations living in areas contaminated by the Chernobyl accident: the ETHOS project. In: NEA (*ed.*): The societal aspects of decision making in complex radiological situations. [1st] Workshop Proc., Villigen, Switzerland, 13–15 January 1998. OECD, Paris:33-42.

[G91] Herzig, E. B. & E. R. Statham (1993): When rationality and good science are not enough: science, politics and the policy process. In: E. B. Herzig & A. H. Mushkatel (*eds.*): Problems and prospects for nuclear waste disposal policy. Greenwood, Westport, CT.

[G92] Hill, M. D. & M. Gunton (2001): A multi-attribute comparison of indefinite storage and geological disposal of long-lived wastes. Pangea Technical Report PTR-01-01. Pangea Resources International, Baden, Switzerland. www.pangea.com.au

[G93] Högberg, L. (1997): Closing session. Summary and conclusions. Regulating the long-term safety of radioactive waste disposal. NEA International Workshop. Córdoba, 20–23 Jan. CSN, Madrid.

[G94] Hooft, E., A. Bergmans, K. Derveaux & L. Vanhoof (2002): Local partnerships: achieving stakeholder consensus on low-level waste disposal? WM '02 Conference. 24–28 Feb, Tucson, AZ.

[G95] HSK (2002): Jahresbericht 2001 der regulatorischen Sicherheitsforschung. HSK-AN-4246. HSK, Villigen-HSK.

[G96] Hurtt, D. *et al.* (2002): Opening and operating a nuclear disposal facility: Lessons learned in public outreach. In: NEA (*ed.*): Better integration of radiation protection in modern society. [2nd] Workshop Proc., Villigen, Switzerland, 23–25 January 2001. OECD, Paris:157-169.

[G97] IAEA (1956): Statute, approved on 23 Oct 1956. International Atomic Energy Agency IAEA, Vienna. www.iaea.org.

[G98] IAEA (1988): Code on the safety of nuclear power plants: governmental organization. Safety Standards. Safety Series No. 50-C-G. A publication within the NUSS Programme. International Atomic Energy Agency IAEA, Vienna.

[G99] IAEA (1989): Safety principles and technical criteria for the underground disposal of high-level radioactive wastes. Safety Standards. Safety Series No. 99.

[G100] IAEA (1991): Safety culture. A report by the International Nuclear Safety Advisory Group. Safety Series No. 75-INSAG-4.

[G101] IAEA (1992): Radioactive waste. IAEA Bulletin. Vol. 34. No. 3.

[G102] IAEA (1993): Report on radioactive waste disposal. Technical Report Series 349. IAEA, Vienna.

[G103] IAEA (1994a): Nuclear Safety Convention. 1994-6-17. IAEA, Vienna.

[G104] IAEA (1994b): Classification of radioactive waste. A safety guide. A publication within the Radwass Programme. Safety Series No. 111-G-1.1.

[G105] IAEA (1994c): Siting of geological disposal facilities. A safety guide. A publication within the Radwass Programme. Safety Series No. 111-G-4.1.

[G106] IAEA (1994d): Safety indicators in different time frames for the safety assessment of underground radioactive waste repositories. First report of the INWAC Subgroup on Principles and Criteria for Radioactive Waste Disposal. TECDOC-767.

[G107] IAEA (1995a): The principles of radioactive waste management. Safety Fundamentals. Safety Series No. 111-F. A publication within the Radwass Programme.

[G108] IAEA (1995b): Establishing a national system for radioactive waste management. A publication within the RADWASS Programme. Safety Standards. Safety Series No. 111-S-1.

[G109] IAEA (1996): Issues in radioactive waste disposal. Second report of the Working Group on Principles and Criteria for Radioactive Waste Disposal. TECDOC-909. October 1996.

[G110] IAEA (1997a): Joint Convention on the Safety of Spent Fuel Management and on the Safety of Radioactive Waste Management (Waste Convention). 1997-9-5. IAEA, Vienna.

[G111] IAEA (1997b): Regulatory decision making in the presence of uncertainty in the context of the disposal of long lived radioactive wastes. Third report of the Working Group on Principles and Criteria for Radioactive Waste Disposal. TECDOC-975. October 1997.

[G112] IAEA (1998a): Technical, institutional and economic factors important for developing a multinational radioactive waste repository. TECDOC-1021. June 1998.

[G113] IAEA (1998b): Meeting on topical issues in nuclear, radiation, and radioactive waste safety. Aug 30–Sep 4, 1998 (summary).
[G114] IAEA (1999a): Topical issues in nuclear, radiation and radioactive waste safety. Proceedings of an International Conference. Vienna, 31 Aug – 4 Sep 1998. IAEA, Vienna:233-255.
[G115] IAEA (1999b): Maintenance of records for radioactive waste disposal. TECDOC-1097. July 1999.
[G116] IAEA (1999c): Assessment of regulatory effectiveness. Peer discussions on regulatory practices. PDRP-4.
[G117] IAEA (1999d): Basic safety principles for nuclear power plants. A report by the International Nuclear Safety Advisory Group. 75-INSAG-3 Rev.1.
[G118] IAEA (2000a): Measures to strengthen international cooperation in nuclear, radiation and waste safety. International Conference on the safety of radioactive waste management. Córdoba, 31 May 2000.
[G119] IAEA (2000b): 3rd Scientific Forum. Radioactive Waste Management – Turning Options into Solutions. Vienna, 19–20 Sep. IAEA, Vienna.
[G120] IAEA (2000c): Retrievability of high level waste and spent nuclear fuel. Proc. of an international seminar organized by the Swedish National Council for Nuclear Waste [KASAM] in co-operation with the [IAEA]. Saltsjöbaden, Sweden, 24–27 Oct 1999. KASAM/IAEA. TECDOC-1187. December 2000.
[G121] IAEA (2001a): Report of Senior Consultants meeting on societal issues. IAEA Headquarters, Vienna. 28 to 30 May 2001. (internal report)
[G122] IAEA (2001b): Measures to strengthen international co-operation in nuclear, radiation, transport and waste safety. Waste safety (Secretariat responses to waste safety issues of Member States). Attachment 1. Report on the safety of radioactive waste management. Board of Governors. General Conference. GOV/2001/31-GC(45)/14. 19 July 2001.
[G123] IAEA (2002a): Ethical considerations in protecting the environment from the effects of ionizing radiation. IAEA-TECDOC-1270. IAEA, Vienna.
[G124] IAEA (2002b): Institutional framework for long term management of high level waste and/or spent nuclear fuel. TECDOC-1323.
[G125] IAEA (2003a): Report of Technical Meeting – Waste Safety Standards – Stakeholder dialogue. 24–27 February 2003. IAEA, Vienna (internal report).
[G126] IAEA (2003b): Safety indicators for the safety assessment of radioactive waste disposal. Sixth report of the Working Group on Principles and Criteria for Radioactive Waste Disposal. TECDOC-1372.
[G127] ICRP (1985): Radiation protection principles for disposal of solid radioactive waste. ICRP-Publication 46. International Commission on Radiological Protection, ICRP, Paris.
[G128] ICRP (1991): 1990 Recommendations of the ICRP. Publication 60. Pergamon Press, Oxford.
[G129] ICRP (1993): Protection from potential exposure: A conceptual framework. ICRP-Publication 64. Annals of the ICRP. Vol. 23. No. 1 1993. ICRP, Paris.
[G130] ICRP (2002): A framework for assessing the impact of ionising radiation on non-human species. ICRP Publication 91, Oct. 2002. Pergamon Press/Elsevier, Amsterdam.
[G131] ILK, International Committee on Nuclear Technology (2001): ILK statement on the potential suitability of the Gorleben Site as a deep repository for radioactive waste. ILK-08 E. ILK, Augsburg.

[G132] ILK (2003): ILK statement on the Recommendations of the Committee on a Selection procedure for Repository Sites (AkEnd). ILK-14 E. ILK, Augsburg. www.ilk-online.org.

[G133] INRA (Europe) (2002): Europeans and radioactive waste. Eurobarometer 56.2. Report. 19 April 2002. DG Energy and Transport. INRA – European Co-ordination Office, Brussels.

[G134] Isaacs, T. (1984): The institutional dimension of siting nuclear waste disposal facilities. Office of Strategic Planning and International programs. Office of Civilian Radioactive Waste Management. US Department of Energy, Washington.

[G135] Jenkins-Smith, H., D. A. Pratt & G. W. Bassett (1994): Analyzing mass perceptions of nuclear politics: puzzling over trust. 5th ICHLRWM, Las Vegas. Vol. 1. ANS, La Grange Park, IL:131-135.

[G136] Jensen, M. (1993, *ed.*): Conservation and retrieval of information. Elements of a strategy to inform future societies about nuclear waste repositories. Final Report of the Nordic Nuclear Safety Research Project KAN-1.3. Nordiske Seminar- og Arbeijdsrapporter 1993:596. SSI, Stockholm.

[G137] Johnson, J. (1998): Nuclear waste cleanup stirs. Chemical & Engineering News. June 8:26-27.

[G138] Kaluzny, Y. (1998): French research on the disposal of high-level long-lived radioactive waste in deep geological formations. 8th ICHLRWM, Las Vegas. ANS, La Grange Park, IL:304-307.

[G139] KASAM (1988): Ethical aspects on nuclear waste. KASAM (Consultative Committee for Nuclear Waste Management). SKN Report 29. SKN (National Board for Spent Nuclear Fuel), Stockholm.

[G140] KASAM (1993): Final disposal of spent nuclear fuel. KASAM's review of the Swedish Nuclear Fuel and Waste Management Co's (SKB's) RD&D Programme 92. Swedish Official Reports Series SOU 1993:67. The Printing Works of the Cabinet Office and Ministries, Stockholm.

[G141] KASAM (1996): Nuclear waste. Disposal technology and site selection. KASAM's review of the Swedish Nuclear Fuel and Waste Management Co's (SKB's) RD&D Programme 95. SOU 1996:101. The Printing Works of the Cabinet Office and Ministries, Stockholm.

[G142] KASAM (1998a): The state of knowledge in the nuclear waste area in 1998. SSI News. No. 2:14-16.

[G143] KASAM (1998b): Nuclear waste. State-of-the-art reports 1998. SOU 1998:68. Norstedts Tryckeri AB, Stockholm.

[G144] KASAM (1999): Nuclear waste. Method – site – environmental impact. KASAM's review of the ... RD&D Programme 98. SOU 1999:67. The Printing Works of the Cabinet Office and Ministries, Stockholm.

[G145] KASAM (2001): Nuclear Waste. State-of-the-art reports 2001. SOU 2001:35. Norstedts Tryckeri AB, Stockholm.

[G146] Kasperson, R., C. Hohenemser, J. X. Kasperson & R. W. Kates (1982): Institutional responses to different perceptions of risk. In: D. L. Sills, C. P. Wolf & V. B. Shelanski (*eds.*): Accident at Three Mile Island: the human dimensions. Westview Press, Boulder, CO:39-46.

[G147] Kasperson, R. E., P. Derr & R. W. Kates (1983): Confronting equity in radioactive waste management: modest proposals for a socially just and acceptable program. In: R. E. Kasperson (*ed.*): Equity issues in radioactive waste management. Oelgeschlager, Gunn & Hain, Cambridge, UK:331-368.

[G148] Keating, M. (1993): The Earth Summit's Agenda for change. A plain language version of Agenda 21 and the other Rio Agreements. Centre for Our Common Future, Geneva.

[G149] King, G. P., J. Katz & J. F. Munro (1991): Public education, public confidence, and public acceptance of radioactive waste management facilities. 2nd ICHLRWM, Las Vegas. Vol. 1. ANS, La Grange Park, IL:470-476.

[G150] Kirchner, G. (1995): Isolationszeiträume für die Endlagerung radioaktiver Abfälle. In: IPPNW (ed.): Die Endlagerung radioaktiver Abfälle. Hirzel, Stuttgart.

[G151] Kühn, K. & W. Hawickhorst (2000): Entsorgung und Endlagerung in Deutschland – inhaltlich gescheitert und ohne sachliche Grundlage? atomwirtschaft. Heft 7. Juli:453-456.

[G152] Kunreuther, H., D. Easterling, W. Desvouges & P. Slovic (1990): Public attitudes toward siting a high-level nuclear waste repository in Nevada. Risk Analysis. Vol. 10. No. 4:469-484.

[G153] Kunreuther, H., K. Fitzgerald & T. D. Aarts (1993): Siting noxious facilities: A test of the facility siting credo. Risk Analysis. Vol. 13. No. 3:301-318.

[G154] LeBars, Y. and C. Pescatore (2003) The NEA Forum on Stakeholder Confidence – its activities and main lessons. in: K. Andersson (ed.): VALDOR 2003 – Proceedings. Stockholm, June 9–13. SCK•CEN, SKI, SSI, NKS, OECD/NEA, UK Nirex, Stockholm:257-261.

[G155] Litmanen, T. (1996): Environmental conflict as a social construction: Nuclear waste conflicts in Finland. Society & Natural Resources. Vol. 9:523-535.

[G156] Little, R. H. et al. (1996): Post-disposal safety assessment of toxic and radioactive waste: Development and testing of the SACO methodology and code. European Commission Report EUR 16871 EN. Luxembourg.

[G157] Littleboy, A. K., P. J. Degnan, R. S. McLeod & S. Norris (1997): Site characterisation strategy and its role in post closure performance assessment. 21st International Symposium on the Scientific Basis for Nuclear Waste Management, Davos. Materials Research Society Symposium Proc. Vol. 506:719-730.

[G158] Lochard, J. (1998): Concluding remarks. In: NEA (ed.): Better integration of radiation protection in modern society. [2nd] Workshop Proc., Villigen, Switzerland, 23–25 January 2001. OECD, Paris:263-264.

[G159] Macy, J. (1989): Nuclear Guardianship Project. Berkeley.

[G160] Marshall, E. (1987): Hanford's radioactive tumbleweed. Science. Vol. 236: 1616-1620.

[G161] Marshall, E. (1991): The geopolitics of nuclear waste. Science. Vol. 251:864-867.

[G162] McCombie, Ch., I. G. McKinley & P. Zuidema (1990): Sufficient validation: the value of robustness in performance assessment and system design. In: OECD/SKI (eds.): Safety Assessment of Radioactive Waste Repositories. GEOVAL-1990. Symposium on Validation of Geosphere Flow and Transport Models. Stockholm, 14–17 May 1990:598-610.

[G163] Meier, D., H. Niedzielski-Eichner & W. C. Probst (1990): Public involvement: keystone to public confidence in the Civilian Radioactive Waste Management Program. 1st ICHLRWM, Las Vegas. Vol. 1. ANS, La Grange Park, IL:220-225.

[G164] Metz, W.C. (1994): Potential negative impacts of nuclear activities on local economies: rethinking the issue. Risk Analysis. Vol. 14. No. 5:763-777.

[G165] Miller, B. & T. Ikeda (1998): Monitoring of deep geological repositories for high-level radioactive wastes. 8th ICHLRWM, Las Vegas. ANS, La Grange Park, IL: 744-746.

[G166] Miller, W. M. & G. M. Smith (1993): Fluxes of elements and radionuclides form the geosphere. Carried out for the Swedish Institute for Radiation Protection, SSI. IE3438-1. INTERA Environmental Division, Melton Mowbray, UK.

[G167] Miller, W. M., G. M. Smith, P. A. Towler & D. Savage (1997): Natural elemental mass movement in the vicinity of the Äspö Hard Rock Laboratory. SKi Report 97:29. Swedish Nuclear Power Inspectorate SKi, Stockholm.

[G168] Mohanty, S. & B. Sagar (2002): Importance of transparency and traceability on building a safety case for high-level nuclear waste repositories. Risk Analysis. Vol. 22. No. 1:7-15.

[G169] Mushkatel, A., K. D. Pijawka & J. M. Nigg (1990): Social impact assessment of siting the high-level nuclear waste repository in Nevada: the use of risk future scenarios in survey research. 1st ICHLRWM, Las Vegas. Vol. 2. ANS, La Grange Park, IL:978-984.

[G170] Nathwani, J. S. (1993): The unintended social risks of nuclear waste disposal. Risk Abstracts. A Quarterly Journal of Abstracts, Reviews, and References. Vol. 10. No. 4. December:1-4.

[G171] NEA, Nuclear Energy Agency (1977): Objectives, concepts and strategies for the management of radioactive waste arising from nuclear power programmes. Report by a Group of Experts. September. OECD, Paris.

[G172] NEA (1982): Disposal of radioactive waste. an overview of the principles involved. OECD, Paris.

[G173] NEA (1991a): Disposal of radioactive wastes. The International Probabilistic System Assessment Group. Background and results. OECD, Paris.

[G174] NEA (1991b): Review of safety assessment methods. OECD, Paris.

[G175] NEA (1995a): The management of long-lived radioactive waste. The environmental and ethical basis of geological disposal of long-lived radioactive wastes. A collective opinion of the Radioactive Waste Management Committee. OECD, Paris.

[G176] NEA (1995b): The disposal of Canada's nuclear fuel waste. Report of the OECD/Nuclear Energy Agency Review Group. Natural Resources Canada, no place.

[G177] NEA (1997a): Disposal of radioactive waste. The Probabilistic System Assessment Group. History and achievements 1985-1994. OECD, Paris.

[G178] NEA (1997b): International peer review of the 1996 performance assessment of the US Waste Isolation Pilot Plant (WIPP). Report of the NEA/IAEA International Review Group. OECD, Paris.

[G179] NEA (1997c): Disposal of radioactive waste. Lessons learnt from ten performance assessment studies. OECD, Paris.

[G180] NEA (1998): Future nuclear regulatory challenges. A report by the NEA Committee on Nuclear Regulatory Activities. OECD, Paris.

[G181] NEA (1999a): Progress towards geologic disposal of radioactive waste: Where do we stand? An international assessment. OECD, Paris.

[G182] NEA (1999b): Confidence in the long-term safety of deep geological repositories. Its development and communication. OECD, Paris.

[G183] NEA (1999c): Geological disposal of radioactive waste. Review of developments in the last decade. OECD, Paris.

[G184] NEA (1999d): Strategic areas in radioactive waste management. The viewpoint and work orientations of the NEA Radioactive Waste Management Committee. OECD, Paris.

[G185] NEA (1999e): The role of the nuclear regulator in promoting and evaluating safety culture. OECD, Paris.

[G186] NEA (2000a): Stakeholder confidence and radioactive waste disposal. Workshop Proc. Paris, 28–31 Aug 2000. OECD, Paris.
[G187] NEA (2000b): Radiological impacts of spent nuclear fuel management options. A comparative study. 2000. OECD, Paris.
[G188] NEA (2001a): Nuclear regulatory challenges arising from competition in electricity markets. OECD, Paris.
[G189] NEA (2001b): Improving regulatory effectiveness. NEA/CNRA/R(2001)3. OECD, Paris.
[G190] NEA (2001c): Collective statement on the role of research in a nuclear regulatory context. OECD, Paris.
[G191] NEA (2001d): Reversibility and retrievability in geologic disposal of radioactive waste. Reflections at the international level. OECD, Paris.
[G192] NEA (2002a): Stepwise decision making in Finland for the disposal of spent nuclear fuel. Workshop Proc. Turku, Finland, 15–16 Nov 2001. OECD, Paris.
[G193] NEA (2002b): Establishing and communicating confidence in the safety of deep geologic disposal: approaches and arguments. OECD, Paris.
[G194] NEA (2002c): The handling of timescales in assessing post-closure safety of deep geological repositories. Workshop Proc. Paris, France, 16–18 Apr 2002. OECD, Paris.
[G195] NEA (2002d): An International Peer Review of the Yucca Mountain Project TSPA-SR. OECD, Paris.
[G196] NEA (2003a) The regulator's evolving role and image in radioactive waste management. Lessons learnt within the NEA Forum on Stakeholder Confidence. OECD, Paris.
[G197] NEA (2003b): Stakeholder participation in decision making involving radiation: Exploring processes and implications. 3rd Villigen Workshop, Villigen, 21–23 October 2003. Announcement. October 2003. OECD, Paris.
[G198] NEA (2003c) Public information, consultation and involvement in radioactive waste management. An international overview of approaches and experiences. OECD, Paris.
[G199] NEA/CSN/enresa (1997): Regulating the long-term safety of radioactive waste disposal. Córdoba, 20-23 Jan 1997. CSN, Madrid. (Annex: Compilation of national regulations on the disposal of long-lived radioactive waste)
[G200] NEA/IAEA (1997): International Peer Review of the 1996 Performance Assessment of the US Waste Isolation Pilot Plant (WIPP). NEA/IAEA Intern. Review Group. OECD, Paris.
[G201] NEA/IAEA/CEC (1991): Disposal of radioactive waste: Can long-term safety be evaluated? An international collective opinion. OECD, Paris.
[G202] Nelkin, D. (1974). The role of experts in a nuclear siting controversy. Bulletin of the Atomic Scientists. Vol. 30. No. 11:29-36.
[G203] Nilsson, A. (2001): Responsibility, equity and credibility – ethical dilemmas relating to nuclear waste. Published at the initiative of the Special Coordinator for Nuclear Waste Disposal (Ministry of the Environment). Kommentus Förlag, Stockholm.
[G204] Nikula, A. (1999): Nuclear energy in Finland. Five new plants are in discussion. atomwirtschaft atw. Heft 8-9. August/September:520-522.
[G205] Nirex *et al.* (2004, co-ordinator): Thematic network on the role of monitoring in a phased approach to geological disposal of radioactive waste. Final report. European Commission. Nuclear Science and Technology. Contract No. FIKW-CT-2001-20130.
[G206] Nörrby, S. (1998): A regulatory authority's needs for R&D to develop competence in assessing nuclear waste management safety. In: I. G. McKinley & Ch. McCombie

(*eds.*): Scientific Basis for Nuclear Waste Management XXI. Sep 28–Oct 3, 1997. Davos. Symposium Proc. Vol. 506. Materials Research Society, Warrendale, PA:11-17.

[G207] Nuclear Engineering International (2002): Bush oks Yucca Mountain. March 2002:10.

[G208] Nuclear Regulatory Commission, NRC (2002): Call for anticipatory research projects [to submit recommendations by June 1, 2002]. www.nrc.gov/what-we-do/regulatory/-research/rsch-projects.html.

[G209] Nucnet (2003). EC Nuclear Directives will "infringe national competencies". The World's Nuclear News Agency. No. 77/03/B. 19 February.

[G210] NUKEM Market Report (1994): Science friction: How technology, time, and conciliation could end the bitter war over nuclear waste. November 1994:4-60.

[G211] NWMO, Nuclear Waste Management Organization (2003): Asking the right questions? The future management of Canada's used nuclear fuel. Nov. 2003. NWMO, Toronto. (papers and extensive discussion on www.nwmo.ca)

[G212] Nygårds, P. (2000): The Swedish way – step-wise approach for public acceptance. International Symposium on Radioactive Waste Management – Sustainable Disposal or Tentative Solutions? March 30, Bern. Forum vera, Zurich.

[G213] Office of Technology Assessment (1982): Managing the nation's commercial high-level radioactive wastes. US Congress, Washington, DC.

[G214] Office of Technology Assessment/Congress of the United States (1983): Managing commercial high-level radioactive waste. Radioactive Waste Management and the Nuclear Fuel Cycle. Vol. 3. Nos. 3-4:279-345.

[G215] Otway, H. J., D. Maurer & K. Thomas (1978): Nuclear power. The question of public acceptance. Futures. April:109-118.

[G216] Palmer, C. (1992): Ecological sustainable development. A place in the sun for nuclear energy? 3rd ICHLRWM, Las Vegas. Vol. 2. ANS, La Grange Park, IL:1470-1477.

[G217] Parker, F. L., R. E. Broshears & J. Pasztor (1984): The disposal of high-level radioactive waste 1984. A comparative analysis of the state-of-the-art in selected countries. Volume I + II. NAK Rapport 11, Swedish National Board for Spent Nuclear Fuel. Beijer Institute, Royal Swedish Academy of Sciences, Stockholm.

[G218] Parker, F. L. (2000): An overview of the world status of radioactive waste management. 44th IAEA General Conference. 18–22 Sept 2000, Vienna. www.f40.-iaea.org/worldatom/About/GC/GC44/Sciprog/parker.pdf

[G219] Pearce, D.W. (1979): Social cost-benefit analysis and nuclear futures. In: G. T. Goodman & W. D. Rowe (*eds.*): Energy risk management. Academic Press, London: 253-267.

[G220] Peters, H. P., G. Albrecht, L. Hennen & H. U. Stegelmann (1990): "Chernobyl" and the nuclear power issue in West German public opinion. Environmental Psychology. Vol. 10:121-134.

[G221] Pijawka, K. D. & A. H. Mushkatel (1991): Public opposition to the siting of the high-level nuclear waste repository: the importance of trust. Policy Studies Review. Vol. 10 (Winter 1991/92), No. 4:180-194.

[G222] Posner, R. (Hg., 1984): Und in alle Ewigkeit ... Kommunikation über 10000 Jahre: Wie sagen wir unsern Kindern, wo der Atommüll liegt? Zeitschrift für Semiotik. Bd. 6. Heft 3:195-330.

[G223] Posner, R. (1990): Das Drei-Kammer-System: Ein Weg zur demokratischen Organisation von kollektivem Wissen und Gewissen über Jahrtausende. In: R. Posner (Hg.): Warnungen an die ferne Zukunft. Atommüll als Kommunikationsproblem. Raben, München:259-304.

[G224] The Radiation Protection and Nuclear Safety Authorities in Denmark, Finland, Iceland, Norway and Sweden (1989): Disposal of high level radioactive waste. Consideration of some basic criteria. A consultative document (called "Nordic Flag Book"). No place.
[G225] Reliability Engineering & System Safety, Vol. 42, 1993 or Vol. 69, 2000.
[G226] Renn, O. (1990): Public responses to the Chernobyl accident. Environmental Psychology. Vol. 10, No. 2:151-167.
[G227] République de France/Ministère de l'Industrie et du Commerce Extérieur (1991): Loi no. 91-1381 du 30 décembre relative aux recherches sur la gestion des déchets radioactifs/Basic Safety Rule. Rule No. III.2.f. Journal officiel de la République de France, 1er janvier 1992.
[G228] Richardson, P. J. (1998): Research priorities in nuclear waste management; a view from the far field. In: I. G. McKinley & Ch. McCombie (*eds.*): Scientific basis for nuclear waste management XXI. Sep 28–Oct 3, 1997. Davos. Symposium Proc. Vol. 506. Materials Research Society, Warrendale, PA:19-31.
[G229] Risk Analysis (1999). Vol. 129. No. 5.
[G230] Roberts, W., H. A. Haerer & D. von Winterfeldt (1992): Decision management for the Hanford environmental dose reconstruction project. 3rd ICHLRWM, Las Vegas. Vol. 2. ANS, La Grange Park, IL:1743-1750.
[G231] Rochlin, G. I. (1977): Nuclear waste disposal: two social criteria. Science. Vol. 195:23-31.
[G232] Roseboom, E. H. (1994): The case for retrievable high-level nuclear waste disposal. 5th ICHLRWM, Las Vegas. Vol. 3. ANS, La Grange Park, IL:1774-1781.
[G233] RWMAC, Radioactive Waste Management Advisory Committee (2001): RWMAC's advice to ministers on the process for formulation of future policy for the long term management of UK solid radioactive waste. Annex 2. Further observations and comments on the proposed five-stage policy development process. Sept. 2001. www.defra.gov.uk/rwmac/reports.htm.
[G234] RWMAC (2001): 23rd annual report of the RWMAC2003. www.defra.gov.uk/rwmac/reports.htm.
[G235] Ryhänen, V. (2000): To implement or not to implement? The Finnish application for final disposal of spent nuclear fuel. International Symposium on Radioactive Waste Management – Sustainable Disposal or Tentative Solutions? March 30, Bern. Forum vera, Zurich.
[G236] Sebeok, T. (1984): Communication measures to bridge ten millennia. Technical report. Atom 01 BMI/ONWI-532. Office of Nuclear Waste Isolation, Batelle Memorial Institute, Columbus, OH.
[G237] Secretary of Energy Advisory Board (SEAB) Task Force of Civilian Radioactive Waste Management: Earning public trust and confidence: Requisites for managing radioactive wastes. Final report. U. S. DOE, Washington.
[G238] Selling, H. A. (2000): Retrievable disposal – opposing views on ethics. In: IAEA: Retrievability of high level waste and spent nuclear fuel. Proc. of an international seminar organized by the Swedish National Council for Nuclear Waste [KASAM] in co-operation with the [IAEA]. Saltsjöbaden, Sweden, 24–27 Oct 1999. KASAM/IAEA. TECDOC-1187. December 2000. IAEA, Vienna:137-144.
[G239] Selwyn, J. (2002): Experience with citizens panels. In: NEA (*ed.*): Better integration of radiation protection in modern society. [2nd] Workshop Proc., Villigen, Switzerland, 23–25 January 2001. OECD, Paris:137-140.

[G240] Shrader-Frechette, K. (1993): Burying uncertainty. Risk and the case against geological disposal of nuclear waste. Univ. of California Press, Berkeley.

[G241] Silini, S. (1992): Ethical issues in radiation protection – the 1992 Sievert Lecture. Health Physics. Vol. 63. No. 2:139-148.

[G242] Sjöberg, L. & B.-M. Drottz-Sjöberg (1994): Risk perception of nuclear waste: experts and the public. Risk Research Report, No. 16. Center for Risk Research, Stockholm School of Economics.

[G243] Sjöberg, L. & B. M. Drottz-Sjöberg (2001): Fairness, risk and risk tolerance in the siting of a nuclear waste repository. Risk Research. Vol. 4. No. 1:75-101.

[G244] SKB (1996): Information, conservation and retrieval. Dec. SKB Technical Report 96-18. Swedish Nuclear Fuel and Waste Management Company SKB, Stockholm.

[G245] SKi, Swedish Nuclear Power Inspectorate (1988): Review of the final repository for reactor waste SFR-1. Division of Nuclear Waste. SKi Technical Report 88:5. SKi, Stockholm.

[G246] SKi/HSK/SSI (1990): Regulatory guidance for radioactive waste disposal – an advisory document. SKi Technical Report 90:15. SKi, Stockholm; SSI, Stockholm; HSK, Würenlingen.

[G247] Slovic, P., J. H. Flynn & M. Layman (1991): Perceived risk, trust, and the politics of nuclear waste. Science. Vol. 254:1603-1607.

[G248] Slovic, P., M. Layman & J. H. Flynn (1991): Lessons from Yucca Mountain. Environment. Vol. 33. No. 3:7-11, 28-30.

[G249] Slovic, P., M. Layman, N. Kraus, J. Flynn, J. Chalmers & G. Gesell (1991): Perceived risk, stigma, and potential economic impacts of a high-level nuclear waste repository in Nevada. Risk Analysis. Vol. 11. No. 4:683-702.

[G250] Slovic, P., S. Lichtenstein & B. Fischhoff (1979): Images of disaster: perception and acceptance of risks from nuclear power. In: G. T. Goodman & W. D. Rowe (*eds.*): Energy risk management. Academic Press, London:223-245.

[G251] SSI (1997): Health, environment and high level waste. The Swedish Radiation Protection Institute's proposed regulations concerning the final management of spent nuclear fuel or nuclear waste. SSI Rapport 97:07. Mai 1997. Swedish Radiation Protection Institute, SSI, Stockholm.

[G252] Sprecher, W., J. Katz & R. J. Redmond (1992): An integrated approach to strategic planning in the Civilian High-Level Radioactive Waste Management Program. 3rd ICHLRWM, Las Vegas. Vol. 2. ANS, La Grange Park, IL:1559-1564.

[G253] Sprecher, W. M. & E. Turner (1991): Risk communication: translation technically complex information to facilitate informed decision making. 2nd ICHLRWM, Las Vegas. Vol. 1. ANS, La Grange Park, IL:766-772.

[G254] Strohl, P. (1995): Notes sur l'information du public relative aux aspects institutionels de la gestion des déchets radioactifs. In: NEA (1995): Informing the public about radioactive waste management: Proc. of an NEA International Seminar, Rauma, Finland, 13–15 June 1995. OECD, Paris:125-131.

[G255] Swahn, J. (1992): The long-term nuclear explosives predicament. The final disposal of militarily usable fissile material in nuclear waste from nuclear power and from the elimination of nuclear weapons. Chalmers Tekniska Högskola, Göteborg.

[G256] Tannenbaum, P. H. (1984): Communication across 300 generations: Deterring human interference with waste deposit sites. Technical report. Atom 01 BMI/ONWI-535. Office of Nuclear Waste Isolation, Batelle Memorial Institute, Columbus, OH.

[G257] Terrell, R., R. Philpott, S. L. Smith & J. Gibson (1991): Building consensus in developing radioactive waste management systems. 2nd ICHLRWM, Las Vegas. Vol. 2. ANS, La Grange Park, IL:1042-1048.
[G258] Thompson, B. G. J. (1999): The role of performance assessment in the regulation of underground disposal of radioactive wastes: an international perspective. Risk Analysis. Vol. 19. No. 5:809-846.
[G259] Thunberg, A.-M. (2000): Retrievability in an ethical perspective. In: IAEA: Retrievability of high level waste and spent nuclear fuel. Proc. of an international seminar organized by the Swedish National Council for Nuclear Waste [KASAM] in co-operation with the [IAEA]. Saltsjöbaden, Sweden, 24–27 Oct 1999. KASAM/IAEA. TECDOC-1187. December 2000. IAEA, Vienna:129-135.
[G260] van der Pligt, J. (1985): Public attitudes to nuclear energy: salience and anxiety. Environmental Psychology. Vol. 5:87-97.
[G261] Ungermark, S. (1991): Communicating confidence and creating credibility. 2nd ICHLRWM, Las Vegas. Vol. 1. ANS, La Grange Park, IL:466-469.
[G262] United Nations, Economic Commission for Europe (1998): Convention on access to information, public participation in decision making and access to justice in environmental matters. ECE/CEP/43. Aarhus, DK. 1998-6-25.
[G263] U. S. DOE (1998): Viability assessment of a repository at Yucca Mountain. Overview. DOE/RW-0508. DOE, Washington, DC.
[G264] U. S. DOE, Department of Energy (1999): A roadmap for developing Accelerator Transmutation of Waste (ATW) technology. A report to Congress. DOE/RW-0519. DOE, Washington, DC.
[G265] U. S. EPA, Environmental Protection Agency (1985): Environmental standards for the management and disposal of spent nuclear fuel, high-level and transuranic radioactive wastes; final rule. Part 11, 40 CFR part 191. US EPA, Washington, DC.
[G266] U. S. USGS, Geological Survey (1999): Yucca Mountain as a radioactive-waste repository. Circular 1184. US Department of the Interior. USGS Information Services, Denver, CO.
[G267] U. S. NAS, National Academy of Sciences (1957): The disposal of radioactive waste on land. Report of the Committee on Waste Disposal of the Division of Earth Sciences. NAS, Washington, DC.
[G268] U. S. NAS, Commission on Geosciences, Environment and Resources (2000): Long-term institutional management of U. S. Department of Energy legacy waste sites. NAS, Washington, DC.
[G269] U. S. NRC, National Research Council (1990): Rethinking high-level radioactive waste disposal. A position statement of the Board on Radioactive Waste Management. National Academy Press, Washington, DC.
[G270] U. S. NRC (1996): Nuclear wastes: Technologies for separations and transmutations. National Academy Press, Washington, DC.
[G271] U. S. NRC (2002) One step at a time. The staged development of geologic repositories for high-level radioactive waste. Committee on Principles and Operational Strategies for Staged Repository Systems, Board on Radioactive Waste Management. The National Academies Press, Washington, DC.
[G272] van de Vate, L. (2001): Retrievable disposal of radioactive waste in the Netherlands. Dec. TNO-NITG-Information:12-16.
[G273] Vonken, F. & P. M. Maurenbrecher (1997): Overview of the feasibility study to hazardous waste in the Netherlands. Proc. 3rd European Engineering Geology Conference, Sep 10–14, 1997. Newcastle upon Tyne:453.

[G274] Wakerley, W. M. & J. Edwards (1986): Long-term storage of radioactive solid waste within disposal facilities. RW/86/080, UK Dept. of Energy, London.
[G275] Weinberg, A. M. (1972): Social institutions and nuclear energy. Science. Vol. 177: 7 July:27-34.
[G276] Wild, D. (2001): The experience of Nirex. IBC's 3rd International Two-day Conference on Managing Radioactive Waste. London, 11–12 Dec, 2001. IBC Global Conferences, London.
[G277] Wickham, S. M., R. D. Wilmot, D. A. Galson & R. Yearsley (1998): Comparing radioactive waste regulatory research and policy in nine countries. 8th ICHLRWM, Las Vegas. ANS, La Grange Park, IL:829-831.
[G278] Wilmot, R. D., D. A. Galson & B. G. J. Thompson (1998): Management of safety assessments. Lessons learned from national projects. 8th ICHLRWM, Las Vegas. ANS, La Grange Park, IL:838-840.
[G279] Wilmot, R. D., D. A. Galson & B. G. J. Thompson (1998): Management of safety assessments. Lessons learned from national projects. 8th ICHLRWM, Las Vegas. ANS, La Grange Park, IL:838-840.
[G280] WISE (2001): Dutch waste disposal commission reports on retrievable waste disposal. World Information Service on Energy, WISE. News Communique on March 2, 2001.
[G281] Wynne, B. (1989): Sheepfarming after Chernobyl. A case study in communicating scientific information. Environment. Vol. 31. No. 2:10-15, 33-39.
[G282] Zio, E. (2000): A formal structure for the treatment of site-representation uncertainty in performance assessments of nuclear waste repositories. Radioactive Waste Management and Environmental Restoration. Vol. 22:135-161.

2.5 Radioactive waste: Meta-analyses of decision processes (MA)

[MA1] Andersson, K. et al. (2004): Transparency and public participation in radioactive waste management. RISCOM II final report. Oct./Dec. 2003. SKI Report 2004:08. www.karinta-konsult.se/RISCOM.htm, www.valdoc.org, www.ski.se.
[MA2] Ballard, K. R. & R. G. Kuhn (1996): Developing and testing a facility location model for Canadian nuclear fuel waste. Risk Analysis. Vol. 16. No. 6:821-832.
[MA3] Barthe Y. & C. Mays (2001): Communication and information in France's underground laboratory siting process: clarity of procedure, ambivalence of effects. Risk Research. Vol. 4. No. 4:411-430.
[MA4] Berkhout, F. (1991): Radioactive waste. Politics and technology. Routledge, London.
[MA5] Blowers, A., D. Lowry & B. D. Solomon (1991): The international politics of nuclear waste. St. Martin's Press, New York.
[MA6] Carter, L. J. (1987): Nuclear imperatives and public trust: dealing with radioactive waste. Resources of the Future, Washington, DC.
[MA7] Carter, L.J. (1993): Ending the gridlock on nuclear waste storage. Issues in Science and Technology. Vol. 2 (Fall). No. 1:73-79.
[MA8] Colglazier, E. W. (1982): The politics of nuclear waste. Pergamon Press, New York.
[MA9] Colglazier, E. W. & R. B. Langum (1988): Policy conflicts in the process for siting nuclear waste repositories. Annual Review of Energy. Vol. 13:317-357.

[MA10] COWAM Network (2003): Nuclear waste management from a local perspective. Reflections for a better governance. Final report. Community Waste Management (COWAM). Nov. 2003. Mutadis, Paris. www.cowam.com/zBoard.asp.
[MA11] de la Bruhèze, A. .A. A. (1992): Political construction of technology. Nuclear waste disposal in The United States, 1945–1972. Univ. Twente, Enschede.
[MA12] Dormuth, K. W., P. A. Gillespie & S. H. Whitaker (1992): Considerations in managing the assessment of the Canadian nuclear fuel waste management disposal concept. 3rd ICHLRWM, Las Vegas. Vol. 2. ANS, La Grange Park, IL:1737-1742.
[MA13] Dormuth, K. W., P. A. Gillespie & S. H. Whitaker (1998): Perspectives on the review of AECL's proposal for disposing of Canada's nuclear fuel waste. In: I. G. McKinley & Ch. McCombie (eds.): Scientific basis for nuclear waste management XXI. Sep 28–Oct 3, 1997. Davos. Symposium Proc. Vol. 506. Materials Research Society, Warrendale, PA:637-648.
[MA14] Dunlap, R.E., M.E. Kraft & E.A. Rosa (1993): Public reactions to nuclear waste: citizens' views of repository siting. Duke Univ. Press, Durham, NC.
[MA15] Easterling, D. & H. Kunreuther (1995): The dilemma of siting a high-level nuclear waste repository. Studies in Risk and Uncertainty (no. 5). Kluwer Academic Publishers, Boston.
[MA16] English, M. R. (1992): Siting low-level radioactive waste disposal facilities. The public policy dilemma. Quorum Books, New York.
[MA17] Gerrard, M. B. (^{2}1995): Whose backyard, whose risk. Fear and fairness on toxic and nuclear waste siting. MIT Press, Cambridge, MA.
[MA18] Gregory, R. & S. Lichtenstein (1987): A review of the high-level nuclear repository siting analysis. Risk Analysis. Vol. 7. No. 2:219-223.
[MA19] Herzik, E. B. & A. H. Mushkatel (1991): Intergovernmental complexity in nuclear waste disposal policy: the indeterminate role of local government. Policy Studies Review. Vol. 10 (Winter 1991/92), No. 4:139-151.
[MA20] High-Level Nuclear Waste Disposal Policy Research Project (1987): High-level radioactive waste and spent nuclear fuel disposal: an assessment of impact evaluations and decisionmaking systems. Lyndon B. Johnson School of Public Affairs. Policy Research Project Report No. 84. Board of Regents, Univ. of Texas, Austin.
[MA21] Hine, D. W., C. Summers, M. Prystupa & A. McKenzie-Richer (1997): Public opposition to a proposed nuclear waste repository in Canada: an investigation of cultural and economic effects. Risk Analysis. Vol. 17. No. 3:293-302.
[MA22] Hinman, G., E. A. Rosa, R. R. Kleinhesselink & T. C. Lowinger (1993): Perceptions of nuclear and other risks in Japan and the United States. Risk Analysis. Vol. 13. No. 4:449-455.
[MA23] Jacob, G. (1990): Site unseen: the politics of nuclear waste repository. Univ. of Pittsburgh Press, Pittsburgh, PA.
[MA24] Kemp, R. (1992): The politics of radioactive waste disposal. Manchester Univ. Press, Manchester.
[MA25] Kraft, M. E. (1991): Public and state responses to high-level nuclear waste disposal: learning from policy failure. Policy Studies Review. Vol. 10 (Winter 1991/92), No. 4:152-166.
[MA26] Kraft, M. E. & B. B. Clary (1993): Public testimony in nuclear waste repository hearings: a content analysis. In: R. E. Dunlap, M. E. Kraft & E. A. Rosa (eds.): Public reactions to nuclear waste: citizens' views of repository siting. Duke Univ. Press, Durham, NC:89-114.

[MA27] Lemons, J., C. Malone & B. Piasecki (1989): America's high-level nuclear waste repository: a case study of environmental science and public policy. International Journal of Environmental Studies. Vol. 34:25-42.

[MA28] Lenssen, N. (1991, German 1994): Atommüll: The problem that won't go away. Worldwatch Institute. Worldwatch Paper, Bd. 6. Wochenschau Verlag, Schwalbach/Ts., BRD.

[MA29] Lidskog, R. (1994): Radioactive and hazardous waste management in Sweden. Movements, politics and science. Acta Universitatis Upsaliensis. Studia Sociologica Upsaliensia 38. Uppsala Univ., Uppsala.

[MA30] Linnerooth-Bayer, J. & K. B. Fitzgerald (1996): Conflicting views on fair siting processes: Evidence from Austria and the U. S. Risk: Health, Safety & Environment 7 (spring):119-134

[MA31] Malone, C. R. (1991): High-level nuclear waste disposal: a perspective on technocracy and democracy. Growth and Change. Vol. 22. No. 2 (Spring):69-74.

[MA32] Munton, D. (1996, *ed.*): Hazardous waste siting and democratic choice. Georgetown Univ. Press, Washington, DC.

[MA33] Mutadis (2000): COWAM Community Waste Management. European Commission – Science, Research, Development. Fifth Framework Programme of the European Atomic Energy Community (EURATOM) for Research and Training in the Field of Nuclear Energy (1998–2002). Mutadis, Paris (outline). www.cowam.org.

[MA34] Mutadis (2004): COWAM 2 (2004–2007). Mutadis, Paris.

[MA35] North, D. W. (1999): A perspective on nuclear waste. Risk Analysis. Vol. 19. No. 4:751-758.

[MA36] NRC, National Research Council (2002): Principles and operational strategies for staged repository systems: Progress report. 20 March, 2002. Committee on Principles and Operational Strategies for Staged Repository Systems, appointed by the National Research Council, Washington, DC.

[MA37] NRC (2003): On step at a time. The staged development of geologic repositories for high-level radioactive waste. Committee on Principles and Operational Strategies for Staged Repository Systems, Board on Radioactive Waste Management. The National Academies Press, Washington, DC.

[MA38] Openshaw, S., S. Carver & J. Fernie (1989): Britain's nuclear waste: Siting and safety. Belhaven Press, London.

[MA39] Rabe, B. G. (1994): Beyond NIMBY. Hazardous waste siting in Canada and the United States. The Brookings Institution, Washington, DC.

[MA40] Richardson, P. J. (1997): Public involvement in the siting of contentious facilities; lessons from the radioactive waste repository siting programmes in Canada and the United States, with special reference to the Swedish repository siting process. Rapport 97-11. Swedish Radiation Protection Institute SSI, Stockholm.

[MA41] Richardson, P. J. (1998): Basic requirements for successful public involvement in siting contentious facilities. 8th CHLRWM, Las Vegas. ANS, La Grange Park, IL:846-848.

[MA42] Söderberg, O. (1998): The role of the decision-maker: Whoever that might be. Experiences from the siting process for a spent nuclear fuel repository in Sweden. In: NEA (*ed.*): The societal aspects of decision making in complex radiological situations. [1st] Workshop Proc., Villigen, Switzerland, 13–15 January 1998. OECD, Paris:115-137.

[MA43] Sundqvist, G. 2002): The bedrock of opinion. Science, technology and society in the siting of high-level nuclear waste. Environment & Policy. Vol. 32. Kluwer Acad. Publ., Dordrecht, NL.
[MA44] van der Pligt, J. (1992): Nuclear energy and the public. Blackwell Publ., Oxford, UK.
[MA45] Vári, A., P. Reagan-Cirincone & J.L. Mumpower (1994): LLRW disposal facility siting. Successes and failures in six countries. Kluwer Academic Publishers, Dordrecht NL.

Index

Aargau 6, 155, 162
Aarhus Convention 270, 325
abbreviations xxiv
Abelson 103, 300
acceptance viii, 12, 26, 43, 44, 48, 61, 80, 83, 101, 116, 139, 150, 171, 178, 179, 188, 193, 195, 197, 202, 210, 218, 222, 232, 251, 256, 259, 264, 309, 315, 319, 322, 324
accountability viii, 110, 167, 251, 255
accountable 244, 253, 271
active 14, 23, 78, 87, 102, 121, 123, 124, 139, 140, 146, 153, 154, 155, 172, 183, 184, 185, 205, 213, 233, 257, 260, 271
activity ix, 15, 58, 60, 63, 90, 105, 106, 107, 126, 161, 237
actors *See* stakeholders
    common ground 210
    constellation 20, 211
    future 227
    individual 255
    information 19
    institutional 36
    interested 270
    interpretation 218
    legitimacy for 187
    local 196
    main 14, 133, 196
    mental models 12
    model 114, 117
    official 152, 205
    other 36
    political 87
    professional 67
    reality models 11
    social 11, 36
    use of terms 34
administration 19, 20, 34, 54, 67, 87, 92, 93, 134, 179, 244, 258
advisory body vi, xxv, xxvi, 20, 91, 93, 164, 242, 258, 269, 285, 314
AEC 224
AECL xxiv, 159, 191, 228, 310, 315, 327
Agenda-21 205, 319
AGNEB xxiv, xxvi, 8, 20, 66, 77, 90, 92, 93, 95, 97, 115, 117, 127, 138, 142, 162, 165, 166, 182, 243, 246, 247, 248, 260, 281, 289, 290, 294
Airolo 154, 245
AkEnd 189, 310, 314, 318
Aktion Mühleberg stillegen 59, 286
Allègre 194
Alps 155
alternatives 58, 66, 102, 103, 107, 108, 139, 142, 168, 201, 229, 307
Altzellen 154
anchor statements 33
ANDRA xxiv, 10, 190, 194, 219, 310
anhydrite 155

anticipatory research 26, 28, 138, 322
applicant 30, 36, 71, 72, 75, 86, 89, 150, 157, 159, 160, 166, 168, 172, 212, 229, 258, 260, 261, 271, 277 *See also* proponent
arena vii, 9, 97, 109, 172, 183
arguments ix, xi, xxvii, 17, 31, 34, 38, 39, 56, 65, 77, 92, 115, 124, 132, 134, 147, 154, 161, 167, 168, 170, 175, 176, 177, 180, 182, 220, 222, 227, 229, 232, 241, 244, 252, 254, 268, 277, 321
Argyris 172, 250, 300
ASK *See* HSK
assumptions 11, 19, 30, 31, 37, 45, 56, 110, 125, 145, 177, 214, 219, 236, 240, 252
asymmetry 89, 108, 254
Atomic Energy Act 3, 5, 6, 7, *8*, 74, 84, 91, 139, 142, 152, 212, 245, 246, 248, 278
Aube 54, 219, 234, 310
authenticity 33, 88, 89
authorities ix, x, xi, 6, 10, 26, 29, 30, 32, 52, 53, 58, 60, 66, 67, 71, 75, 89, 92, 94, 95, 97, 113, 114, 116, 117, 120, 121, 127, 131, 132, 140, 144, 154, 161, 163, 164, 168, 171, 172, 179, 180, 181, 186, 187, 188, 192, 210, 211, 215, 218, 234, 237, 241, 245, 247, 249, 250, 254, 255, 258, 260, 266, 268, 269, 270, 271, 274, 276, 277, 278

back end 17, 80, 82, 216
background of safety 72
backlog 91, 150, 176
Baden-Württemberg 167
Baer 272, 310
Ballard 67, 70, 326
barriers
 concept 233, 264, 277, 301
 geological 55, 225, 235
 institutional 259
 multiple viii, 240
 overbuilding 233
 primary 147, 224
 safety 71, 145, 211, 225, 238

technical ix, 18, 40, 64, 70, 147, 221, 224, 232, 237, 259
basic underlying assumptions 11
Beck 45, 121, 204, 305
Beierle 198, 295
Belgium 60, 65, 196, 261, 280, 314
benefit 22, 46, 48, 53, 58, 61, 80, 90, 107, 108, 116, 118, 148, 159, 170, 204, 254, 267, 268, 299, 305, 309, 322
Benken 6, 10, 16, 20, 155, 165, 167
best site 167
Bex 154
Beznau 3, 4, 62, 151, 154
BFE xxiv, 280, 281, 282, 283, 294, 295, 297
biases x, 30, 33, 34, 43, 44, 106, 109, 110, 116, 123, 128, 129, 306, 309
Bijker 34, 295, 299
biodiversity 58
BNFL xxiv, 3, 4, 189, 312
Board on Radioactive Waste Management ix, 144, 311, 313, 325, 328
Bohnet 108, 301
Bois de la Glaive 155, 157, 291 *See also* Ollon
Böschen 178, 296
Böttstein 6, 16, 162
Brehmer 46, 305
Breitschmid 148, 224, 225, 226, 282
bribery 169, 188
Brundtland 13, 304
Bullard 120, 311
burden [on future generations] 57, 71, 141, 207, 208, 211, 216, 221, 222, 228
Bure 10, 190
Burmaster 146, 302
Buser 4, 60, 92, 95, 234, 283

Canada xxiv, 67, 77, 159, 164, 187, 191, 201, 203, 228, 310, 311, 315, 320, 322, 326, 327, 328
Cantonal Council 7, 20, 150, 156, 161, 245, 247, 275
Cantonal Government 7, 143, 153, 158, 159, 160, 161, 166, 172, 184, 248, 261 *See also* Cantonal Council
Carter 70, 210, 252, 326

Index

case study ix, 21, 27, 190, 196, 297, 299, 313, 326, 328
CASS 27, 305
catch-22 situation 159
categories x, 14, 15, 33, 34, 39, 40, 45, 64, 102, 158, 164, 209
Cayford 198, 295
CC 52, 54, 61, 62, 72, 176, 203
 Concentrate and Confine
CEC xxiv, 219, 296, 321
Central Interim Storage xxvi, 6, 63, 165, 170, 273 *See also* ZWILAG
Centre de la Manche 69
chemical 54, 55, 62, 66, 69, 125, 196 *See also* toxic
Chernobyl 48, 66, 92, 120, 198, 299, 313, 315, 322, 323, 326
China 93
Cinderella effect vii, 82
citizens' groups 10, 66, 83, 190, 212
Clarke 186, 199, 311
Clary 32, 327
climate change 17
closed circle 182, 186
closed fuel cycle 63
closed issue 9, 34, 40, 178, 179, 188, 274
closure 22, 34, 52, 79, 80, 86, 88, 123, 126, 129, 168, 209, 211, 212, 213, 219, 227, 235, 236, 237, 243, 252, 258, 267, 268, 271, 272, 277, 292, 319, 321
Cogéma xxiv, 3, 4, 63
cognitive biases 43, 116
coherence viii, 136, 251
coherent 88, 136, 254
Colglazier 70, 71, 86, 89, 326
common ground 168, 210, 213, 252
communication x, 12, 30, 32, 34, 36, 39, 44, 53, 57, 106, 107, 116, 117, 161, 251, 278, 306, 307, 308, 320, 324
comparability 36, 142, 152, 155, 156, 223
compensation 90, 159, 167, 169, 170, 188, 189, 190, 215, 268
 environmental 170, 268
 factual 170
 financial 169, 268
 locally concentrated risk 189
 not bribery 188
 to overcome NIMBY 170

complex goals 107, 124, 138
complexity vi, x, 19, 29, 30, 34, 40, 54, 65, 105, 110, 120, 126, 132, 137, 142, 153, 178, 241, 253, 268, 299, 327
compromise 213
concept of robustness 58
Confederation xxiv, 4, 22, 90, 92, 93, 94, 95, 115, 128, 130, 158, 166, 206, 214, 238, 245, 279
confidence xi, 23, 34, 48, 65, 87, 111, 120, 139, 147, 187, 201, 231, 232, 239, 256, 259, 260, 275, 277, 306, 307, 308, 311, 319, 321, 323, 325
confirmatory research 28
conflict 33, 45, 81, 118, 139, 143, 144, 170, 177, 210, 244, 269, 283, 302, 307, 309, 319
consensus 31, 34, 78, 139, 143, 168, 189, 190, 194, 196, 213, 223, 233, 238, 244, 247, 249, 255, 298, 304, 316, 325
consensus conference 31, 143, 189
consent 104, 143, 159, 162, 189, 195, 207, 253, 254, 301
consequence 11, 45, 51, 54, 56, 70, 81, 83, 84, 95, 96, 101, 104, 106, 114, 116, 120, 130, 133, 134, 149, 156, 165, 169, 178, 191, 204, 205, 216, 220, 237, 251, 254, 278
consequence analysis 51, 56
conservatism 146, 236, 240
conservative 56, 145, 232, 272
consistent 88, 89, 96, 113, 136, 137, 148, 172, 184, 241, 243, 253, 257, 259
constraint xi, 69, 76, 90, 101, 105, 129, 178, 179, 189, 203, 263
content analysis x, xi, 30, 31, 32, 33, 37, 39, 40, 54, 102, 124, 151, 175, 185, 222, 244, 276, 277, 298, 327
context viii, x, 3, 12, 27, 29, 34, 39, 46, 48, 53, 64, 68, 80, 90, 92, 93, 104, 107, 110, 127, 139, 141, 142, 146, 159, 167, 168, 170, 175, 176, 181, 186, 191, 194, 195, 196, 206, 233, 238, 245, 246, 249, 250, 255, 264, 268, 274, 303, 304, 306, 308, 312, 316, 321
continuity 82, 91, 92, 187, 193, 208, 220, 255, 264, 274

continuous ix, 55, 68, 147, 172, 175, 184, 191, 208, 217, 227, 254, 258, 261, 272
contradiction vii, xi, 21, 88, 107, 120, 121, 123, 132, 223, 278
control ix, xi, 13, 18, 23, 48, 49, 52, 64, 69, 70, 71, 73, 75, 76, 77, 78, 79, 86, 94, 110, 119, 124, 126, 127, 129, 132, 139, 140, 146, 147, 157, 181, 186, 187, 191, 193, 195, 202, 205, 207, 209, 211, 213, 214, 216, 217, 219, 220, 221, 223, 225, 226, 227, 229, 232, 234, 235, 236, 238, 240, 250, 252, 256, 257, 258, 259, 262, 264, 267, 271, 272, 273, 274, 276, 277, 278, 279, 300, 301, 302
controllability xi, 6, 7, 28, 33, 44, 46, 48, 52, 61, 70, 74, 75, 77, 78, 79, 80, 95, 153, 158, 176, 180, 182, 183, 184, 207, 211, 215, 222, 223, 227, 239, 244, 253, 277, 314
controversial vi, 34, 70, 203, 227, 261
controversy 34, 66, 83, 223, 321
Córdoba Conference[s] 9, 192, 194, 316, 317, 321
CORE xxiv, 28, 138, 228, 283
core beliefs 213, 214
core principles 210, 252
Courvoisier 245
Covello 113, 296, 305, 309
COWAM xxiv, 40, 167, 190, 196, 280, 283, 285, 314, 315, 327, 328
credibility 33, 53, 56, 87, 88, 134, 187, 256, 321, 325
credible 49, 69, 88, 89, 120, 240, 254
cross-cutting vii, 54
crystalline 5, 6, 15, 65, 90, 130, 155, 161, 162, 163, 164, 166, 172, 182, 183, 184, 187
Cyert 149, 255, 300, 303

damage 17, 33, 43, 45, 47, 48, 52, 53, 55, 56, 59, 61, 64, 70, 77, 78, 144, 147, 205, 226
Damveld 141, 217, 223, 225, 226, 227, 283
danger vii, 45, 47, 48, 52, 57, 67, 80, 84, 88, 139, 170, 205, 224
DD 52, 72, 73, 151, 176, 203   Dilute and Disperse

debate
  in reverse order xi, 76, 181, 202, 276
Decide–Announce–Defend 114, 179, 182, 186, 198, 260, 263
decision
  "co-decision" 13
  "good" 65, 103, 105, 110, 253, 263, 300
  acceptance 116, 139
  affected 45
  and control 69
  and organisations 303
  and risk 66
  and technical progress 101
  anomalies 11, 29, 43, 101, 113, 123, 128, 313
  aspects 177
  attributes 253
  authority 36
  base 114, 189, 201, 242, 249, 260, 276
  basis xi
  Bayes 304
  behaviour 111
  choice 149, 168, 302
  collective 101, 103
  complex 110
  concepts 113
  conceptual 139, 229, 268, 271, 279
  conditions 178, 180
  consequences 254, 276
  consequential xi
  criteria 106, 110, 249
  cut-off 163, 164, 255
  cyclic process 103
  definition 103
  environmental 295
  for action 101
  framing 302, 304
  future generations 225, 228, 267
  goal-oriented 103
  impacts 263, 271
  implementation 103
  in principle 218
  in radioactive waste governance 123
  incremental 257
  informed consent 254
  instrument 58
  interim 168, 223, 253, 275
  irrevocable 238

Index

issue 102, 113, 253
issues 57
legitimacy 111, 251
maker xi, 29, 43, 45, 56, 78, 84, 119, 192, 204, 209, 214, 220, 221, 225, 253, 261, 269, 272, 277, 328
model 44, 86, 114, 121, 144, 161, 198, 243, 260, 263, 274, 278
monitoring 238
non-decision 227, 228
on action 102
on information 102
open 242
optimum 103, 138
option 201, 253
organisational 34
partial 103, 255
paths 218, 257
patterns 178, 179
periods 63
phases 104, 126
points 188, 251, 266
policy 33, 177
political xi, 53, 94, 218, 226, 277
postponed 86, 88, 108, 124, 146, 149, 221, 228, 229 *See* delay
power of 213
preceding 105
precedures 71
pre-decisions xi, 277
procedures 118, 303
qualification 106
reasons 159
responsibility 84
reversibility 46, 79, 225, 256
robust 265
rules 111
scientific 13
siting 89
situation x, 11, 29, 139
societal 13, 194, 253
status quo 303
steps 114, 116
stepwise 277, 321
strategic process 104
strategy 25, 26, 29, 177, 185, 187
structural 120
sub-optimum 171

subsequent 63
support 131, 222, 223, 278
sustained 104
trade-offs 232
type 179
under uncertainty *See* uncertainty
unpopular 83
weaknesses 271
decision making 12, 19, 44, 76, 102, 109, 113, 114, 117, 123, 149, 150, 151, 161, 191, 196, 199, 229, 238, 246, 249, 250, 278, 294, 295, 296, 300, 303, 305, 314, 315, 316, 321, 328
  broadly based 249
  criteria 123
  democratic 246
  dynamic and pluralistic 250
  linear 114, 161, 278
  regulatory 229
  sophisticated linear 117
  sub-optimum 151
  transparency 274
decision process *See* process
decision research 25, 101, 102, 103, 105, 111, 175
decision science x, xxiv, 21, 29, 31, 33, 34, 102, 106, 111, 233, 274, 300, 302
decisionistic models 182, 198
decision-making process *See* process
decomposition 34, 107, 110, 124, 136, 137
defence in depth viii, 65, 233, 256, 277
definitory power viii, 180
delay x, 26, 74, 127, 150, 166, 181, 183, 190, 195, 209, 229, 236, 274, 275
democratic deficit 117, 272
demonstrate 4, 7, 62, 72, 145, 163, 166, 191, 209, 245, 267
demonstration 7, 34, 38, 40, 52, 58, 60, 61, 76, 86, 88, 92, 123, 126, 127, 129, 137, 147, 152, 156, 162, 163, 164, 165, 172, 180, 184, 209, 222, 236, 237, 238, 239, 240, 248, 254, 279
design ix, x, 18, 22, 32, 51, 52, 59, 63, 71, 78, 103, 153, 161, 176, 187, 195, 202, 226, 231, 233, 234, 235, 236, 237, 240, 243, 252, 258, 272, 302, 311, 319

DETEC xxiv, xxvi, 6, 7, 19, 20, 27, 141, 152, 165, 166, 217, 228, 245, 273, 284, 294
deterministic approach 51, 55, 56, 108, 114, 116, 145, 233
dilemma 63, 84, 128, 148, 170, 226, 282, 321, 327
dimension 201
   constraint 203
   debate xi, 13, 76, 169, 276
   decisional 107
   discourse 102, 170, 202, 213
   ecological 202
   economic 202, 215
   ethical 204
   general risk 44
   historical 33
   institutional vi, 132, 211, 257, 269, 318
   institutional time 64
   long-term 13, 55, 252 *See* safety
   objective time 64
   political 202, 209, 232
   project-related 73
   risk 306
   risk analysis 307
   social 76, 202
   societal 194, 209, 232
   strategies 193
   technical 107, 202, 218
   temporal 202, 203
   time xi, 64, 123, 132, 137, 176, 223, 224, 259, 278
   variety 202
direct disposal 58, 62, 63
discipline vii, 35, 115, 218
discounting 70, 84, 90, 170, 221, 312
discourse vii, xi, 38, 63, 76, 94, 102, 113, 124, 140, 142, 148, 152, 170, 176, 178, 180, 188, 202, 203, 213, 224, 226, 249, 253, 254, 262, 264, 268, 269, 275, 277, 307
disposal viii, ix, x, xi, xxvii, 4, 5, 6, 7, 8, 9, 14, 16, 17, 18, 22, 25, 26, 28, 31, 32, 51, 52, 54, 55, 56, 57, 58, 59, 61, 62, 63, 64, 69, 70, 71, 73, 74, 75, 76, 77, 78, 79, 80, 81, 85, 86, 88, 89, 90, 91, 92, 93, 94, 95, 96, 97, 117, 118, 121, 124, 125, 126, 127, 132, 133, 134, 136, 137, 138, 139, 140, 143, 144, 145, 147, 149, 151, 152, 153, 154, 158, 159, 161, 162, 163, 164, 165, 166, 168, 169, 170, 171, 172, 177, 178, 179, 180, 183, 184, 185, 186, 187, 188, 190, 192, 194, 195, 197, 201, 203, 205, 207, 209, 211, 216, 217, 218, 219, 220, 223, 224, 226, 227, 228, 231, 232, 233, 234, 235, 236, 237, 238, 239, 240, 241, 245, 246, 247, 248, 253, 254, 259, 260, 265, 268, 269, 270, 272, 276, 277, 278, 279, 282, 283, 287, 288, 290, 291, 292, 310, 311, 312, 313, 315, 316, 317, 318, 319, 320, 321, 322, 323, 324, 325, 326, 327, 328, 329
disposal concept 4, 7, 8, 14, 73, 75, 76, 88, 91, 95, 125, 127, 140, 143, 152, 158, 164, 165, 166, 184, 207, 218, 223, 224, 226, 236, 248, 265, 327
disposal philosophy 18
disposition viii, ix, x, 3, 5, 6, 19, 21, 22, 27, 54, 58, 59, 61, 62, 63, 69, 74, 77, 81, 88, 90, 91, 93, 115, 121, 128, 130, 136, 139, 140, 141, 149, 152, 153, 165, 176, 177, 180, 185, 201, 202, 206, 209, 220, 221, 225, 228, 232, 246, 250, 254, 262, 266, 267, 269, 278, 279
disposition philosophy 91
dissent 130, 213
distrust 159, 306
Dobson 208, 296
document analysis x, 21, 31, 37
DOE xxvi, 10, 66, 67, 69, 88, 120, 219, 267, 323, 325
Dörner 138, 300
dose xxv, 17, 55, 57, 58, 59, 64, 69, 70, 120, 234, 235, 239, 323
Dounreay 68
dread vi, 47, 48, 52
Drigg 69, 234
Drottz-Sjöberg 45, 46, 308, 324
dumping 3, 63, 68, 72, 73, 89, 96, 151, 170, 171
Dürrenmatt 83

Easterling 190, 191, 313, 319, 327
ecology xi, 76, 215, 276

Index

economy xi, 76, 169, 215, 263, 276, 313
Edwards, J. 236, 326
Edwards, W. 118, 300, 305, 309
effective 44, 46, 70, 78, 107, 110, 140, 217, 226, 252, 265
effectiveness viii, 25, 32, 33, 64, 109, 244, 251, 256, 272, 317, 321
efficiency 32, 49, 67, 109, 134
efficient 106, 214, 215, 224, 252, 254, 256
EIA 187, 298, 304, 311 = environmental impact assessment
Eisikovic 83, 101, 113, 254, 292, 307
EKRA xxiv, 6, 7, 76, 78, 79, 91, 130, 141, 142, 143, 153, 158, 166, 179, 182, 185, 195, 217, 219, 223, 225, 227, 234, 235, 236, 237, 238, 239, 240, 242, 248, 266, 270, 283, 284, 286
El Cabril 234
Elander 121, 307
Elster 56, 300, 305
empirical findings 59, 72, 90, 127, 129, 132, 137, 140, 142, 147, 149, 150, 177
energy vi, vii, x, xxvii, 4, 9, 26, 27, 40, 44, 53, 61, 62, 70, 80, 81, 82, 83, 86, 88, 93, 94, 95, 118, 119, 123, 127, 128, 130, 133, 134, 138, 139, 140, 141, 149, 152, 159, 171, 177, 181, 183, 188, 195, 197, 204, 206, 208, 216, 217, 226, 227, 228, 245, 246, 247, 260, 276, 277, 289, 294, 309, 312, 321, 322, 325, 326, 329
Energy Dialogue 6, 142, 143, 144, 179, 212, 223, 247
energy policy 9, 40, 82, 83, 128, 133, 140, 177, 188, 228
engineering x, 64, 73, 78, 85, 104, 109, 111, 115, 152, 171, 183, 184, 194, 207, 215, 238, 240, 257, 276, 277, 299, 303
Environment Action Programme 26, 197
environmental organisations *See* NGO
EPA 58, 325
equal opportunities 71, 207, 208
equal treatment 156, 157, 176, 227, 246
equity viii, 26, 44, 53, 67, 83, 86, 107, 124, 148, 149, 197, 203, 207, 208, 221, 232, 254, 318, 321
ERDA 224

error 56, 65, 110, 111, 120, 172, 183, 302
error tolerance 56, 65
Espejo 187, 299, 313
Establish criteria–Consult–Filter–Decide 186
ethical ix, xi, xxvii, 13, 26, 27, 48, 71, 76, 149, 181, 191, 193, 201, 202, 204, 207, 208, 213, 225, 227, 228, 260, 276, 283, 310, 315, 320, 321, 325
ethics 48, 167, 203, 205, 206, 207, 283, 296, 298, 323
  consequential 205
  deontological 205
  egalitarian 206, 315
  responsibility 205
  teleological 205
  utilitarian 205, 315
  various principles 207
ETHOS 198, 315
EU xxiv, 26, 40, 87, 89, 167, 187, 196, 197, 205, 238, 244, 280, 285, 313, 314
Eurobarometer 71, 87, 149, 192, 197, 318
European Communities ix, 195, 296, 298, 311, 312, 313, 315
European Framework Programme viii, 26, 197, 232, 312, 313
European Union xxiv, 21, 59, 87, 96, 149, 196, 197, 205, 243, 313
evaluation viii, 32, 103, 104, 114, 121, 153, 157, 163, 172, 184, 196, 202, 212, 216, 219, 231, 233, 241, 250, 251, 258, 267, 275, 301, 302, 305, 309
Evers 66, 72, 83, 301
evidence 29, 36, 49, 72, 73, 79, 86, 88, 89, 104, 106, 119, 130, 148, 153, 157, 158, 159, 162, 163, 177, 191, 238, 241, 244, 264, 267, 308
exclusion criteria 53, 61, 86, 153, 154, 158, 159, 160, 161, 172, 176, 182, 183, 184, 242, 246, 248, 271, 278
experience 33, 39, 45, 47, 48, 49, 52, 65, 67, 68, 69, 75, 84, 145, 160, 177, 191, 198, 232, 255, 256, 311, 312, 326
expert 6, 9, 12, 39, 49, 66, 69, 75, 81, 94, 95, 115, 119, 128, 130, 146, 152, 158, 159, 160, 162, 167, 178, 179, 180, 182, 194, 198, 199, 215, 231, 239, 243, 246, 247, 248, 261, 262, 263, 264

expertise 67, 82, 156, 162, 187, 196, 198, 231, 245, 249, 260, 261, 262, 263
experts ix, x, xxvii, 8, 10, 22, 26, 32, 36, 39, 44, 45, 47, 48, 53, 61, 73, 75, 80, 84, 85, 86, 94, 97, 116, 117, 118, 119, 120, 128, 131, 139, 143, 147, 152, 156, 159, 162, 163, 167, 171, 178, 179, 182, 183, 184, 187, 188, 196, 198, 201, 210, 212, 215, 227, 242, 245, 248, 249, 260, 261, 262, 263, 274, 275, 276, 309, 321, 324
explanatory power 32, 33, 37, 87
explorability 156, 160, 162
exploratory gallery 6, 7, 150, 155, 156, 157, 158, 159, 242, 247, 248
extended final disposal xi, 185, 277

facilitation 144
fair viii, 26, 53, 55, 72, 84, 154, 157, 195, 196, 197, 232, 254, 328
fallacies x, 30, 34, 106, 123, 128, 129, 130
familiar 48, 72, 105, 183, 232
familiarity 33, 46, 48, 53, 72, 307
FDI xxiv, 273
feasibility 4, 7, 60, 73, 86, 118, 129, 137, 152, 153, 154, 156, 162, 163, *164*, 165, 166, 187, 188, 239, 248, 282, 292
Federal Council 4, 5, 6, 7, 9, 19, 20, 73, 90, 91, 93, 94, 95, 128, 129, 130, 133, 156, 157, 161, 162, 163, 165, 179, 210, 212, 217, 226, 248, 273, 284, 285, 293
Federal Government xxiv, 7, 92, 152, 156, 157, 160, 164, 189, 209, 216, 268, 273, 310 *See* also Federal Council
Federal Interim Storage 14, 273
final disposal viii, ix, xi, 3, 4, 6, 7, 8, 14, 16, 21, 54, 60, 61, 75, 77, 88, 91, 105, 115, 121, 128, 130, 140, 141, 142, 143, 151, 152, 153, 165, 166, 177, 178, 179, 182, 186, 187, 216, 217, 219, 222, 223, 224, 225, 226, 228, 229, 234, 235, 236, 240, 247, 273, 277, 279, 313, 323, 324
Finland 82, 164, 187, 192, 218, 228, 315, 319, 321, 323, 324

Fischhoff 46, 67, 113, 301, 305, 306, 307, 309, 324
Flüeler *No special entry*
FOE xxiv, 4, 19, 67, 135, 166, 248, 269, 273
FOPH xxv, 3, 4, 14, 20, 245, 273
Forsmark 71, 156
Forum vera 88, 141, 225, 238, 285, 286, 289, 295, 322, 323
framework xi, 40, 59, 107, 121, 142, 164, 177, 194, 195, 196, 202, 207, 249, 251, 258, 259, 261, 268, 277, 279, 296, 306, 313, 317
framing 104, 110, 111, 125, 161, 178, 218, 302, 304
France xxiv, 10, 62, 65, 68, 69, 73, 77, 80, 82, 171, 187, 189, 190, 196, 225, 234, 268, 287, 308, 318, 321, 323, 326
Franzen 108, 148, 301
Freiburghaus 109
Frey 108, 170, 292, 297, 301
FSC xxv, 192, 193, 196, 231
fuel cycle 17, 70, 228
Funke 105, 301
future generations 13, 48, 57, 71, 72, 78, 79, 83, 84, 85, 108, 123, 128, 139, 140, 141, 144, 146, 148, 170, 192, 206, 207, 208, 211, 217, 221, 222, 223, 224, 225, 226, 227, 228, 244, 254, 256, 270, 271, 272, 296, 298, 313

Garrick 115, 298
Germany 10, 62, 69, 70, 84, 89, 120, 167, 171, 189, 195, 196, 197, 248, 268, 310, 322, 328
Gibbons 25, 255, 256, 297, 299
Gigerenzer 11, 301
Gill 187, 313
Global Approach ix, 195, 296
GNW xxv, 6, 7, 10, 94, 130, 142, 144, 150, 157, 159, 160, 172, 178, 179, 184, 225, 226, 235, 236, 237, 242, 247, 248, 261, 286, 289
goal conflict 27, 79, 141, 206
goal relation x, 30, 34, 107, 124, 138, 140, 148, 177, 210, 256
goals 23, 31, 34, 51, 52, 54, 57, 58, 59, 64, 65, 78, 79, 82, 90, 103, 105, 106, 107, 109, 124, 138, 139, 140, 141,

Index 341

153, 160, 175, 176, 177, 189, 209, 210, 213, 214, 216, 218, 220, 225, 227, 234, 241, 250, 252, 253, 256, 264, 270, 271, 272, 275, 279
Goldstone 272, 301
Gomez 106, 301
Gorleben 10, 189, 317
Gösgen 62, 143
governance viii
government vi, xxiv, xxvii, 3, 5, 7, 8, 20, 73, 76, 77, 87, 92, 130, 143, 152, 153, 156, 157, 158, 159, 160, 161, 164, 187, 188, 189, 190, 191, 209, 215, 216, 225, 248, 258, 261, 268, 273, 308, 310
Graf 60, 286
Great Britain *See* UK
Greber 228, 315
Greenpeace 61, 75, 77, 128, 130, 224, 286, 287, 293, 294
Grisons 155, 246
groupthink 45, 107, 123, 131, 226, 301, 302
Guarantee 4, 5, 8, 15, 16, 64, 74, 88, 92, 93, 94, 117, 127, 129, 137, 152, 156, 162, 164, 165, 166, 170, 171, 179, 187, 190, 246, 247
Guba 38, 301
Guideline R-21 54, 57, 59, 61, 64, 71, 211, 270
Gunderson 257, 297

Habermas 11, 182, 198, 302
Hammond 219, 315
Hanford 68, 319, 323
Hanson 108, 126, 144, 302
Harrisburg *See* Three Mile Island
hazard vi, x, 13, 23, 29, 33, 44, 45, 47, 52, 53, 55, 61, 63, 70, 71, 72, 73, 74, 75, 84, 85, 121, 125, 203, 228, 229, 236, 252, 253, 306
hazard potential 45
Heierli 219, 287
Helton 146, 302
Hériard Dubreuil 219, 315
heuristic 43, 44, 48, 68, 116, 250, 303, 306
heuristics 309
hidden agenda 78, 139, 140, 222, 261

high-level xxv, 9, 176, 327, 328
HLLW xxv, 91, 93, 129
HLW xxv, 5, 6, 7, 14, 15, 16, 20, 85, 87, 89, 92, 93, 96, 118, 150, 152, 155, 156, 161, 163, 164, 165, 166, 171, 175, 176, 187, 190, 192, 193, 195, 197, 217, 239, 240, 248, 264, 265, 273, 292
Hohenemser 55, 306, 309, 318
host rock 5, 6, 15, 153, 156, 157, 160, 161, 162, 164, 166, 175, 176, 187, 218, 236, 237
HSK xxiv, 6, 7, 14, 20, 28, 56, 61, 67, 82, 90, 92, 95, 96, 115, 129, 133, 134, 135, 143, 152, 156, 158, 160, 162, 163, 164, 165, 171, 231, 242, 245, 269, 273, 280, 281, 283, 287, 290, 291, 293, 295, 314, 316, 324
Hug 119, 287, 288
Hunzinger 245

IAEA vi, viii, xxv, 9, 19, 59, 64, 69, 73, 82, 85, 91, 95, 119, 136, 139, 161, 171, 184, 185, 192, 193, 194, 195, 201, 203, 206, 208, 212, 216, 219, 229, 231, 233, 234, 238, 241, 257, 268, 270, 272, 280, 288, 290, 306, 308, 310, 314, 316, 317, 320, 321, 322, 323, 325
ICHLRWM xxv
ICRP xxv, 58, 86, 186, 194, 311, 317
IDHEAP 143, 212
ill-defined 104, 105, 125, 303
ill-prepared 95
ill-structured 34, 105, 109, 301
image 27, 68, 81, 84, 119, 297, 321, 324
 external (Nagra) 94, 176
 self-image (Nagra) 94, 134, 176
Imboden, M. 127, 181, 245, 288
immediacy 47
impacts xi, 17, 47, 52, 64, 69, 118, 119, 120, 124, 132, 190, 191, 193, 205, 206, 207, 223, 228, 244, 263, 271, 278, 319, 321, 324
implementation vi, x, 28, 68, 71, 72, 74, 76, 78, 91, 94, 102, 103, 106, 136, 161, 168, 171, 184, 190, 195, 196, 214, 218, 228, 236, 238, 253, 255, 257, 258, 262, 265, 277, 278, 279, 290, 311

implementer ix, xxvi, 9, 10, 28, 32, 53, 80, 81, 83, 84, 114, 116, 134, 190, 209, 228, 250, 258, 266, 268
incertitude 146
incoherent 89
inconsistent 92, 96, 157, 217, 227
independent 38, 53, 59, 61, 66, 138, 258
indeterminacy 65, 146
indeterminate 47, 90, 327
indicator 33, 52, 55, 58, 59, 61, 153, 182, 183, 234, 235, 241, 243, 244, 271, 278, 316, 317
inevitability 45, 48, 52, 69
inevitable 62, 72, 168, 181, 232
inference 37, 39
information 19, 28, 32, 34, 37, 39, 45, 48, 49, 53, 56, 69, 86, 97, 102, 105, 106, 107, 108, 110, 111, 115, 117, 119, 120, 124, 125, 130, 143, 144, 145, 160, 161, 166, 176, 177, 184, 187, 189, 191, 193, 212, 215, 219, 226, 228, 241, 245, 249, 251, 260, 263, 269, 272, 282, 306, 318, 321, 324, 325, 326
Inglestam 211, 297
insecurity 146
insights vii, ix, x, 26, 29, 33, 35, 54, 106, 111, 124, 148, 154, 177, 224, 226, 233, 274, 278
institutional vi, ix, x, xi, xxvii, 10, 11, 18, 19, 21, 22, 25, 26, 28, 29, 34, 35, 36, 43, 49, 64, 65, 71, 73, 74, 78, 82, 91, 93, 101, 115, 120, 123, 125, 129, 132, 140, 147, 159, 172, 177, 182, 185, 186, 193, 202, 203, 207, 208, 211, 213, 220, 233, 244, 252, 256, 257, 258, 259, 265, 267, 268, 269, 271, 272, 273, 277, 279, 300, 301, 304, 313, 316, 318, 325
institutional constancy 211, 270
interdisciplinarity 262
interdisciplinary 53
interests 31, 66, 82, 85, 94, 101, 146, 148, 167, 208, 215, 221, 223, 244, 254, 255, 256
intergenerational issues 26, 52, 53, 124, 148, 208, 221, 254, 272
intermediate-level ix, xxv, 5, 7, 14, 16, 21, 32, 60, 70, 74, 88, 95, 153, 154, 156, 160, 163, 164, 165, 176, 203, 228, 273, 292
interpretative flexibility 34, 218
intervention 13, 78, 124, 130, 132, 140, 148, 207, 213, 218, 219, 221, 223, 232, 253, 272
intragenerational issues 52, 53, 83, 84, 124, 148, 203, 207, 208, 221
intransparent 105, 107, 125, 145
intrusion
    human 17, 55, 126, 207, 257, 258
    water 55
involuntary 44, 47, 70
IRGC 297
irreversibility 45, 119, 204
irreversible 79, 227, 237
ISO 9000 255, 297
isolation 15, 23, 59, 64, 75, 81, 146, 160, 175, 176, 202, 203, 245, 252, 267, 312
Issler 74, 127, 134, 165, 166, 217, 224, 288, 289, 295
iterative 23, 36, 103, 129, 167, 219, 244, 250, 253

Jacob 186, 203, 327
Jaeger 95
Janis 103, 302
Japan 62, 164, 327
Jasanoff 184, 264, 298, 302, 309
Jonas 56, 203, 205, 298
judgement 11, 44, 48, 74, 102, 103, 167, 193, 231, 253, 260, 261, 278
Jülich 120, 311
Jungermann 69, 111, 119, 302, 303, 304, 305, 306, 307
justification viii, 311

Kahneman 43, 104, 302, 304, 306, 309
Kaplan 115, 298
KARA xxv, 9, 60, 75, 115, 151, 288
Karlsson 84, 85, 309
KASAM xxv, 71, 76, 105, 115, 158, 187, 188, 207, 238, 278, 300, 317, 318, 323, 325
Kasperson 84, 87, 142, 208, 306, 309, 312, 318
Kasser 60, 289
Keeney 139, 306, 307
Kemp 82, 119, 186, 187, 327

Index

KFW vii, xxv, 7, 20, 39, 95, 143, 153, 158, 159, 161, 179, 182, 237, 239, 242, 243, 248, 261, 280, 285, 289, 314
Kiener 74, 269, 289
King 120, 319
Kirchner 59, 319
Kissling-Näf 172, 263, 302
Kleindorfer 109, 302
KNE xxv, 6, 8, 20, 66, 79, 92, 129, 157, 158, 162, 163, 164, 166, 182, 227, 247, 248, 265, 289
Knoepfel 35, 263, 289, 300, 302
knowledge
    "co-production" of 199
    "trading zones" of 214
    absence 130, 226, 260
    access to 261
    action related or transformation 21
    and "vagueness" 144
    and action 27
    and control 78
    and risk perception 48
    and uncertainty 47, 124
    and understanding 46, 48, 53, 86
    application of 27
    assumptions 125
    base 251
    -based society 26, 197, 244, 313
    basic underlying 37
    body 274
    context-sensitive 264
    degree of 48
    delegation 261
    exchange 184
    expert 263
    extension 65
    from other fields 68
    generation 167, 178, 214, 215, 223, 255, 263, 270
    goal-oriented 260
    insufficient 146, 229
    integration 21, 23
    interest-bound 260
    lack 147
    limited 57
    limits 308
    local 275
    management 271
    mode-2 255
    new production 297
    partial 263
    personal 48
    production 13
    purpose-oriented 102
    relevant 25
    reliable 264
    scientific 12
    secured 65
    socio-technical 314
    state 19, 163
    sufficient 102, 138, 260
    system 20, 214
    systems 21, 25, 27, 145, 209, 210, 253
    target 21, 25, 27
    technological 66, 83
    transfer 184, 214, 255, 260, 262, 267, 272, 310
    transformation 25, 27
    types 262
KORA xxv, 45, 131, 179, 212, 284, 289, 296
Kraft 32, 307, 327
Krasnoyarsk 68
Kreuzer 77, 141, 225, 289, 290
Krippendorff 38, 298
Krohn 45, 119, 306, 307
Krücken 45, 119, 302, 306, 307
Krugman 45, 307
KSA vi, xxv, 20, 28, 143, 158, 164, 166, 245, 280, 287, 290
Küffer 74, 133, 290
Kühn 319
Kuhn, R. G. 67, 70, 326
Kuhn, T. S. 12, 298
Kunreuther 85, 190, 191, 302, 319, 327

La Hague 62, 63
Langum 70, 71, 89, 326
latency periods x, 30, 34, 67, 109, 124, 149, 150, 191
latent content 37
Latour 34, 298
laypeople 43, 44, 47, 48, 68, 72, 82, 86, 87, 94, 116, 143, 167, 250, 255, 263
learning vii, x, 25, 30, 34, 67, 86, 94, 95, 106, 109, 172, 183, 188, 191, 197, 199, 217, 233, 249, 250, 254, 263,

272, 274, 277, 285, 296, 297, 299, 300, 302, 303, 313, 327
legacies 54, 82, 267
legal means 177
legislation 34, 57, 81, 130, 211, 247, 249, 269, 296
Leibstadt 62, 225
Leuenberger 143
Leuggern 6, 162
Levi 103, 300
liability 45, 82, 137, 204, 209, 215, 216, 238
liberalised market 95, 134, 165, 216, 283
licence 4, 7, 14, 68, 71, 74, 126, 130, 150, 154, 157, 169, 224, 235, 241, 246, 247, 248
Lidskog 121, 307, 328
life-cycle analysis 17, 80
Lincoln 38, 301
LLILW xxv, 7, 16, 155, 160
LLW xxv, 5, 6, 15, 16, 39, 69, 74, 87, 90, 92, 94, 95, 120, 143, 150, 156, 157, 161, 170, 175, 176, 188, 197, 235, 239, 248, 273, 285, 314
local 9, 10, 47, 85, 97, 119, 121, 154, 159, 160, 166, 169, 182, 187, 188, 189, 190, 191, 195, 196, 245, 248, 258, 261, 267, 275, 283, 294, 310, 311, 319, 327
Lochard 198, 319
London Dumping Convention 73
long-lived xxv, 5, 7, 16, 70, 74, 125, 160, 163, 164, 165, 193, 203, 228, 235, 273, 292, 311, 315, 316, 318, 320, 321
long-lived intermediate-level 16, 70
long-term i, iii, vi, viii, ix, x, xi, 6, 7, 8, 13, 17, 18, 19, 21, 22, 23, 26, 34, 40, 47, 49, 51, 52, 54, 55, 57, 60, 61, 64, 65, 67, 70, 73, 75, 76, 77, 78, 79, 80, 82, 84, 86, 88, 107, 110, 118, 123, 124, 125, 126, 129, 130, 132, 134, 136, 137, 140, 141, 142, 143, 145, 146, 147, 150, 152, 153, 158, 162, 171, 177, 178, 179, 180, 181, 182, 183, 185, 186, 189, 191, 194, 195, 196, 199, 203, 204, 206, 208, 209, 211, 215, 216, 219, 222, 224, 225, 226, 227, 228, 229, 232, 233, 234, 235, 236, 237, 238, 239, 240, 241, 245, 250, 252, 253, 257, 264, 265, 266, 267, 269, 271, 272, 274, 276, 277, 278, 279, 287, 305, 309, 311, 314, 316, 320, 321, 324
Love Canal 66
low-level ix, 4, 5, 7, 16, 17, 21, 47, 55, 57, 64, 74, 88, 95, 120, 125, 126, 154, 160, 172, 183, 184, 249, 252, 316, 327
Lucens 3, 4, 245
Luhmann 11, 45, 57, 120, 205, 302, 307
LULU 85, 306
Lutz 96, 141, 290

Macy 141, 219, 227, 319
Mag 103, 138, 303
Majone 202, 298, 303
manifest text 37
Mann 103, 302
March 62, 149, 255, 300, 303
Marks 307
marl 155
Mayak 68
Mays 192, 326
Mazmanian 119, 303, 307
McCombie 6, 67, 81, 164, 285, 290, 311, 319, 321, 323, 327
mechanism 17, 48, 55, 125, 130, 132, 168, 194, 237, 252
media 5, 7, 36, 48, 89, 94, 147, 153, 165, 166, 179, 184, 224
mediation xxv, 31, 45, 131, 178, 212, 214, 244, 247, 255
Meet–Understand–Modify 186, 199
Menkes 33, 177, 298, 305, 309
mental models 11, 12, 114, 116, 213, 250, 251
Mettau 6, 10, 155, 163
Midlands 95, 154, 155, 156, 157, 160
milestones 3, 97, 171, 218, 244, 253, 267, 272
Miller, W. M. [B.] 59, 319, 320
minimum regret 71, 186, 237
Ministry of Energy 66, 144
Minsch 189, 215, 298
Mintzberg 103, 104, 126, 303
MIR xxv, 3, 5, 14, 15, 16, 20, 151, 273
mistrust 49, 81
mitigation 59, 71
mixed oxide 62

Index

MNA xxv, 7, 130, 161, 246, 293, 294
Molasse 129, 155, 162, 282
MONA 261
monitoring 54, 71, 76, 78, 79, 123, 124, 152, 157, 193, 195, 202, 204, 206, 213, 217, 219, 220, 234, 235, 236, 237, 238, 239, 240, 243, 249, 257, 258, 264, 267, 270, 271, 279, 310, 311, 312, 313, 321
Mont Terri 155
moratorium 8, 62, 73, 74, 143, 178, 179, 183, 228, 248
Morell 119, 307
Morone 119, 303
Morsleben 69
MO_x 62
Mühleberg 3, 59, 62, 96, 151, 286
multidimensional 30, 117
multidimensionality 13, 205, 250
multinational 134
multinational facility 91, 96, 165, 193, 316
Mumpower 113, 296, 305, 309, 329
Mushkatel 87, 315, 320, 322, 327
mutual learning 25, 39, 214, 252, 253, 254

Naegelin 92, 143, 291
Nagra x, xxv, 4, 5, 6, 7, 8, 10, 14, 15, 28, 37, 56, 60, 61, 63, 64, 67, 69, 71, 73, 74, 76, 77, 81, 82, 83, 88, 89, 90, 91, 92, 93, 94, 95, 96, 115, 118, 121, 127, 129, 130, 132, 133, 134, 137, 138, 140, 143, 144, 147, 150, 151, 152, 153, 154, 155, 156, 157, 158, 159, 160, 161, 162, 163, 164, 165, 166, 171, 172, 175, 176, 177, 179, 180, 184, 187, 214, 219, 224, 226, 235, 236, 238, 239, 240, 242, 245, 246, 247, 248, 268, 273, 282, 283, 284, 285, 287, 288, 289, 290, 291, 292, 293, 294
Nathwani 216, 320
National Academy of Sciences 70, 85, 87, 325
national task 6, 91, 92, 94, 135, 138, 172, 185, 195, 209, 211, 264, 268

NEA viii, ix, xxv, 22, 28, 56, 62, 63, 75, 85, 118, 139, 146, 147, 163, 180, 184, 192, 193, 194, 196, 201, 206, 207, 212, 219, 231, 233, 238, 240, 244, 256, 278, 292, 295, 296, 305, 310, 311, 312, 314, 315, 316, 319, 320, 321, 323, 324, 328
Netherlands 80, 96, 141, 186, 205, 297, 310, 312, 325, 326
network 238, 255, 283, 310, 327
Nevada 85, 87, 234, 313, 319, 320, 324
New Approach ix, 195
NGO vi, xxv, xxvi, 6, 10, 19, 36, 73, 77, 97, 117, 120, 127, 130, 140, 141, 143, 147, 150, 152, 157, 181, 188, 190, 212, 215, 223, 224, 226, 238, 247, 258, 260, 261, 273
NIABY 85
Nidecker 60, 292
Nidwalden xxv, 6, 7, 20, 74, 96, 130, 143, 144, 150, 152, 153, 156, 157, 158, 161, 169, 172, 179, 184, 212, 224, 242, 246, 247, 248, 249, 261, 294
NIMBY 47, 48, 82, 85, 94, 119, 120, 121, 149, 169, 170, 176, 307, 328
NIMTOO 82, 209
Ninck 109, 131, 303
Nirex 66, 81, 188, 189, 311, 312, 314, 319, 321, 326
North, D. W. 9, 328
Northern Atlantic Ocean 4, 14, 60
novelty 47
Nowotny 25, 66, 72, 83, 101, 113, 214, 254, 264, 292, 299, 301, 307
NPP xxv, 4, 5, 6, 10, 15, 62, 73, 81, 90, 94, 96, 131, 138, 143, 151, 152, 154, 164, 166, 170, 187, 192, 216, 225, 245, 246, 261
NRC xxvi, 28, 195, 322, 325, 328
Nuclear Energy Act 7, *62*, 73, 96, 139, 153, 165, 212, 217, 228, 239, 248, 270, 285, 294
nuclear guardianship 88, 96, 141, 152, 227
Nuclear Regulatory Commission xxvi, 28, 89, 322
Nuclear Safety Convention 82, 90, 91, 216, 316
NWMO 191, 322

Oberbauenstock 5, 15, 155, 156, 291
Oberholzer 87, 89, 292
objectives ix, 11, 21, 26, 35, 54, 109, 197, 210, 219, 220, 244, 287
Odermatt 150
OECD xxv, xxvi, 27, 127, 193, 195, 292, 295, 296, 299, 305, 310, 311, 312, 314, 315, 316, 319, 320, 321, 323, 324, 328
Office of Energy xxiv, 4, 19, 37, 74, 90, 96, 151, 273, 280, 284
Ogi 144, 294
Ollon 246, 284, 291
ONDRAF-NIRAS 261
Opalinus Clay 6, 155, 162, 163, 164, 166, 240, 292
openness viii, 55, 107, 178, 193, 251, 256, 267
opposition 3, 7, 9, 10, 73, 75, 84, 94, 95, 113, 116, 119, 120, 129, 130, 132, 134, 139, 144, 150, 154, 159, 166, 179, 180, 182, 190, 242, 245, 246, 247, 248, 255, 288, 315, 322, 327
optimum solution 105
option 23, 52, 60, 62, 70, 77, 89, 90, 93, 96, 102, 108, 111, 119, 128, 130, 139, 146, 158, 163, 164, 166, 172, 176, 184, 220, 222, 227, 244, 247, 253, 260, 267, 268, 275
Oskarshamn model 187, 311
Otway 113, 261, 308, 309, 322
overbuilding 56, 233
overconfidence 48
overestimation 44, 47, 48, 88
overoptimistic 56, 150, 188
overshooting 69

paradigm 12, 58, 105, 117, 180, 196, 198, 249, 263, 304, 308
para-public 32, 94
Parker 130, 194, 195, 213, 246, 268, 322
participation viii, 28, 34, 46, 48, 53, 61, 66, 67, 68, 69, 83, 117, 120, 139, 143, 159, 182, 191, 196, 198, 209, 210, 212, 213, 215, 241, 242, 244, 245, 249, 251, 252, 260, 263, 267, 272, 275, 279, 295, 306, 307, 310, 311, 312, 321, 325, 326

participatory 94, 111, 140, 184, 187, 190, 196, 215, 244, 249
partitioning *See* transmutation
passive xi, 23, 71, 78, 102, 117, 119, 139, 140, 172, 185, 189, 198, 199, 205, 211, 213, 222, 226, 232, 235, 236, 237, 238, 240, 244, 256, 257, 262, 270, 272, 277
Paul Scherrer Institute xxvi, 14, 20, 28, 138, 184, 214, 273
peer-review 66, 146, 234, 256, 258, 275
perceivability 33, 46, 48, 53, 72
perceivable 48, 53, 72
performance assessments xi, 54, 56, 140, 145, 232, 260, 277, 302, 326
Perrow 11, 303
perspectives vii, x, 8, 9, 12, 13, 29, 30, 31, 32, 33, 44, 48, 58, 61, 63, 75, 80, 81, 83, 101, 107, 110, 113, 121, 124, 128, 130, 131, 133, 141, 142, 153, 154, 178, 191, 202, 209, 210, 213, 216, 227, 232, 244, 251, 252, 253, 260, 261, 263, 274, 276, 283, 295, 302, 306, 310
  from above 29, 30, 99, 212, 274
  from below 29, 30, 196, 274
pessimistic 56
Peters 66, 322
Petitpierre 127, 292
phases 258, 266
Pijawka 87, 320, 322
Piller 128, 308
pilot facility 76, 79, 227, 237, 239, 266
Piz Pian Grand 155, 291
planning horizon 84, 85
players ix, 108, 148, 177, 183
plutonium 16, 62, 68, 78, 125, 186
policy making 33, 177, 244
polls xxvii, 89
polluter pays
  causality 48, 52, 53, 222
polytely 105, 138, 139
Posner 84, 208, 322
postponement *See* delay
power utilities 4, 133, 178, 179, 188
Pre-alps 155
predictability 129, 130, 146, 155, 156
predictable 55, 162, 207, 220
Prêtre 82, 165, 292

Index

prevention 59, 95
principles
   "chain of obligation" 227, 267
   "rolling present" 227, 267
   causality 146, 165, 177, 221, 228,
      229, 244, 261, 270, 274, 276, 279
   concentrate and confine xxiv
   core (value) 210, 252
   decision theory 189
   defence in depth viii, 233
   dilute and disperse xxiv
   ethical xi, 71, 207, 208
   final disposal 311, 316, 317, 320
   fundamental 223
   management 257
   methodological 31, 38, 110
   multi-barrier 277
   political 76, 276
   polluter pays- *See* causality
   precautionary 205, 296, 307
   problem solving 301
   protection 206
   radiation protection 317
   radioactive waste management safety
      316
   regret principle 237
   robust systems 314
   safety assessment 229
   safety for nuclear power plants 317
   siting 190
   staged final disposal 325, 328
   stakeholder dialogue 309
   sustainability 229, 253
   sustainable development 223
   transdisciplinarity 299
prisoner's dilemma 57, 108, 148
probabilistic approach 47, 51, 55, 56,
   108, 116, 124, 145, 146, 255, 298
probability 43, 45, 47, 48, 51, 55, 56,
   107, 144, 145, 157, 309
problem 180
   "free rider" 57
   and goals 109
   and solution xi
   approach 110
   communication 304, 322
   complex ix, 10, 26, 105, 107, 123,
      138, 301
   complexity 30

   co-operation 108, 124, 148
   decision 102, 103, 105, 107
   definition 102, 105, 106, 109, 128,
      141, 175, 178, 180, 253
   development 103
   -driven 25, 255
   emerging 217
   energy shortage 130
   environment 106
   environmental 68, 105
   ethical 204
   expert 180
   explicit 121, 131
   finding 109
   formulation 103
   future 269
   horizon 178, 180, 215
   identification 106, 109, 258
   ill-defined 104, 107, 109, 303
   ill-structured 109, 301, 303
   immanent 61
   implicit 105, 123, 128, 130
   informational 102
   judicial 293
   long-term 134
   modelling 299
   not solved 309, 328
   of comparability 155
   of representativeness 295
   old 109
   or risks 66
   ownership 83, 208, 214, 255, 267
   partial 107
   perception 290
   perception and communication 34,
      106
   perpetuation 40
   perspectives 29
   political x, 6, 7, 57, 81, 131, 176, 180,
      181, 205
   pressure 131
   psychological 245
   range 131, 132, 278
   recognition 103, 105, 213
   relevance 107, 110
   situation 131
   socio-technical 260
   solution 7, 34, 105, 223
   solvable 176

solving 8, 103, 106, 107, 115, 136, 260, 285, 296, 299, 301, 302, 303
solving behaviour 107
solving capacity 217
solving cycle 104
solving procedure 107
solving process 106
solving strategy 103, 295
structure x, 30, 34, 105, 107, 123, 130, 132
surrounding 124
technical 6, 18, 176, 178, 232, 245
technological 120
trust 87, 307
type 177, 180
understanding 126, 147
unexpected 156
unsolvable 9
view 176
waste 4, 9, 23, 96, 128, 148, 151, 245, 246, 277, 313, 315
Probst 106, 301, 319
procedural aspects ix, x, xi, 18, 22, 53, 107, 171, 172, 175, 176, 183, 213, 222, 244, 271, 276, 277, 278, 279
procedural problem 6
procedure ix, x, xi, 5, 7, 18, 22, 31, 39, 48, 53, 57, 60, 67, 76, 83, 104, 107, 109, 110, 118, 129, 147, 148, 150, 154, 155, 158, 160, 161, 164, 167, 168, 171, 172, 182, 183, 184, 187, 188, 189, 190, 212, 218, 226, 240, 241, 242, 243, 244, 247, 250, 253, 254, 255, 256, 258, 260, 262, 270, 271, 275, 277, 278, 304, 310, 318, 326
process
　alternatives 103
　and delays 168
　and dispute resolving 275
　and errors 172
　and goals 253, 275
　and negotiation 215
　and networks 255
　and perspectives 252
　and results 149, 168
　approach 252
　-based 49, 147, 240, 260, 278
　choice 110
　communication 36, 39, 305
　complex 37, 253
　convergence 256
　decision viii, x, 26, 30, 31, 32, 99, 182, 197, 232, 242, 248, 299, 303, 310
　decision-making ix, 10, 18, 21, 36, 48, 64, 83, 106, 170, 177, 183, 186, 192, 196, 199, 207, 215, 232, 238, 243, 253, 277, 315, 300
　development 139
　dialogue 196, 309
　dynamic 250
　EIA 187, 311
　good 253
　governance 66, 251
　guardian 273
　historical 33, 177
　implementation 257
　improvement 110, 158, 228
　integrated 179
　iterative 23
　known 86
　learning vii, 106, 183, 217, 254
　level 196
　licensing 19
　long-term 110, 279
　mediation 214
　mutual learning 297
　of negotiation 83
　on track 190
　ongoing vii, 34, 177
　open 256
　orientation ix, 30, 35, 38, 67, 111, 148, 149, 161, 172, 184, 189, 226, 233, 243, 255, 256, 314
　ownership 240
　paradigm 263
　partial 177
　participation 120, 321
　participatory 111, 190
　phase 242
　planning 187
　policy 298, 303, 315, 323
　political 35, 109
　problem-solving
　qualification 159
　regulatory 302
　regulatory decision making 314
　-related 241, 253

Index

reliable 271
requirements 195
research 38
reviewing xi, 52, 123, 132, 223, 257, 275, 277, 278, 279
risk analysis 114, 116, 117, 249, 308
role in 159
selection 89
single-option 158
siting 53, 61, 66, 159, 160, 167, 182, 187, 190
stepwise ix, 18, 150, 243, 248
Swiss radioactive waste management 244
transferable 33, 177
under control 186
understanding 27
utilities 107, 124, 139, 255
volunteer 190
vs. outcome 256
products 258
profit xi, 34, 57
proliferation 119, 217, 221
proof 17, 53, 60, 65, 74, 86, 88, 89, 127, 147, 148, 181, 182, 232, 238, 252
proponent ix, x, xxiv, xxv, 6, 8, 20, 22, 25, 26, 28, 29, 48, 56, 69, 76, 96, 107, 114, 116, 117, 127, 128, 129, 134, 140, 159, 171, 178, 181, 184, 187, 216, 223, 242, 245, 247, 248, 249, 258, 269, 276, 278
Propose–Learn–Share–Decide 274
protection 4, 13, 20, 51, 52, 54, 57, 58, 59, 60, 61, 64, 65, 70, 85, 86, 89, 118, 124, 126, 139, 140, 141, 153, 160, 172, 176, 184, 193, 194, 197, 198, 202, 206, 211, 213, 218, 220, 223, 227, 233, 234, 239, 241, 244, 256, 264, 270, 272, 278, 296, 305, 310, 311, 312, 313, 316, 317, 319, 323, 324
protection goal 54, 57, 58, 59, 175, 211, 213, 239
protest 3, 121, 166, 245
PSI xxvi, 14, 20, 115, 138, 273, 297
PSR 60, 292, 293, 294
psychometric approach 44, 306, 309
public ix, 26, 28, 29, 41, 57, 59, 60, 61, 63, 67, 69, 72, 74, 82, 83, 84, 85, 87, 88, 94, 96, 113, 116, 117, 118, 119, 120, 135, 138, 146, 147, 149, 154, 161, 164, 169, 177, 179, 180, 182, 183, 185, 186, 187, 189, 190, 191, 194, 195, 196, 199, 204, 208, 209, 210, 212, 215, 216, 231, 234, 238, 240, 242, 244, 246, 248, 249, 250, 251, 253, 256, 258, 260, 265, 266, 269, 270, 271, 275, 278
public defender of the future 84, 208

quality
  of decision 263
  of expertise 260, 262
  of knowledge 260
  of speediness 275
quality assurance 18, 63, 124, 125, 182, 213, 220, 234, 235, 258, 264, 273, 275, 279, 297
quality management 271

R&D xxvi, 82, 105, 187, 216, 218, 228, 273, 290, 321
radiation xxv, 44, 46, 58, 59, 64, 69, 85, 130, 170, 192, 193, 194, 197, 198, 211, 223, 225, 233, 249, 252, 290, 295, 296, 305, 306, 308, 310, 312, 315, 316, 317, 319, 321, 323, 324
Radiation Protection Act 58, *125*, *294*
Radiation Protection Article 9
radionuclide vi, 15, 58, 64, 85, 138, 145, 153, 160, 161, 193, 228, 239, 241
radon 63, 249
rationality
  "technical" 115
  absolute 114, 116, 250
  amplification 25
  bounded 110, 116, 129, 250, 301
  communicative 11
  concept 34
  concepts 11, 12
  enlarged x, 29
  impositions of 205
  issue 102
  model 115
  notion 251
  objective-oriented 11
  perception 250
  perspectives 113, 121, 177
  purpose-oriented 113

restricted 30
risk and 299
social 118, 119, 232, 250, 251
systems 11
types 251
Rausch 245, 246, 293
Ravetz 66, 308
Rawls 208, 217, 299
realistic 56, 190, 267
Rechsteiner 225
reciprocity 57, 148
reconstruction ix, x, 21, 22, 23, 31, 33, 323
recovery 78, 79, 222
recycling 8, 62, 125, 285
reflection vii, 9, 10, 11, 120, 204, 207, 223
reflexivity vii, 159, 215, 263
regional development 189
regulatory bodies 58, 82, 204, 223, 269, 272, 274
re-integration 34, 107, 124, 136, 137
reliability
    content analysis 38
    instrumental 38
    intercoder 38, 40
    internal 38
    intersubjective 38
    notion 240, 256
    organisational culture 304
    performance assessment 306
    stability 38
    temporal 38
reliable 253
remediation 78, 79, 170, 267
Renn 120, 299, 306, 307, 323
repositories 96, 121, 125, 126, 129, 130, 139, 145, 148, 150, 153, 158, 160, 163, 166, 169, 170, 176, 178, 182, 189, 192, 193, 203, 207, 211, 216, 222, 224, 228, 233, 234, 235, 236, 237, 238, 247, 262, 264, 266
repository 55, 60, 64, 69, 74, 75, 96, 138, 142, 144, 166, 170, 171, 193, 195, 216, 225, 236, 238, 264, 310, 315, 316, 318, 319, 320, 321, 325, 326, 328
reprocessing xxiv, 3, 4, 6, 8, 16, 58, 62, 63, 64, 70, 73, 80, 89, 93, 96, 119, 125, 127, 136, 143, 151, 165, 170, 171, 176, 228
research political embedding 25
resilience xi, 142, 240, 257, 277, 297
resource concept 34
resources vi, xi, 12, 25, 34, 38, 52, 78, 79, 80, 90, 92, 93, 107, 115, 124, 131, 132, 137, 138, 139, 140, 143, 144, 166, 169, 176, 178, 179, 184, 186, 202, 212, 215, 216, 217, 221, 222, 227, 228, 233, 261, 268, 270, 272, 277
responsibility 46, 48, 53, 54, 57, 80, 81, 83, 84, 85, 89, 90, 91, 138, 177, 187, 194, 195, 199, 205, 208, 209, 211, 214, 216, 228, 238, 240, 255, 258, 266, 268, 269, 271, 277, 278, 298, 321
responsibility principle 205
retention 59, 64
retrievability viii, xi, 6, 7, 22, 28, 52, 53, 61, 71, 74, 75, 77, 78, 79, 80, 95, 96, 127, 141, 153, 157, 158, 176, 180, 182, 183, 184, 195, 202, 207, 211, 215, 220, 222, 223, 225, 227, 236, 237, 239, 244, 246, 253, 262, 264, 271, 277, 279, 311, 315, 317, 321, 323, 325
retrieval ix, 54, 61, 71, 77, 78, 79, 125, 139, 201, 207, 211, 227, 236, 268, 318, 324
reversibility viii, 22, 33, 46, 48, 53, 71, 72, 75, 77, 79, 202, 220, 222, 225, 226, 256, 321
review xi, 7, 18, 46, 53, 61, 63, 86, 92, 95, 106, 110, 114, 115, 116, 123, 127, 132, 159, 164, 166, 172, 176, 184, 187, 189, 191, 196, 219, 223, 231, 234, 246, 249, 253, 257, 269, 275, 278, 279, 292, 302, 304, 318, 320, 327
Rio Conference *See* Agenda 21
Rip 244, 252, 303
RISCOM 75, 117, 196, 274, 310, 312, 313, 326
risk vii, 47
    acceptability 48, 308
    acceptance 44, 116, 271, 324
    analysis ix, x, 12, 13, 22, 30, 32, 33, 43, 44, 45, 47, 51, 55, 59, 60, 65, 86, 102, 113, 116, 119, 121, 175,

176, 233, 238, 249, 250, 251, 296, 298, 307, 308
analysis, integrated 231
and benefits 306, 309
and costs 299
and decisions 305
and fairness 327
and rationality 299
and systems theory 297
and uncertainty 303
approach 48
aspects 177
assessment 34, 35, 44, 59, 64, 67, 121, 146, 181, 297, 300, 308, 309
attributes 119
awareness 119
based decision making 296, 298
based regulation 39, 280, 297, 315
bearers 86, 118, 208, 209, 221
chemical 196
chronic 17, 47, 48, 52, 55, 64, 125, 252
classification 307
communication 44, 53, 116, 117, 161, 251, 278, 305, 306, 308, 324
comparisons 63
concentrated 189
concept 12, 87
concern 57, 65
control 71, 236
cost benefit 118
crisis 49
criteria 46, 234, 235, 250
debate 44, 66, 181, 254
decision 205
decisional vi
decisions under 108, 144, 302
definition 33, 47, 51, 54, 249, 298, 309
dimensions 44, 46, 306
distribution 34, 48, 53, 54, 83, 84, 86, 118, 268
emerging 127, 299
environmental 304, 305, 308, 309
evaluation 250, 305, 309
evolutionary 45, 119
export 83
function 51
future 85

goals 176
governance 30, 251, 296, 315
hypothetical 47
knowledge 115, 215
learning 67
level 120
limit 57
limits 312
long-term 118, 309
management ix, x, 43, 86, 175, 176, 181, 198, 251, 256, 292, 296, 297, 298, 299, 301, 305, 308, 312, 322
methodology 39, 113
model x, 29, 30
notion x, 29, 44, 47, 51, 54, 251
objective 44, 45, 116
of diluting engineering practice 238
perceived 46, 48, 309, 324
perception x, 21, 25, 33, 34, 35, 41, 44, 45, 46, 47, 48, 51, 53, 55, 56, 61, 101, 113, 114, 116, 175, 233, 244, 251, 270, 274, 304, 305, 308, 309, 313, 318, 324, 327
policy 312
potential 47, 120
premium 90, 170, 268
probabilistic assessment 298
profile 84
psychology 305
psychometric study 306
quantitative 55
radiation 130, 170, 249, 290, 306, 308
rating 309
reporting 307
research 43, 47, 67
residual 59, 181
sensitivity 46
situation 203
smeared 17, 55, 59
social 320
social amplification 54, 306
societal 271
society 305
source 44, 46
subjective 43
systemic 27, 127, 274
target 33, 47, 52, 59, 114, 116, 251
technical 17, 47, 195, 297
technological *See* technical

technologies 120, 303, 305, 307, 309
tolerance 324
uncertainties 186
understanding 34, 249
voluntary 47
zero 167
Ritschard v, 9, 73, 151, 245
robustness
  and ethics 208
  approach 121
  concept ix, 18, 21, 23, 149, 233
  decision *See* societal
  defence in depth 233
  definition 18, 238, 239
  disposal design 51, 52
  engineered 233
  institutional 259
  integral 35 *See* overall system
  integrated 232, 233
  intrinsic 233
  notion 233, 239
  of decisions 256
  of repository system 233
  of safety demonstration 61
  of solution 123
  overall 22, 23, 244, 268, 274, 277
  overall system x, 22, 23, 231, 243, 257, 264, 265
  performance 234, 235
  performance assessment 319
  radioactive waste management 314
  repository system viii
  scenario 221
  social *See* societal
  societal 243, 259, 263
  system viii
  technical 18, 56, 232, 233, 235, 240, 265
  technical and societal 233
  technical-system 234, 235
Rocky Flats 68
Rohrmann 10, 34, 303, 304
role ix, 44, 48, 64, 66, 68, 92, 93, 94, 95, 101, 104, 119, 159, 172, 176, 184, 192, 193, 195, 202, 209, 214, 238, 248, 255, 268, 269, 270, 308, 310, 312, 313, 319, 320, 321, 325, 327, 328
role of state 176
Rometsch 5, 127, 131, 137, 236, 246, 293

Ropohl 19, 298, 299
Roseboom 219, 323
Rowe 65, 299, 322, 324
Ruh 143
rules viii, 43, 64, 66, 83, 109, 111, 143, 213, 238, 251, 253, 261, 271, 304
Ryhänen 228, 323

SAEFL xxvi, 27, 73, 273
safeguards 220
safety viii, ix, x, xi, 5, 10, 12, 13, 14, 17, 18, 20, 23, 26, 28, 34, 36, 40, 45, 48, 49, 52, 53, 54, 55, 58, 59, 60, 61, 65, 66, 69, 71, 72, 74, 76, 77, 78, 79, 80, 82, 86, 88, 90, 91, 92, 95, 96, 97, 107, 108, 114, 115, 116, 123, 126, 129, 130, 133, 134, 135, 136, 137, 138, 139, 140, 142, 145, 146, 147, 148, 149, 152, 153, 156, 158, 161, 162, 164, 166, 167, 170, 171, 178, 179, 180, 181, 182, 183, 188, 189, 191, 192, 194, 197, 204, 205, 209, 210, 211, 212, 216, 219, 220, 221, 222, 223, 225, 226, 227, 229, 231, 232, 233, 234, 235, 236, 237, 238, 239, 240, 241, 242, 243, 245, 248, 250, 252, 253, 258, 260, 262, 264, 265, 266, 267, 268, 269, 270, 271, 273, 274, 276, 277, 278, 292, 298, 309, 310, 311, 314, 316, 317, 319, 320, 321, 326, 328
  "organised" 52, 72, 90, 260
  long-term ix, x, xi, 17, 18, 23, 34, 40, 52, 73, 76, 77, 78, 79, 86, 88, 123, 129, 137, 145, 153, 162, 171, 178, 180, 182, 183, 194, 204, 209, 220, 222, 225, 226, 227, 232, 233, 235, 238, 240, 252, 253, 265, 266, 267, 269, 271, 276, 277
safety analysis 146, 180, 212, 236, 237
safety assessment 18, 26, 197, 232, 236, 291
safety case 18, 78, 153, 239, 241
  models 236
safety indicator *See* indicator
Samuelson 108, 303
Sandman 44, 308
Savannah River 68

scenario 17, 45, 55, 56, 60, 64, 65, 126, 129, 144, 145, 147, 189, 221, 275, 315, 320
Schaltegger 170, 297
Schlumpf 91
Scholz 26, 214, 254, 258, 259, 285, 296, 299, 314
Schön 250, 300
Schweizerhalle 66
scope of action 79, 227
Seaborn 191
sealing 88, 126, 137, 221, 225, 235, 237, 238, 239, 243, 258, 267, 271
security 78, 79
sediments 6, 155, 161, 162
Seiler 4, 117, 217, 247, 293, 297
Sellafield 62, 188
Selten 11, 301
separation *See* transmutation
SES vi, xxvi, 37, 130, 141, 282, 283, 290, 293, 294
short-lived xxv, 5, 70, 158, 160, 161, 203, 208, 224
Siblingen 118, 283
signal value 69
Silini 85, 324
Simon 11, 105, 296, 301, 303
site selection 32, 39, 53, 56, 61, 66, 153, 158, 160, 167, 171, 172, 176, 187, 188, 190, 218, 240, 252, 291, 318
siting ix, 7, 21, 32, 36, 53, 66, 69, 75, 83, 84, 86, 87, 88, 89, 90, 120, 121, 129, 132, 143, 151, 152, 156, 162, 163, 165, 167, 168, 169, 172, 182, 184, 188, 189, 190, 218, 219, 233, 234, 242, 243, 244, 248, 261, 267, 285, 306, 307, 311, 314, 318, 319, 320, 321, 322, 324, 326, 327, 328, 329
siting procedure ix, 21, 167, 189
Sjöberg 45, 46, 308, 313, 324
SKB xxvi, 71, 76, 84, 85, 105, 187, 188, 238, 271, 318, 324
SKi xxvi, 56, 97, 187, 311, 313, 320, 324
Slovic 69, 87, 119, 296, 305, 306, 307, 308, 309, 315, 319, 324
society vi, xi, 9, 13, 25, 26, 27, 57, 69, 70, 72, 76, 81, 83, 94, 108, 110, 120, 123, 129, 188, 194, 197, 198, 199, 202, 203, 208, 210, 211, 213, 214, 215, 216, 221, 222, 244, 251, 253, 254, 256, 259, 263, 268, 271, 272, 275, 276, 279, 285, 290, 296, 298, 299, 302, 305, 306, 308, 309, 310, 312, 313, 316, 319, 323, 329
socio-technical i, viii, ix, xi, 19, 21, 22, 106, 123, 125, 130, 181, 232, 244, 260, 270, 274, 276, 277, 314
source analysis 32
South Africa 68
Sowden 255, 304
Spain 164, 196, 234, 308
spent fuel 7, 8, 16, 62, 64, 96, 125, 151, 163, 164, 193, 228, 264, 292
Sprecher 65, 324
SSI xxvi, 56, 58, 187, 300, 311, 312, 313, 314, 318, 319, 320, 324, 328
stakeholders
 "official" 75, 117, 133
 additional 183, 263
 advisory board 195
 arenas 305
 as clients 256
 attitudes 214
 belief systems 252
 common ground 168
 concerned x, 22, 29, 89, 167, 254
 concerns 196
 confidence 311
 consensus 316
 consent 244
 constellation 210, 212, 214
 contact 107
 core beliefs 213
 critical topics 244
 definition ix, 192
 dialogue 189, 309, 317
 experts 210
 Forum on Stakeholder Confidence (FSC) xxv, 231
 FSC 319, 321
 goals 140, 210, 252
 groupthink 131
 inclusive involvement 83, 250
 individual 131
 influence 189
 information 269
 instruments 212
 interaction 184

involvement viii, ix, xi, 8, 26, 34, 40, 89, 97, 113, 118, 121, 161, 164, 171, 177, 194, 196, 197, 198, 199, 208, 222, 243, 249, 252, 255, 257, 269, 270, 277, 278
isolation 226
knowledge 125
main xi, 6, 34, 36, 37, 80, 96, 111, 114, 116, 126, 131, 132, 136, 168, 177, 179, 183, 223, 257, 278
motives 80
mutual learning 254
notions 240
origin 36
participation 295
passive 199
perspectives 133, 134, 135, 209, 210, 253
political blocks 95
positions 109, 115, 147, 215
primary and secondary 109
problem and solution ranges 278
problem definition 132
proxies for 271
range 39, 183
regulators 270
relevant 231, 268, 279
responsibility 209, 251, 259
roles 158, 279
situation 125
strategies 81
Switzerland 210, 223, 245
traditional 83
trust 111, 167, 262
types ix, 26, 179, 258, 261
values 110
stalemate 9, 10, 11, 275
Stallén 72, 309
Starr 46, 48, 309
status quo 108, 190, 303
Steinmann 166, 294
stepwise 253 *See* process
stewardship 267, 274, 312
STOLA 261
Stopp Wellenberg 60
storage ix, 3, 4, 6, 8, 21, 47, 60, 67, 73, 74, 75, 77, 78, 79, 80, 91, 95, 96, 125, 126, 127, 130, 136, 139, 140, 141, 143, 144, 146, 147, 150, 164, 165, 169, 170, 171, 176, 186, 201, 202, 205, 206, 216, 217, 222, 223, 224, 225, 226, 227, 228, 229, 234, 247, 254, 257, 258, 273, 310, 311, 315, 326
strategic planning 67, 303, 324
strategy ix, 21, 22, 25, 26, 27, 29, 35, 49, 54, 56, 63, 70, 78, 80, 81, 89, 96, 103, 105, 106, 108, 110, 111, 121, 129, 136, 139, 141, 148, 150, 153, 160, 161, 165, 176, 177, 179, 185, 186, 187, 188, 193, 202, 203, 206, 213, 215, 218, 225, 226, 228, 232, 238, 239, 241, 242, 243, 255, 260, 278, 318, 319
Strohl 256, 267, 324
STS 13, 34, 40, 184, 211, 233 Science and technology studies
STUK xxvi, 82, 192
Subgroup xxvi, 8, 65, 77, 92, 93, 95, 97, 117, 128, 160, 162, 182, 243, 246, 247, 248, 294, 316
subsystem x, 19, 30, 34, 40, 106, 107, 110, 124, 125, 126, 136, 257
Sundquist 105, 182, 188
supervision 91, 92, 119, 176, 238
surveillance 60, 71, 76, 78, 88, 125, 139, 142, 193, 211, 217, 220, 225, 227, 235, 236, 237, 252, 264, 267, 279
survey 234, 312, 325
sustainability ix, x, xi, 13, 21, 22, 23, 27, 28, 30, 34, 35, 48, 58, 71, 82, 124, 139, 140, 141, 202, 205, 206, 207, 209, 210, 211, 214, 215, 217, 218, 219, 220, 223, 229, 232, 253, 256, 262, 277, 305, 313
sustainable development xxvii
SVA xxvi, 4, 19, 37, 68, 74, 286, 290, 292, 294, 295
Svenson 84, 85, 309
Swahn 217, 324
Sweden 46, 58, 60, 71, 76, 82, 84, 85, 92, 97, 105, 128, 156, 164, 169, 182, 185, 187, 188, 196, 203, 207, 217, 261, 268, 269, 271, 278, 282, 297, 307, 308, 309, 311, 313, 317, 318, 320, 322, 323, 325, 328
"Swedish model" 92, 164
Swiss Energy Foundation *See* SES 73, 141

Index 355

system
  behaviour 18, 236, 239
  belief 210, 254
  change 119, 220
  characteristics ix, 17, 79, 125, 232, 252, 255
  complex 27, 64, 86, 106, 175, 183, 189, 275
  control 18, 140, 225, 227, 232
  degradation 55, 144, 145
  destabilisation 106
  disposal viii, 32, 56, 153, 186, 209, 219, 232, 233, 237, 239
  disposition 202, 278
  dynamic 106, 125
  environment 106, 123
  failure 78, 79, 119, 209, 222, 236, 237
  flow 220
  geological 55, 75, 145
  human 214
  impacts xi, 132, 278
  improvement 181
  industrial 47
  information 56
  institutional 159
  long-term vi, xi, 54, 277
  modelling 103, 123, 126
  multi-barrier 54
  multidimensional 117
  national 91
  open 17, 19, 55, 252
  partial 123, 124, 136, 267
  performance 124, 138
  predictability 123, 129, 130
  radioactive waste xi, 19, 20, 194, 256, 277
  ranges 132
  reference 205, 253
  robust 18, 145, 232, 238, 239
  roles 187
  socially robust 244
  socio-technical 19, 21, 25, 232, 270, 274
  stable 232, 264, 272
  structure 8, 106, 125
  Swiss radioactive waste 210, 267
  technical 11, 18, 55, 62, 64, 106, 107, 113, 231, 232
  thinking 256
  trust 255
  understanding x, 30, 34, 56, 106, 123, 125, 127, 276
  unique 52
  units 131
systems engineering 104, 109, 111, 299, 301
systems theory x, 29, 297

target conflict 78, 107
targets 23, 59, 76, 107, 114, 116, 209, 239, 251
test facility 237, 266
third parties xi, 38, 52, 61, 74, 97, 117, 161, 182, 188, 223, 245, 247, 248, 250, 278
Thomas 113, 308, 309, 322
Thompson 32, 325, 326
Three Mile Island 66, 92, 302, 303, 311, 318
Thunberg 71, 325
Ticino 154, 155, 245
timetables 188, 190
Tomsk 68
toxic waste 13, 17, 18, 60, 62, 66, 67, 82, 207, 252, 262, 311, 312, 319, 327
toxicity 13, 14, 59, 62, 125
traceability xi, 56, 61, 158, 167, 171, 176, 182, 183, 222, 244, 277, 320
traceable x, 17, 39, 57, 86, 129, 153, 161, 171, 241, 254, 262, 271, 276
transdisciplinarity 25, 39, 214, 254, 285, 296, 299
transdisciplinary vii, 25, 26, 28, 124, 202, 214, 254, 260, 263, 285, 296, 297
transmutation 52, 63, 70, 195, 202, 221, 227, 228, 254
transparency vi, xi, 33, 61, 107, 117, 129, 143, 153, 158, 167, 171, 176, 183, 193, 196, 197, 209, 222, 240, 241, 243, 276, 277, 296, 299, 306, 312, 320
transparent ix, x, 19, 21, 22, 26, 51, 53, 56, 61, 66, 83, 86, 126, 129, 154, 158, 159, 161, 165, 171, 172, 184, 186, 196, 244, 248, 249, 254, 262, 271, 276
trans-science 299
trans-scientific 25, 194, 263
trial and error 172

trust 23, 34, 46, 49, 53, 54, 57, 72, 82, 87, 88, 89, 90, 111, 120, 124, 134, 138, 148, 165, 167, 190, 192, 196, 197, 245, 247, 254, 255, 256, 259, 260, 262, 263, 305, 306, 307, 308, 315, 318, 322, 323, 324, 326
TRUSTNET viii, 117, 196, 296, 315
trustworthy 89
Turner 65, 324
Tversky 43, 104, 302, 304, 306, 309

UK 3, 69, 73, 77, 80, 89, 171, 185, 187, 188, 196, 234, 299, 302, 305, 306, 308, 309, 311, 312, 314, 318, 319, 320, 323, 326, 329
uncertainty
  aleatory and epistemic 146
  analysis ix, 18, 51
  and "proof" of long-term safety 60
  and ignorance 146
  and knowledge 299
  and risk in final disposal 324
  and variability 145
  as a factor 65
  banning vii
  conditions 254
  conditions of 83
  decision under 56
  decisional 187, 229, 316
  decisions and 102
  decisions and risk 301
  decisions under 107, 108, 124, 144, 147
  in performance assessment 302, 326
  insecurity 146, 221
  irreducible 213, 228
  judgement under 306
  key aspect vii
  key notion 146
  limits 300
  model 144
  notion 12
  parametric 147
  reduced by information 260
  reduction of 124
  safety indicators 234
  scenario 147
  societal 147
  sources 233

  stochastic and statistical 144
  strategy to deal with trust 49
  structural 145
  subjective 120
  temporal 145
  types 56, 119, 124, 145, 147, 232, 244
  understanding 299
  vagueness 221
uncontrollable 44
understanding viii, 26, 27, 34, 44, 46, 48, 53, 56, 73, 86, 106, 115, 123, 125, 126, 127, 128, 142, 147, 178, 194, 196, 197, 215, 232, 233, 238, 244, 249, 262, 271, 275, 276, 277, 279, 304, 309
unidimensional 115
United Kingdom *See* UK
United States *xxvi*, *9*, *322*, *327*, *328*
unpredictable 27
unrealistic 21, 48, 73, 132, 150
uranium 62, 68, 78, 125, 164, 186
Uri 75, 246, 275, 290
US ix, xxvi, 10, 27, 28, 32, 56, 58, 66, 67, 69, 70, 71, 84, 85, 87, 88, 89, 120, 144, 195, 202, 219, 224, 229, 234, 300, 318, 320, 321, 322, 325
USA 10, 65, 66, 67, 80, 84, 88, 89, 120, 171, 185, 186, 190, 267, 268, 302
USA/U. S. *See* US
UVEK *See* DETEC

Vaalputs 68, 69
Val Canaria 155
validation
  communicative 26
  content analysis 29, 37, 38, 186
  performance assessment 136, 211, 216, 236, 237, 257, 258, 265, 270, 271, 319
validity
  assumptions 219
  communicative 40
  construct 39
  correlational 39
  data-related 38
  external 38
  periods 107
  predictive 39
  prerequisite 38

process-oriented 39
product-oriented 38
sampling 39
semantical 39
value 9, 12, 27, 32, 44, 45, 47, 48, 55, 58, 69, 70, 81, 84, 85, 87, 118, 120, 127, 141, 153, 194, 206, 216, 251, 261, 272, 319
van den Berg 141, 217, 223, 225, 226, 227, 283
van Dorp 21, 55, 62, 65, 146, 314
variability 56, 65, 107, 123, 145
variant xi, 76, 154, 158, 207, 220, 222, 276
Vatter 268, 294
Vaud 3, 154, 155, 246
verification 216, 258
veto principle 119, 188
Vlek 72, 309
voluntariness 48, 52, 69, 143
voluntary 69, 188, 190
volunteerism 167, 188, 189
Von Foerster 113
von Winterfeldt 118, 139, 305, 307, 308, 309, 323
VSE 15, 74, 133, 288, 295
vulnerability 240, 272, 313

Wackersdorf 62
Wakerley 236, 326
Wälti 94, 214, 289, 295
waste
    definition 80, 125
    from industry 3, 14, 16, 20, 70, 80, 130, 273, 279
    from medicine 3, 14, 16, 20, 70, 80, 130, 273, 279
    from research 3, 14, 16, 20, 80, 130, 273, 279

Waste Convention 91, 179, 185, 195, 207, 209, 212, 219, 264, 268, 270, 276, 278, 316
Weber, M. 11, 304
Weinberg 25, 70, 227, 299, 326
Weinstein 48, 309
Wellenberg vii, xxv, 6, 7, 10, 16, 20, 39, 60, 70, 74, 76, 88, 90, 95, 130, 143, 144, 150, 153, 154, 155, 156, 157, 158, 159, 160, 161, 172, 178, 179, 182, 183, 184, 185, 203, 224, 226, 235, 236, 237, 238, 239, 242, 247, 248, 275, 280, 281, 284, 285, 286, 287, 288, 289, 291, 292, 294, 314
Wiedemann 10, 34, 303, 304
Wild 81, 188, 326
Wildi 60, 95, 283, 290, 294, 314
willingness to participate
    as a contrast to volunteerism 189
Windhoff-Héritier 35, 304
WIPP xxvi, 32, 71, 120, 186, 320, 321
Woodhouse 119, 303
worst case 56, 270
Wynne 89, 119, 143, 167, 199, 251, 304, 308, 309, 326

Yucca Mountain 10, 85, 87, 115, 186, 191, 219, 234, 240, 311, 313, 315, 321, 322, 324, 325

Zeckhauser 108, 303
zero alternative 228
Zillessen 119, 304
Zimmermann 109, 301
Zio 65, 300, 314, 326
Zuidema 143, 166, 217, 264, 295, 319
Zünd 60, 286
Zürcher Weinland 6, 85, 155, 163, 165, 166, 167, 292
ZWILAG xxvi, 14, 63, 73, 133, 273

ENVIRONMENT & POLICY

1. Dutch Committee for Long-Term Environmental Policy: *The Environment: Towards a Sustainable Future.* 1994   ISBN 0-7923-2655-5; Pb 0-7923-2656-3
2. O. Kuik, P. Peters and N. Schrijver (eds.): *Joint Implementation to Curb Climate Change. Legal and Economic Aspects.* 1994   ISBN 0-7923-2825-6
3. C.J. Jepma (ed.): *The Feasibility of Joint Implementation.* 1995   ISBN 0-7923-3426-4
4. F.J. Dietz, H.R.J. Vollebergh and J.L. de Vries (eds.): *Environment, Incentives and the Common Market.* 1995   ISBN 0-7923-3602-X
5. J.F.Th. Schoute, P.A. Finke, F.R. Veeneklaas and H.P. Wolfert (eds.): *Scenario Studies for the Rural Environment.* 1995   ISBN 0-7923-3748-4
6. R.E. Munn, J.W.M. la Rivière and N. van Lookeren Campagne: *Policy Making in an Era of Global Environmental Change.* 1996   ISBN 0-7923-3872-3
7. F. Oosterhuis, F. Rubik and G. Scholl: *Product Policy in Europe: New Environmental Perspectives.* 1996   ISBN 0-7923-4078-7
8. J. Gupta: *The Climate Change Convention and Developing Countries: From Conflict to Consensus?* 1997   ISBN 0-7923-4577-0
9. M. Rolén, H. Sjöberg and U. Svedin (eds.): *International Governance on Environmental Issues.* 1997   ISBN 0-7923-4701-3
10. M.A. Ridley: *Lowering the Cost of Emission Reduction: Joint Implementation in the Framework Convention on Climate Change.* 1998   ISBN 0-7923-4914-8
11. G.J.I. Schrama (ed.): *Drinking Water Supply and Agricultural Pollution.* Preventive Action by the Water Supply Sector in the European Union and the United States. 1998   ISBN 0-7923-5104-5
12. P. Glasbergen: *Co-operative Environmental Governance: Public-Private Agreements as a Policy Strategy.* 1998   ISBN 0-7923-5148-7; Pb 0-7923-5149-5
13. P. Vellinga, F. Berkhout and J. Gupta (eds.): *Managing a Material World.* Perspectives in Industrial Ecology. 1998   ISBN 0-7923-5153-3; Pb 0-7923-5206-8
14. F.H.J.M. Coenen, D. Huitema and L.J. O'Toole, Jr. (eds.): *Participation and the Quality of Environmental Decision Making.* 1998   ISBN 0-7923-5264-5
15. D.M. Pugh and J.V. Tarazona (eds.): *Regulation for Chemical Safety in Europe: Analysis, Comment and Criticism.* 1998   ISBN 0-7923-5269-6
16. W. Østreng (ed.): *National Security and International Environmental Cooperation in the Arctic – the Case of the Northern Sea Route.* 1999   ISBN 0-7923-5528-8
17. S.V. Meijerink: *Conflict and Cooperation on the Scheldt River Basin.* A Case Study of Decision Making on International Scheldt Issues between 1967 and 1997. 1999   ISBN 0-7923-5650-0
18. M.A. Mohamed Salih: *Environmental Politics and Liberation in Contemporary Africa.* 1999   ISBN 0-7923-5650-0
19. C.J. Jepma and W. van der Gaast (eds.): *On the Compatibility of Flexible Instruments.* 1999   ISBN 0-7923-5728-0
20. M. Andersson: *Change and Continuity in Poland's Environmental Policy.* 1999   ISBN 0-7923-6051-6

ENVIRONMENT & POLICY

21. W. Kägi: *Economics of Climate Change: The Contribution of Forestry Projects.* 2000
    ISBN 0-7923-6103-2
22. E. van der Voet, J.B. Guinée and H.A.U. de Haes (eds.): *Heavy Metals: A Problem Solved?* Methods and Models to Evaluate Policy Strategies for Heavy Metals. 2000
    ISBN 0-7923-6192-X
23. G. Hønneland: *Coercive and Discursive Compliance Mechanisms in the Management of Natural Resourses.* A Case Study from the Barents Sea Fisheries. 2000
    ISBN 0-7923-6243-8
24. J. van Tatenhove, B. Arts and P. Leroy (eds.): *Political Modernisation and the Environments.* The Renewal of Environmental Policy Arrangements. 2000
    ISBN 0-7923-6312-4
25. G.K. Rosendal: *The Convention on Biological Diversity and Developing Countries.* 2000
    ISBN 0-7923-6375-2
26. G.H. Vonkeman (ed.): *Sustainable Development of European Cities and Regions.* 2000
    ISBN 0-7923-6423-6
27. J. Gupta and M. Grubb (eds.): *Climate Change and European Leadership.* A Sustainable Role for Europe? 2000
    ISBN 0-7923-6466-X
28. D. Vidas (ed.): *Implementing the Environmental Protection Regime for the Antarctic.* 2000
    ISBN 0-7923-6609-3; Pb 0-7923-6610-7
29. K. Eder and M. Kousis (eds.): *Environmental Politics in Southern Europe: Actors, Institutions and Discourses in a Europeanizing Society.* 2000   ISBN 0-7923-6753-7
30. R. Schwarze: *Law and Economics of International Climate Change Policy.* 2001
    ISBN 0-7923-6800-2
31. M.J. Scoullos, G.H. Vonkeman, I. Thornton, and Z. Makuch: *Mercury - Cadmium-Lead: Handbook for Sustainable Heavy Metals Policy and Regulation.* 2001
    ISBN 1-4020-0224-6
32. G. Sundqvist: *The Bedrock of Opinion.* Science, Technology and Society in the Siting of High-Level Nuclear Waste. 2002   ISBN 1-4020-0477-X
33. P.P.J. Driessen and P. Glasbergen (eds.): *Greening Society.* The Paradigm Shift in Dutch Environmental Politics. 2002   ISBN 1-4020-0652-7
34. D. Huitema: *Hazardous Decisions.* Hazardous Waste Siting in the UK, The Netherlands and Canada. Institutions and Discourses. 2002   ISBN 1-4020-0969-0
35. D. A. Fuchs: *An Institutional Basis for Environmental Stewardship: The Structure and Quality of Property Rights.* 2003   ISBN 1-4020-1002-8
36. B. Chaytor and K.R. Gray (eds.): *International Environmental Law and Policy in Africa.* 2003   ISBN 1-4020-1287-X
37. F.M. Brouwer, I. Heinz and T. Zabel (eds.): *Governance of Water-Related Conflicts in Agriculture.* New Directions in Agri-Environmental and Water Policies in the EU. 2003   ISBN 1-4020-1553-4

ENVIRONMENT & POLICY

38. G.J.I. Schrama and S. Sedlacek (eds.): *Environmental and Technology Policy in Europe. Technological Innovation and Policy Integration.* 2003
    ISBN 1-4020-1583-6
39. A.J. Dietz, R. Ruben and A. Verhagen (eds.): *The Impact of Climate Change on Drylands.* With a focus on West Africa. 2004   ISBN 1-4020-1952-1
40. I. Kissling-Näf and S. Kuks (eds.): *The Evolution of National Water Regimes in Europe.* Transitions in Water Rights and Water Policies. 2004
    ISBN 1-4020-2483-5
41. H. Bressers and S. Kuks (eds.): *Integrated Governance and Water Basin Management.* Conditions for Regime Change and Sustainability. 2004   ISBN 1-4020-2481-9
42. T. Flüeler: *Decision Making for Complex Socio-Technical Systems.* Robustness from Lessons Learned in Long-Term Radioactive Waste Governance. 2006
    ISBN 1-4020-3480-6
43. E. Croci (ed.): *The Handbook of Environmental Voluntary Agreements.* Design, Implementation and Evaluation Issues. 2005   ISBN 1-4020-3355-9
44. X. Olsthoorn and A.J. Wieczorek (eds.): *Understanding Industrial Transformation.* Views from Different Disciplines. 2005   ISBN 1-4020-3755-4
45. H. Aiking, J. de Boer and J. Vereijken (eds.): *Sustainable Protein Production and Consumption: Pigs or Peas?* 2006   ISBN 1-4020-4062-8
46. F. Brouwer and B.A. McCarl (eds.): *Agriculture and Climate Beyond 2015.* A New Perspective on Future Land Use Patterns. 2006   ISBN 1-4020-4063-6

www.springer.com